Pesticide Resistance in Arthropods

Contributors

Richard T. Roush
Department of Entomology
Cornell University
Ithaca, NY 14853-0999

Bruce E. Tabashnik
Department of Entomology
University of Hawaii
Honolulu, HI 96822

Ralph D. Bagwell
Route 6, Box 50B
Paris, TX 75640

Jeffrey R. Bloomquist
Department of Entomology
Virginia Polytechnic Institute and State
  University
Blacksburg, VA 24061

Clayton Campanhola
EMBRAPA/CNPDA
Caixa Postal 69
Jaguariuna, SP, Brasil

Brian A. Croft
Department of Entomology
Oregon State University
Corvallis, OR 97311

Joanne C. Daly
Division of Entomology, CSIRO
Black Mountain
Canberra, ACT 2601
Australia

Richard H. ffrench-Constant
Department of Entomology
Cornell University
Ithaca, NY 14853-0999

George P. Georghiou
Department of Entomology
University of California
Riverside, CA 92521

Marjorie A. Hoy
Department of Entomological Sciences
University of California
Berkeley, CA 94720

Billy F. McCutchen
Department of Entomology
University of California
Davis, CA 95616

Frederick W. Plapp, Jr.
Department of Entomology
Texas A&M University
College Station, TX 77843

David J. Pree
Agriculture Canada, Res. Sta.
Vineland Station, Ontario LOR 2EO
CANADA

Jeffrey G. Scott
Department of Entomology
Cornell University
Ithaca, NY 14853-0999

David M. Soderlund
New York State Agricultural Station
Cornell University
Geneva, NY 14456

Pesticide Resistance in Arthropods

Edited by
Richard T. Roush
and
Bruce E. Tabashnik

Chapman and Hall
New York and London

First published in 1990 by
Chapman and Hall
an imprint of
Routledge, Chapman & Hall, Inc.
29 West 35 Street
New York, NY 10001

Published in Great Britain by

Chapman and Hall
11 New Fetter Lane
London EC4P 4EE

Library of Congress Cataloging in Publication Data
Pesticide resistance in arthropods / edited by Richard T. Roush and
    Bruce Tabashnik.
        p.   cm.
    Includes bibliographical references.
    ISBN 0-412-01971-X
    1. Insecticide resistance. 2. Arthropod pests—Control.
I. Roush, Richard Tyrone. II. Tabashnik, Bruce.
SB951.5.P48
632'.951—dc20   89-70875

British Library Cataloguing in Publication Data
Pesticide resistance in arthropods.
    1. Pesticides. Resistance of pests
    I. Roush, Richard T. *1954–* II. Tabashnik, Bruce E. *1954–*
    632.95
    ISBN 0-412-01917-X

To
The Memory of
Roman Sawicki, F.R.S.

Contents

1

Introduction

Bruce E. Tabashnik and *Richard T. Roush*

Pesticide resistance is an increasingly urgent worldwide problem. Resistance to one or more pesticides has been documented in more than 440 species of insects and mites. Resistance in vectors of human disease, particularly malaria-transmitting mosquitoes, is a serious threat to public health in many nations. Agricultural productivity is jeopardized because of widespread resistance in crop and livestock pests. Serious resistance problems are also evident in pests of the urban environment, most notably cockroaches.

Better understanding of pesticide resistance is needed to devise techniques for managing resistance (i.e., slowing, preventing, or reversing development of resistance in pests and promoting it in beneficial natural enemies). At the same time, resistance is a dramatic example of evolution. Knowledge of resistance can thus provide fundamental insights into evolution, genetics, physiology, and ecology.

Resistance management can help to reduce the harmful effects of pesticides by decreasing rates of pesticide use and prolonging the efficacy of environmentally safe pesticides. In response to resistance problems, the concentration or frequency of pesticide applications is often increased. Effective resistance management would reduce this type of increased pesticide use. Improved monitoring of resistance would also decrease the number of ineffective pesticide applications that are made when a resistance problem exists but has not been diagnosed. Resistance often leads to replacement of one pesticide with another that is more expensive and less compatible with alternative controls. For example, some pesticides are considered "selective" because they are especially toxic to certain pests but are less toxic to beneficial biological control agents. When resistance to a selective pesticide occurs, it may be replaced by less selective chemicals that kill beneficials and disrupt biological control, which ultimately increases dependence on pesticides.

We hope that this book will provide a useful reference to the current state of knowledge and will stimulate further advances in the understanding and management of pesticide resistance. The idea that resulted in this book originated with John R. Leeper of E.I. duPont de Nemours & Company. John conceived a list of topics and assembled a team of contributors for the book. Because of change in his assignment at duPont, he lacked the time needed for working on the book and he relinquished editorial responsibility. We felt that the project was worthy and should be carried through to completion. Thus, we jointly accepted the editorial duties.

After we accepted editorship, we developed our own concept of the book. After much discussion and consultation with "How to Write a Review Paper" (Chapter 19 in R. A. Day, *How to Write a Scientific Paper* [Philadelphia: ISI Press, 1979), we sent guidelines to each of the contributors. We urged them to provide comprehensive, "state-of-the-art" reviews in each of 10 major areas in the field of pesticide resistance. We asked them to include authoritative and critical evaluations of the literature, seeking to achieve new perspectives and broad understanding.

Our goal is to reach a broad audience, including students and professionals from such areas as entomology, toxicology, evolutionary biology, genetics, ecology, and public policy. Each chapter is designed to have value as a teaching tool; the entire book is suitable for an advanced graduate course on pesticide resistance. The book includes introductory and advanced information on each topic.

To avoid duplication of other books on resistance, we encouraged contributors to emphasize the latest developments, to stress areas not adequately covered in previous work, and to strive for breadth and synthesis. To promote high quality, each chapter was reviewed by at least one outside expert.

The book begins with a review of the latest technical and statistical advances in resistance monitoring (Chapter 2). Chapter 3 gives an overview of practical and relatively simple methods for identifying resistance mechanisms. Chapter 4 provides a comprehensive and critical review of resistance mechanisms. Genetic investigations have played and will continue to play a major role in understanding resistance (Chapter 5). Models provide an essential tool in evaluating resistance management strategies because field experiments are so difficult (Chapter 6). Chapter 7 reviews the impact of agricultural sprays on resistance in insect vectors of human disease, a serious and controversial issue. Considerable effort is now being devoted to increasing resistance of beneficial insects and mites (Chapter 8). The next two chapters provide specific examples of the design and implementation of resistance management programs. Chapter 9 describes management of resistance in *Heliothis virescens,* one of the most important crop pests in the United States. Although most models and genetic investigations focus on individual species, real-world problems often involve multiple-pest complexes. Chapter 10 describes one of the more successful examples of resistance management in such a difficult and complex agroecosystem. Finally, Chapter 11 outlines a philosophy for resistance management based on a worldwide review of resistance management programs.

Several general themes emerge from the book. The most important is that management of pesticide resistance is a component of integrated pest management (IPM), which seeks to minimize pesticide use through application of alternative tactics such as biological and cultural controls. Another theme is that greater emphasis on field experiments and evaluation of resistance in field populations is needed to advance resistance management from theory to practice. Research on genetics, mechanisms, and models will facilitate better monitoring and management of resistance, but social factors cannot be ignored. Successful implementation of resistance management requires an understanding of the legal, political, and economic context of pest management.

Each new generation of pesticides has been heralded by some as the long-term answer to pest control. By now, however, arthropod pests have provided ample evidence of their ability to overcome virtually any pesticide that can be devised or even imagined. Examples of pest resistance to conventional neurotoxic pesticides abound; resistance has also developed to juvenile hormone mimics, chitin inhibitors, and microbial pesticides.

The last example is especially timely because the next generation of pest control may rely heavily on microbial toxins, particularly through expression of *Bacillus thuringiensis* toxin genes in genetically engineered crop plants and microorganisms. The *B. thuringiensis* toxins are exceptionally useful because they readily kill certain pests yet are virtually harmless to most other organisms, including humans. The greatest threat to the effectiveness of these extraordinarily selective pesticides is development of pest resistance, which has already occurred in some cases. We hope that application of the principles expressed in this book will help to preserve the efficacy of microbial toxins and other pesticides, including those yet to be discovered.

We thank John Leeper for starting this project and our many colleagues, friends and family members who helped us finish it. In particular, we thank Greg Payne, Science Editor at Chapman and Hall; the chapter contributors; and the reviewers for their patience and diligence in the project. Special thanks go to our spouses, Anne Frodsham and Rowena Krakauer, for their invaluable editorial assistance, intellectual contributions, and moral support.

2

Resistance Detection and Documentation: The Relative Roles of Pesticidal and Biochemical Assays

Richard H. ffrench-Constant and *Richard T. Roush*

I. Introduction

The last decade has witnessed significant changes in both the philosophy and methods used for the monitoring of insecticide and acaricide resistance. The traditional emphasis has been on the development of precise but artificial techniques that measure change only in the physiological resistance of a strain under laboratory conditions (Busvine 1957). These techniques commonly use topical application of technical-grade insecticide in a suitable solvent and the calculation of median lethal dose estimates (e.g., LD_{50} or LC_{50}) on a per-body-weight basis.

In an effort to coordinate and unify data collection and interpretation, many of these techniques were adopted as standards by such organizations as the United Nations Food and Agriculture Organization (FAO), the World Health Organization (WHO), and the Entomological Society of America (ESA). Although advantageous in many ways, standardization should not be allowed to inhibit improvement of existing techniques or the introduction of new ones, especially where standard techniques have been adopted without the benefit of supporting research. For example, the standard technique used to test for resistance in *Heliothis* spp. (Anonymous 1970) failed to define unambiguously the appropriate weight class for bioassay of *H. virescens* (Mullins and Pieters 1982), apparently because this had not been studied.

Many of the standard approaches adopted during the 1970s were abandoned in the 1980s in favor of faster and more efficient techniques that more closely correlate with field control. Among the first techniques to be challenged was the standard slide dip method for bioassay of spider mites (especially *Tetranychus* spp.). Although widely used in the 1970s (e.g., Roush and Hoy 1978), the slide dip was found to be relatively ineffective for detecting resistance to dicofol, propargite, and cyhexatin because it did not correlate well with field control

We thank W.G. Brogdon, T.J. Dennehy, E.E. Grafton-Cardwell, and B.E. Tabashnik for supplying preprints of unpublished papers and reviews of earlier drafts.

(Dennehy et al. 1983, 1987a; Dennehy and Granett 1984a; Keena and Granett 1985; Welty et al. 1987) and was poor for slow-acting compounds (See Section II.A). Thus, few recent studies on resistance in spider mites have used the slide dip technique (e.g., Chapter 10). Another example, which is one of the most geographically widespread and labor-intensive insecticide resistance monitoring programs yet performed, involved the testing of more than 40,000 *Heliothis virescens* from across the southern United States for pyrethroid resistance. This survey virtually ignored the topical assay (Riley 1989) that had been the official standard for *Heliothis* (Anonymous 1970) and relied instead on the use of adult exposure in treated glass vials (Section II.A). Even where topical assays are still in use for *Heliothis* spp., the emphasis is on the use of diagnostic doses rather than on the calculation of LD_{50}s and resistance ratios (Forrester and Cahill 1987).

Much of the new diversity in resistance monitoring methods has been the result of improved knowledge of resistance mechanisms and expression, as well as a change in the goals and philosophy of monitoring. This new philosophy places an increasing emphasis on relating laboratory data to field control (Ball 1981) in an effort to apply theory to practical problems. Standardization will always be important, but as examples cited in this chapter demonstrate, no one technique is likely to be adequate for all pesticides used against any given species.

Even the traditional definition of resistance, usually paraphrased as the "development of a strain capable of surviving a dose lethal to a majority of individuals in a normal population," has been criticized because of its lack of relevance to field control and the ambiguity of defining a "normal" population. Sawicki (1987) noted these problems and proposed a considerably improved definition: "Resistance (is) a genetic change in response to selection by toxicants that may impair control in the field." In addition, some researchers are calling for reexamination of possible genetic changes in behavior that may effect resistance, requiring the use of techniques not limited to strictly measuring physiological responses (Sparks et al. 1989).

Recent reviews have examined both improvements in insecticide resistance detection methods (Brown and Brogdon 1987) and our current understanding of the biochemistry (Chapter 4) and ecological genetics of resistance (Roush and McKenzie 1987, Chapter 5). However, the advantages and disadvantages of new biochemical (in vitro) monitoring techniques in relation to more traditional insecticide and acaricide (in vivo) bioassays in studies of resistance remain largely unassessed. Thus, the purpose of this chapter is to examine the present and potential contributions of new biochemical techniques and to place these into context with more traditional insecticide and acaricide bioassays. Beforehand, however, we shall discuss how in vivo bioassays may be improved, especially in those situations where biochemical techniques may not easily be developed or applied.

The applicability of the techniques available for resistance monitoring relies primarily on the aims of any given monitoring program and knowledge of the target insect. Thus, the approach to monitoring will differ considerably, depending on whether it is desirable (a) merely to document the efficacy of an insecticide, (b) to determine the frequency of resistant genotypes with accuracy, or (c) to have a high chance of detecting a given frequency of resistant genotypes.

All too often resistance monitoring has been used only to document resistance or, perhaps more generally, to confirm whether or not control failure was caused by resistance. Brent (1986) notes that this is only one of seven possible aims of resistance monitoring. (The other six are numbered and follow here.) If resistance management is to reach its full potential, greater emphasis must be placed on other objectives, including (1) to measure and identify resistant genotypes accurately. The latter is fundamental whether the aim is (2) to provide early warning of an impending resistance problem, (3) to determine changes in the distribution or severity of resistance, (4) to make recommendations for pesticides least affected by resistance, (5) to measure the biological characteristics of genotypes under field conditions, or (6) to test the effectiveness of resistance management tactics.

One of the most promising aspects of resistance management from the perspective of reducing pesticide use is the development of rapid methods that allow users to choose the most effective, safe, or inexpensive pesticides from among those affected by resistance. For some pests resistance to the majority of available pesticides may be found over a large area, but resistance to any given pesticide is localized or sporadic (e.g., Dennehy and Granett 1984b, Dennehy et al. 1987a, Forrester and Cahill 1987, Grafton-Cardwell et al. 1987, Tabashnik et al. 1987, Welty et al. 1987, Dennehy et al. 1988, ffrench-Constant and Devonshire 1988, Chapters 9 and 10). Developing a resistance profile immediately before choosing the pesticide to be applied reduces the frequency of ineffective, wasted applications. Further, resistance detection efforts should be initiated when resistance is still rare, thus providing an early warning or some indication of the genetic potential for resistance to occur.

The currently available test techniques divide broadly into two categories: in vivo assays on intact individuals, which usually involve pesticide exposure, and in vitro biochemical techniques that assess enzyme activity or the nature or quantity of DNA coding for specific resistance genes in insect preparations. An unusual in vivo test is the "hot needle assay" for the direct determination of *kdr*-type nerve insensitivity (Chapters 3 and 4) in house flies, *Musca domestica* (Bloomquist and Miller 1985), which does not involve pesticide exposure, but only behavioral response to physical stress. Such techniques have not been widely used and deserve further experimental attention; readers are encouraged to see the original paper for further details. The former two categories of test techniques are discussed first in light of the types and choice of assay methods available and second in relation to their present and potential contributions to monitoring programs. Although this review is designed to compare the two classes of techniques in order to aid future choices, the two approaches should be seen as complementary and in no way exclusive of each other.

Whereas in vivo bioassay techniques will always be instrumental in the initial detection or documentation of resistance (objectives a and c cited earlier), biochemical techniques should offer the ability to identify and monitor resistant genotypes accurately at low frequencies, an essential attribute for the appraisal of resistance management strategies and improving pesticide recommendations. However, the final choice of monitoring technique may be largely determined by the level of technical support available to any monitoring program.

II. Types and Choice of Detection Methods

A. Pesticide Bioassay

1. Dose–Response and Diagnostic Dose

Traditionally, monitoring has involved comparisons of LD_{50}s, LD_{90}s, or the slopes of dose–response curves between field and laboratory populations (e.g., Staetz 1985). Until recently, little attention has been given to the efficiency of dose or concentration–mortality data in resistance monitoring. However, particularly in view of statistical considerations in sampling (Section IV), the proper choice of test techniques and objectives in data collection can be the most important decision in a monitoring program. Although adequate for documenting resistance at high frequencies, both LD and slope estimates are insensitive to small changes in resistance frequency, particularly when resistance is first appearing in the population (Roush and Miller 1986). Diagnostic tests (such as a fixed dose and exposure period) are more efficient for detecting low frequencies of resistance because all individuals are tested at an appropriate dose and none are wasted on lower doses, where percentage mortality is not informative. (An excellent example of this is given in Figure 8.2, where the presence of resistance is not detectable at concentrations less than the LC_{95}.)

As will be discussed in Section IV, test techniques that accurately discriminate between resistant and susceptible genotypes (i.e., kill >99.9% of susceptibles but <0.1% of resistant individuals—for example, a dose of 2 in Fig. 2.1A) are significantly more efficient for detecting resistance than techniques that misclassify more frequently. A diagnostic dose that kills a large fraction of individuals that can survive field exposure will lower the efficiency of monitoring because some resistant individuals will be killed rather than detected, as illustrated by the 30% mortality of resistant individuals in Figure 2.1B. On the other hand, allowing too many susceptibles to survive a diagnostic dose (such as a dose of 1.0 in Fig. 2.1A) increases the sample size needed to distinguish chance susceptible survivors statistically (false positives) from resistance (Section IV).

In most practical situations a perfectly discriminating test is unknown, either because resistance has not yet been documented or because the expression of resistance has not yet been examined in sufficient detail to facilitate the choice

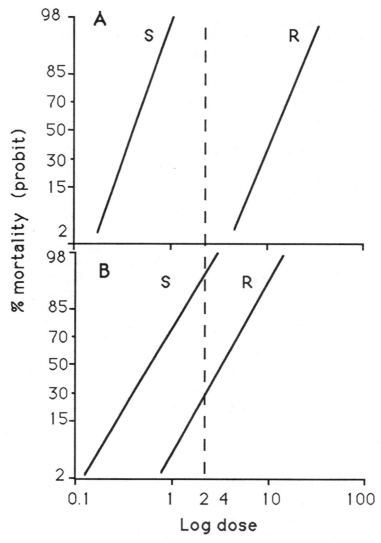

Figure 2.1. Hypothetical dose-mortality regressions illustrating discrimination between resistant and susceptible strains. In A, a dose of 2 nearly perfectly discriminates between the resistant (RR) and susceptible phenotypes (SS and RS if resistance is monogenic and recessive). In B, no dose perfectly discriminates, but a dose of 2 is best for resistance monitoring since it is the LD_{99} for susceptibles and kills no more than 30% of the resistant individuals. Arbitrarily doubling the dose to 4 kills more than 50% of the resistant individuals and is a poor choice for resistance monitoring.

of an appropriate method, particularly if resistance "intensity" (see later) is low. In such circumstances, the best monitoring method for resistance detection will be the use of a dose that kills about 99% of susceptible individuals (a susceptible LD_{99}, or similarly high dose, as discussed in Section IV), which provides a compromise between allowing few susceptible survivors yet does not risk killing as many resistant individuals as a higher dose might (Roush and Miller 1986), as shown in Figure 2.1B. One should not simply estimate an LD_{99} for a susceptible strain and then use a diagnostic dose two- or three-fold greater than this (See Fig. 2.1), as used to be common practice. As illustrated by slide dip assays of dicofol-resistant spider mites, a dose two- to three-fold greater than the susceptible LD_{99} would have killed >98% of the resistant strain (Dennehy et al. 1983). Robertson et al. (1984) have described approaches that maximize the efficiency and statistical accuracy of LD estimates; these may help to estimate precisely the diagnostic dose. In contrast to the LD_{50}, where the best estimates are achieved by an even distribution of doses, the best estimates of an LD_{99} probably require one or two responses <10% and most responses between 75% and 100%. However, as discussed later, selection of the appropriate bioassay technique is at least as critical as the dose used.

One problem with any monitoring program that relies on LD determinations, even for diagnostic doses, is that large unexplained variations in the LD values of unexposed field populations can occur (Sawicki 1987). The responses of laboratory standard strains may also vary from generation to generation (Wolfenberger et al. 1982), and susceptible strains held for long periods in the laboratory may bear little resemblance to susceptible strains currently found in the field. Caution should therefore be taken in the adoption of laboratory susceptible strains as standards, particularly since fully susceptible strains may no longer exist in the field for most major pests (Brown and Brogdon 1987, Sawicki 1987). It is probably more useful to determine the appropriate LD_{99} on the basis of field strains studied before wide commercial introduction of the specific pesticide. In the case of monitoring for pyrethroid resistance in Australian *Helicoverpa* (=*Heliothis*) *armigera*, the diagnostic dose (which killed 99% of susceptible larvae) was calibrated against the mean of 34 susceptible strains from both laboratory and field (Forrester and Cahill 1987). This approach is a good one, particularly since it provides confidence in the LD_{99} needed for the statistical test mentioned earlier (Roush and Miller 1986). Other researchers suggest using the modal response of "susceptible" field strains to define susceptibility, emphasizing the dynamic nature of a "normal response" (Sawicki 1987).

Although a diagnostic dose is the most efficient way to detect resistance, analysis of variance (ANOVA) is an alternative method for comparing the susceptibilities of populations and documenting the presence of resistance. This requires the testing of a number of doses but may be especially useful if the toxicological responses of susceptible or resistant individuals are poorly known (Tabashnik et al. 1987).

The diagnostic dose approach has been criticized in the past because it does not give an estimate of the "magnitude" of resistance, most commonly represented as a resistance ratio, found by dividing an LD value for the resistant strain by that for the susceptible strain. However, the magnitude of a resistance problem is really a function of two factors, the *frequencies* of resistant genotypes and the *intensity* (strength) of resistance associated with each genotype (terminology of Dennehy 1987). Ideally, the resistance ratio should only be used to compare strains of known genotype composition and is confusing if used in reference to field samples, which are usually a heterogeneous mix of genotypes.

Where information on the magnitude of resistance is desirable or relevant, a solution is to use more than one diagnostic dose (e.g., Scott et al. 1989 and references cited therein). This might be particularly appropriate where resistance appears to be polygenic (Chapters 5 and 6) or does not easily conform to a monogenic model. Although a diagnostic dose approach makes no particular assumption about the inheritance of resistance (Roush and Miller 1986), no test technique will perfectly discriminate between resistant and susceptible genotypes if the genotypes have a wide range of overlapping responses. An accurate discriminating dose is only possible if the susceptibilities of different genotypes do not overlap. In some cases, all three genotypes may be distinguished (Fig. 2.2A). With such doses available, one may need only one dose to estimate genotype frequency at the most important locus; continuous resistance may require a series of such "bench marks" (Fig. 2.2B).

In this context it should be noted that the actual *intensity* of resistance cannot in most cases be accurately measured by laboratory assays, as it is often a function of test technique. Thus, Dennehy et al. (1983) found only a sixfold resistance to dicofol by slide dip assay but more than a 500-fold resistance in residual leaf dip assay. Similar examples will be mentioned elsewhere in this chapter. Far more important than calculating the fold level of resistance is to relate mortality in laboratory assays to field control (Ball 1981, Sawicki 1987), as discussed later in this section.

The ideal test technique would thus be fast and efficient and would correlate closely with field control, even if the stage tested or mode of exposure differs from that under field conditions. Examples that approach this ideal include monitoring of adult *H. virescens* in glass vials (Roush and Luttrell 1987, 1989, Chapter 9) and spider mites in plastic petri dishes (Grafton-Cardwell et al. 1989).

2. Choice of Exposure Method

Many resistance management tactics attempt to minimize the discrimination between genotypes, often through the use of pesticides and formulations that kill resistant individuals more effectively (Roush 1989, Chapter 5). The objective in resistance monitoring is just the opposite: to choose test methods and pesticides that exaggerate the differences between susceptible and resistant individuals such

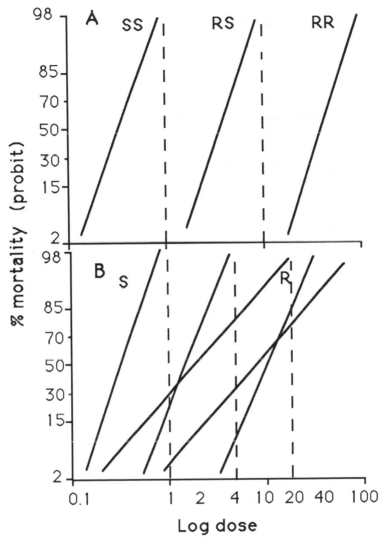

Figure 2.2. Hypothetical dose-mortality regressions illustrating: (A) discrimination between all three genotypes when resistance is monogenic, and (B) absence of complete discrimination where resistance is polygenic and continuous. For (A), the best doses for resistance monitoring are 1 and 10. For (B) the best doses are probably 1, 4 and 20, since the LD_{99} for susceptibles detects resistance and 4 and 20 are approximately logarithmically spaced doses covering a wide range of mortalities among resistant strains.

that the frequency of misclassification is greatly reduced. More specifically, desirable test methods increase the magnitude of the difference between resistant and susceptible genotypes (i.e., resistance ratio) and/or increase the slopes of the dose–response lines, as illustrated in Figure 2.3.

A great many standard insecticide bioassays exist that are capable of detecting resistance (Busvine 1957; WHO 1970, 1976, 1980). However, for purposes of the present discussion, most insecticide and acaricide bioassays can be classified via the manner in which the pesticide is applied: (a) immersion, whereby appropriate life stages are dipped or washed in solutions of known concentration or, alternatively, larvae reared in diet-containing insecticide; (b) residue or surface contact, whereby individuals are exposed to a dry residue of pesticide on a natural (e.g., leaf) or artificial (e.g., glass) substrate; and (c) topical, where a known dose of insecticide is applied directly to individual insects commonly via a microsyringe. Because many currently used pesticides act on contact with terrestrial lifestages, resistance monitoring techniques have been primarily based on the latter two methods of exposure.

A further category, which may become more prominent with increasing emphasis on the development of biological insecticides, is ingestion or feeding techniques (e.g., Hughes et al. 1986). However, since significant detection of resistance to biological insecticides is relatively recent, few bioassay techniques have been tried on both resistant and susceptible populations (e.g., Ignoffo et al. 1985, McGaughey 1985). Thus, we cannot yet make any salient suggestions on choice of exposure method for these types of compounds. Testing for resistance to fumigants also poses special problems, as discussed by Winks (1986a,b) and Winks and Waterford (1986).

Choice of the appropriate exposure method and insecticide can be problematic, especially if resistance is poorly defined, but can often improve discrimination between genotypes. Different exposure routes should be tested. For example, a strain of German cockroaches (*Blattella germanica*) resistant to the pyrethroids permethrin and fenvalerate applied by both topical and residual techniques showed significant resistance to cypermethrin and deltamethrin *only* by topical application (Scott et al. 1986). Thus, permethrin or fenvalerate, the compounds to which resistance is broadly expressed, may be preferable for resistance monitoring. In contrast, if the lack of resistance to cypermethrin and deltamethrin residues is confirmed under field conditions, these compounds would be preferable for resistance management. On the other hand, a leaf residue assay discriminated between the resistant and susceptible strains of spider mites, even though a standard dip assay had failed to show significant resistance (Dennehy et al. 1983).

Although not intuitively obvious, residual tests may produce as little variability as topical exposure, as shown by studies of pyrethroid-resistant house flies (Hinkle et al. 1985) and diamondback moths, *Plutella xylostella* (Tabashnik and Cushing 1987). They may also allow the handling of much larger sample sizes and thereby

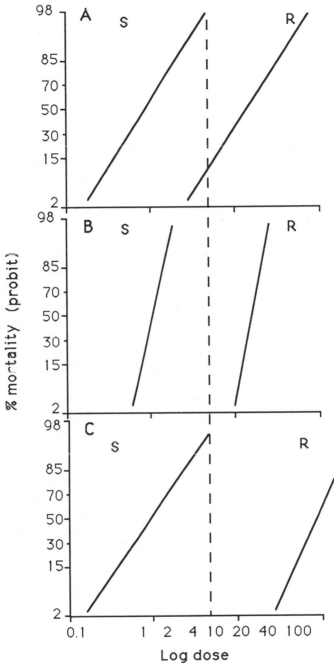

Figure 2.3. Hypothetical illustration of how increases in slopes (B) or resistance ratios (C) can improve discrimination between genotypes in traditional bioassays for resistance monitoring.

compensate for any loss of accuracy on a per-insect basis. Tests that more closely simulate exposure as it occurs in the field, such as leaf residue assays for mites (Hoyt and Harries 1961, Walker et al. 1973, Dennehy et al. 1983) and aphids (Sawicki and Rice 1978), not only can improve the accuracy of resistance detection but also can help to establish the relationship between laboratory bioassays and field control failures, as discussed later. Residual tests using plastic substrates have been developed for spider mites. These correlate closely with residual assays on foliage but are quicker, giving results in 24 hours, and do not require untreated leaves (Dennehy et al. 1987b, Welty et al. 1988, Grafton-Cardwell et al. 1989). The results of residual assays have also been used to make testable predictions about rates of selection in the field (see Section V.C).

Although some researchers have expressed concern that residual tests leave an untreated surface from which the insect can escape exposure (Brown and Brogdon 1987), it is possible to design test containers that can be treated equally on all surfaces. An untreated surface would, however, allow the opportunity to test for some kinds of behavioral resistance (Sparks et al. 1989). One disadvantage of residual assays is that the exact dose accumulated by the insect is not known (Brown and Brogdon 1987), but topical assays may also give little idea of the amount of pesticide penetrating the insect; compound may be lost via a number of routes or not absorbed (e.g., Schouest et al. 1983). However, topical bioassays can be valuable for resistance monitoring. At least for some organophosphorus insecticides, topical assays appear to be superior to residual assays for the German cockroach (Milio et al. 1987).

As a general rule, exposure methods that give steeper dose–response curves (with lower variation and standard deviations) tend to give better discrimination between genotypes because fewer susceptible individuals will survive relatively high doses. Thus, if resistant genotypes are not yet available, a general rule may be to choose exposure methods that give steeper dose—mortality lines (Fig. 2.3). In this regard, immersion techniques may deserve further attention. Although often dismissed as unrealistic or imprecise (results cannot be expressed in terms of toxicant per gram of body weight), immersion tests ensure uniform contact and often give steep concentration mortality curves, with slopes of approximately 4 to 10 on a probit scale (e.g., Hemingway et al. 1984, Halliday and Georghiou 1985), and lower standard deviations than residual or topical tests (e.g., Brindley et al. 1982). Immersion of field-collected lepidopterous larvae in dilute solutions of formulated insecticide has proved to be quick and practical for monitoring resistance in areas where laboratory facilities are unavailable, and it may have potential for use on other insects, such as aphids (Watkinson et al. 1984). A similar approach appears promising for Colorado potato beetle (*Leptinotarsa decemlineata*) larvae (R.T. Roush and N. Carruthers unpublished). However, the inadequacies of slide dip tests for mites, as discussed earlier, suggest that not even immersion tests are a panacea.

3. Life Stages Tested

In some cases the life stage tested may be critical, as resistance can be expressed mainly in certain stages, such as organophosphorus resistance in the larvae of the Australian sheep blowfly, *Lucilia cuprina* (Arnold and Whitten 1975). Alternatively, resistance may vary even within a given life stage; for example, malathion resistance declines with age in adults of the mosquito *Anopheles stephensi* (Rowland and Hemingway 1987) In other examples, larval and adult mortality correlate well (Roush and Luttrell 1987, 1989), which provides great advantages in resistance monitoring (Chapter 9).

4. Response Criteria

Choice of response criteria may also be important, particularly where knockdown resistance to pyrethroids is involved. In residual tests with permethrin on German cockroaches, higher intensities of resistance were displayed when knockdown was used instead of death as a response criterion (Scott et al. 1986). Assuming dose–response slopes are unchanged, knockdown should give better discrimination between resistant and susceptible insects. In contrast, scoring mortality either as dead (no movement after prodding) or as dead plus moribund (unable to walk) made little difference for cyhexatin resistance in European red mite, *Panonychus ulmi* (Welty et al. 1988). Scoring moribund individuals as dead rather than alive in residual assays reduced the resistance ratio at the LC_{50} from >29-fold to 5-fold, but it also increased the slopes of the concentration–response lines, such that there was no change in the discrimination between genotypes. Because "dead" is a more precise category than "moribund," death was recommended as the criterion for mortality (Welty et al. 1988). Scoring moribund *Heliothis* larvae as "dead" may give more reliable and biologically meaningful results than only scoring those actually dead at a given time. However, even by these methods, scoring *Heliothis* larvae reliably for pyrethroid resistance can still take at least three days, in contrast to the scoring of adults in 24 hours (Roush and Luttrell 1989).

5. Choice of Pesticide for Monitoring

The best pesticide for resistance monitoring may not actually be the one used in the field. For example, carbaryl-resistant predatory mites, *Metaseiulus* (*=Galendromus*) *occidentalis,* are cross-resistant to propoxur. Whereas carbaryl gives shallow dose–response regressions and there is overlap between resistant and susceptible strains, propoxur discriminates completely between susceptible (SS) and resistant (RS and RR) individuals (Chapter 5, Fig. 5.5). Dieldrin confers much higher resistance and discriminates better between genotypes than many other compounds thought to act at the same target site, such as lindane (Oppen-oorth 1985) and endosulfan. Thus, even though banned from field use, dieldrin may still be useful in monitoring for cyclodiene resistance. The Australian *Heliothis* monitoring program uses both endosulfan and dieldrin to monitor endosulfan resistance (Forrester and Cahill 1987). Similarly, cypermethrin appears to be more effective for monitoring pyrethroid resistance in the horn fly, *Haematobia irritans*, than does permethrin (Roush et al. 1986).

A survey of the expression of resistance to pyrethroids in a *Culex* mosquito species showed that some gave much higher resistance ratios, especially in heterozygotes, with relatively steep slopes (Halliday and Georghiou 1985). Although these compounds differed in relative dominance, discrimination between RS and SS is more important than dominance per se. If nerve insensitivity (*kdr* type) is the suspected mechanism of pyrethroid resistance, DDT may yield better results than pyrethroids because it often provides higher resistance levels (e.g., Sawicki 1978, Scott et al. 1986). Interestingly, because there are several mechanisms that can confer resistance to DDT (Oppenoorth 1985, Chapter 3), this insecticide can be used to detect resistance to several compounds, even though establishing the precise mechanism would require further investigation.

Pesticide formulation can also be important; thus, an emulsifiable concentrate of cyhexatin gives steeper slopes than a wettable powder formulation in monitoring for resistance in spider mites (Edge and James 1986, Hoy et al. 1988). Finally, while developing assay methods, safety should also be a factor. Thus, when monitoring for resistance, closely related compounds may be nearly as effective and considerably safer to the user (e.g., methomyl or oxamyl rather than aldicarb).

6. Standardization

Many insecticide exposure methods suffer from problems of poor standardization if used by many operators at a number of locations. Although some problems can be overcome by rigorously standardizing techniques (Hassan et al. 1985) and by centralized production, such as for the distribution of insecticide-coated glass vials for the assay of adult *Heliothis virescens* in the cotton-growing regions of the United States (Riley 1989, Chapter 9), insecticides may degrade on glass or on treated filter paper. Such methods should therefore be thoroughly tested before extensive use. Controlling temperature during pesticide exposure may also be a potential problem, but techniques are available to standardize temperature, even in the field (Brindley et al. 1982). However, biochemical methods outlined in the following section offer the potential advantage of more rapid and reproducible assays. These could be used more uniformly at a number of laboratories or field stations, provided that they are developed as easily distributed kits with reagents of reasonable shelf life.

The investigations summarized in this section emphasize the value of testing a number of different lifestages and creative exposure techniques to find the resistance test method that discriminates best between resistant and susceptible genotypes. The best method will often vary with the pesticide of interest. For example, the residual assay for dicofol resistance in spider mites (Dennehy et al. 1983) had to be modified for more repellent acaricides (Dennehy 1987). Nonetheless, the residual assay technique led to the development of even simpler and more rapid assays (Dennehy et al. 1987b). This and other examples suggest that investigation of a number of different exposure routes and pesticides is well worthwhile at the outset of any monitoring effort. Unfortunately, most detection methods cannot be refined to the point of providing a discriminating dose until resistant populations are identified. Novel approaches for doing this are discussed in Chapter 5.

B. Biochemical Assay

1. Type of Biochemical Assay

Currently available biochemical assays (for review see Brown and Brogdon 1987) include (a) detection of enzyme activity in unprocessed insect homogenates using model substrates, (b) use of enzyme-specific antisera to isolate the activity of resistance-conferring enzymes, and (c) detection of specific DNA sequences.

Good examples of the first category come from the study of elevated carboxylesterases and insensitive acetylcholinesterases. Increased hydrolysis of 1-napthyl acetate has been widely used for the detection of elevated carboxylesterase or total esterase activity, such as in individual *Myzus persicae* (Sawicki et al. 1980) and various mosquitoes (Pasteur and Georghiou 1981, Brogdon and Dickinson 1983), often in convenient microtiter plates (Brogdon et al. 1988a,b) or on filter papers (Pasteur and Georghiou 1989). Such tests may have widespread usefulness because similarly high levels of esterase activity have been reported in more than 15 species of insects, ticks, and mites (Pasteur and Georghiou 1989). However, it should be stressed that the finding of elevated hydrolysis of such a substrate alone is not sufficient evidence of resistance, unless it can be proved that the altered (elevated or catalytically more efficient) enzyme responsible also confers resistance (Chapter 3).

Acetylthiocholine iodide has been widely used as a model substrate for monitoring acetylcholinesterase activity in the presence and absence of insecticides, in a variety of individual insects such as house flies (Devonshire and Moores 1984a). Recently, the advent of kinetic microplate readers has led to refinement of this technique, allowing the rapid and accurate identification of genotypes in large numbers of different house fly acetylcholinesterase variants, and even the corresponding combinations of different resistance alleles in their progeny (Moores et al. 1988a). This technique has also been adapted for mosquitoes (ffrench-Constant and Bonning 1989), aphids, and whiteflies (Moores et al. 1988b).

As an example of the second category, an enzyme-specific polyclonal antiserum has been used to detect elevated quantities of the carboxylesterase named esterase-4 (E4), the sole enzyme responsible for resistance in *M. persicae* (Devonshire et al. 1986). Antisera raised against elevated esterase in *Culex* mosquitoes did not cross-react with the elevated esterase conferring resistance in *M. persicae* and showed a limited cross-reaction with other mosquito species (Mouches et al. 1987). Thus, it will probably be necessary to develop such assays on a species-by-species basis. Potential exists for the use of antisera to detect elevated quantities of certain cytochromes P450 following successful purification and cloning of an isozyme from a resistant strain of house flies (Feyereisen et al. 1989). If mono-clonal antibodies were made for such detection methods, they would allow the production of large amounts of pure antisera for use in a number of laboratories or field stations.

Examples of the third category, detection of specific DNA sequences to identify resistance genotypes, are currently limited to the probing of individual aphids for esterase-4 in a dot–blot format, which quantify the extent of E4 gene amplification (Field et al. 1989b). The use of this technique in combination with the E4 immunoassay for resistance monitoring will be discussed further in Section V.C. However, enormous scope for DNA detection mechanisms exists following the cloning of resistance genes in other insects (Section VI).

The use of biochemical monitoring techniques in the field may be limited by their susceptibility to variations in temperature. However, this problem can be overcome by adjusting incubation times at different temperatures (Beach et al. 1989) or through field incubators (Brindley et al. 1982).

2. Choice of Biochemical Assay and Species Confirmation

The choice of assay depends not only on detailed knowledge of biochemistry but also on more practical factors such as resource availability. Thus, less sophisticated assays, including total esterase assays measuring the compound activity of all esterases in an aphid such as *M. persicae* (Sawicki et al. 1980), can be used on a wider basis and by less skilled personnel but may not provide information as accurate or as detailed as the enzyme-specific immunoassay for E4 activity alone, as described earlier.

A further factor often overlooked in the use of all resistance tests is the confirmation that a sample is composed entirely of the correct species. Many pest populations may be composed of sibling species complexes that are difficult to distinguish. In this context, electrophoresis can be extremely useful. For example, three species of *Tetranychus* spider mites are found together on cotton in California and are difficult to distinguish. However, identification is desirable, as each species may respond differently to pesticides. Of these, *Tetranychus pacificus,* in which resistance is most serious, can be unambiguously identified using cellulose acetate electrophoresis (Grafton-Cardwell et al. 1988). Similarly, of two *Helicoverpa* (=*Heliothis*) species found on cotton in Australia, only *H. armigera* shows significant resistance. Although traditional criteria for distinguishing immature stages are unreliable, both eggs and larvae can be distinguished electrophoretically (Daly and Gregg 1985). Thus, the absence of significant numbers of potentially resistant species may make resistance testing unnecessary.

The immunoassay developed to quantify E4 in the aphid *M. persicae* will classify misidentified aphids as susceptibles. In view of the recent description of *M. antirrynhii*, an aphid of extremely similar morphology to *M. persicae* and uncertain host–plant distribution (Blackman and Paterson 1986), it may be necessary to check species composition of samples regularly. This can be done by using a fraction of homogenate from each individual, analyzed by immunoassay, for species confirmation by electrophoresis (ffrench-Constant et al. 1988a).

C. Relationship Between Assay and Field Control

In addition to the development of improved test techniques, one of the major themes of recent research on resistance monitoring has been to relate the results of bioassays to mortality in the field. (The term *field* here incorporates both public health and forest uses.) Laboratory assays are often far too different from field exposure to predict mortality in the field (Ball 1981, Denholm et al. 1984). Even when monitoring assays do match field exposure and predict mortality, prediction of control achieved is still difficult because control failures are a function of both pest density and resistance frequency (Daly and Murray 1988) at the time of treatment. Thus, a high frequency of resistant individuals may not always cause control failure if pest densities are low prior to treatment. Alternatively, even a low frequency of resistance can cause control failure at higher pest densities. Except in those cases where control failures are dramatic, it is often difficult, because of the complex of factors affecting pest densities and crop yields, to prove whether control has been impaired.

In some cases, laboratory assays exaggerate the potential importance of resistance. In *H. virescens*, resistance ratios to pyrethroids of more than 50-fold by topical assay (Martinez-Carrillo and Reynolds 1983) were never associated with poor control. In a more extreme example from the same species, resistance ratios of more than 850-fold to methomyl in topical assays of third-instar larvae apparently only reduced field mortality from 64% to 39% (Roush and Luttrell 1987).

In other cases, standard resistance tests have underestimated the economic significance of resistance. There is often as much as a 10-fold difference in LD values among strains that are considered susceptible (Wolfenbarger et al. 1982, Staetz 1985, Sawicki 1987). Consequently, resistance ratios of less than 10-fold have often been arbitrarily assumed to be irrelevant to control (as reviewed by Denholm et al. 1984). However, as noted earlier, the traditional standard bioassay for spider mite resistance, the slide dip, showed only about a sixfold difference between populations successfully and poorly controlled by dicofol (Dennehy et al. 1983). A sixfold resistance also allows *M. persicae* to survive field rates of demeton-S-methyl (Devonshire et al. 1975). In contrast to the cases cited earlier for *H. virescens*, populations showing resistance ratios to pyrethroids of sixfold or less were associated with control failures (Roush and Luttrell 1989). At least part of the problem in these examples resulted from inadequacies of the standard bioassay techniques; in the cases of dicofol for spider mites and pyrethroids for *Heliothis*, alternative test techniques more clearly demonstrated resistance, as was already apparent by poor field control (Dennehy et al. 1983, Roush and Luttrell 1989).

On the basis of these examples, there is clearly a need to calibrate the results of monitoring techniques with loss of efficacy under field conditions. This does not necessarily require that the mode of exposure, pesticide, or criterion of mortality used in the bioassays must mimic field exposure directly. As already emphasized, exaggerating the differences between resistant and susceptible individuals improves the resolution between genotypes. The amount of activity in at least some biochemical test techniques correlates significantly with resistance ratios in laboratory in vivo bioassays (Devonshire et al. 1986, Pasteur and Georghiou 1980), providing a potential method for roughly estimating both the intensity and frequency of resistance in the field. Such techniques are extremely useful but must still be calibrated against field control to achieve their full potential (ffrench-Constant et al. 1987).

Calibrating the biological relevance of both in vivo and biochemical assays requires establishing a correlation between bioassay results and efficacy. Although the ideal correlation might be one to one (Fig. 2.4), such that bioassay results directly predict the level of control achieved, even a curvilinear relationship can be useful as long as the variance of the best fit relationship is low in the values of toxicological response that predict desired control levels (e.g., 90–100%, see Fig. 2.5). Confidence intervals are less critical when control is much less than this, because control may already be unacceptable and an alternative compound should be sought.

Two methods have been used to establish the relationship between monitoring results and field efficacy: (a) comparing results directly with the field and (b) mimicking field applications under more standardized conditions. The second approach offers significant advantages because far greater numbers of trials can be run and because factors other than pesticides that might reduce or increase population growth, such as predators (Dennehy et al. 1987a, Harrington et al. 1989) or physical conditions, can be controlled experimentally. For example, of the initial six attempts to establish the relationship between resistance tests and control failures with propargite in *Tetranychus* spider mites on cotton, populations were too low in all but three of the trials to produce convincing results (Dennehy et al. 1987, Grafton-Cardwell et al. 1987). Although Grafton-Cardwell et al. (1987) showed that more than 80% mortality at the appropriate diagnostic dose appeared to give acceptable field control, this critical frequency was based on only three fields with a wide gap in test mortalities (between 56% and 83%).

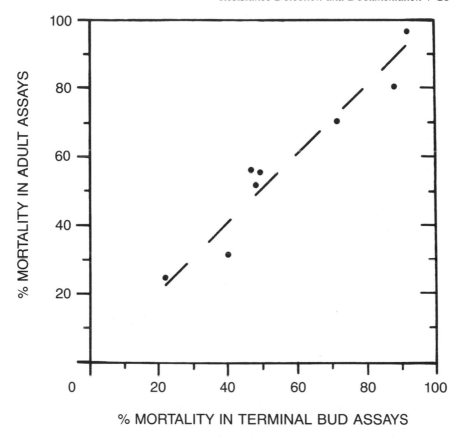

Figure 2.4. Example of a linear correlation between a resistance monitoring technique and apparent field control. Mortality of *Heliothis virescens* moths in glass scintillation vials treated with 5 μg cypermethrin compared to mortality of third-instar larvae placed on cotton terminal foliage treated with recommended field application rates of cypermethrin and fenvalerate. Each data point refers to a single colony collected from a different geographical location (redrawn from Roush and Luttrell 1987).

Although later efforts improved estimates of the correlation between resistance frequencies and control (Grafton-Cardwell et al. 1989), these studies illustrate the difficulty of arranging full-scale field efficacy trials, including untreated plots, of the kind traditionally used to compare pesticides. Even when a large number of populations showing a range of resistance frequencies can be sampled, such factors as crop variety, pest population age structure, and plant size can introduce variability into the results (e.g., Welty et al. 1989). Although these factors may be important to evaluating a technique, in practical terms the most efficient way to evaluate their impact would be to study each of the more important factors separately or in an experimental design allowing an analysis of variance (ffrench-Constant et al. 1987, 1988c).

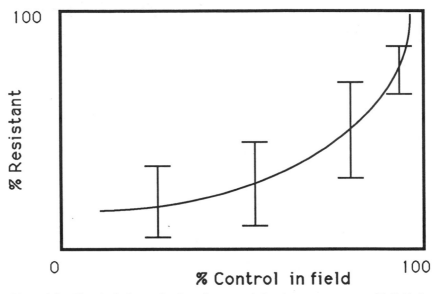

Figure 2.5. Hypothetical example of non-linear correlations between percentage of individuals identified as resistant by monitoring techniques and control observed. Error bars show 95% confidence intervals.

In other studies in which the correlations are made directly to normal field use, it is more often difficult to provide suitable untreated comparisons that quantify efficacy (e.g., Denholm et al. 1984, Farnham et al. 1984). In cases of multiple pesticide resistance (e.g., Grafton-Cardwell et al. 1987), the objective will often be to choose the pesticide that will least select for resistance (ffrench-Constant et al. 1987, Roush 1980), not merely to identify those that show a loss of efficiency.

An alternative to direct field studies in a standard efficacy trial format is to develop application methods for laboratory or field use that duplicate field exposure, or at least study field exposure under more controlled conditions. For example, Luttrell et al. (1987) constructed spray equipment that duplicates actual field applications on "bouquets" of cut apical cotton foliage. This method was then used to compare the pyrethroid resistance of several geographical strains of *H. virescens* under controlled laboratory conditions at one location. Similar techniques have been developed using sprayed panels in cages for house flies and in large-scale, self-contained simulation cages enclosing cotton plants and populations of tobacco (or sweetpotato) whitefly, *Bemisia tabaci* (Denholm et al. 1990). In other cases, different genotypes or strains might be caged or implanted on normally treated surfaces in the field to evaluate mortality where exposure can be controlled. Such techniques have already been used to evaluate genotypic fitnesses under field selection (Roush and McKenzie 1987, Chapter 5). As discussed later, the application of biochemical monitoring to such studies may allow the direct analysis of mixtures of competing genotypes (see Section V.B).

It may always be difficult to predict precisely efficacy by resistance bioassays. Although it is desirable to define a set of frequencies of resistance that would predict control failure at various pest densities and/or trigger the selection of a resistance management tactic or use of an alternative pesticide, it must be emphasized that control failure is a function of both pest density and resistance frequency, as discussed earlier, and perhaps other factors (e.g., Welty et al. 1989). Nonetheless, it is useful to develop some rough estimates that would help guide pest management decisions (Dennehy and Granett 1984b).

Dennehy (1987) has defined the frequency at which resistance becomes an economic problem as the "critical frequency," which seems roughly analogous to the concept of an economic injury level, the pest population density at which damage occurs (although Dennehy and Granett [1984b] suggested that it parallels the economic threshold concept). Dennehy (1987) also proposed the term *action threshold* to specify the resistance frequencies that warrant implementing resistance management procedures. The action threshold has been described as analogous to the economic threshold concept in pest management (Sawicki 1987) but was in fact used very differently. Although the economic threshold concept has evolved, it has always been assumed to be a mean value lower than the economic injury level, so that action can be taken to prevent the population from reaching the economic injury level (Onstad 1987). The action threshold envisioned by Dennehy is higher than the critical frequency in at least one example (Dennehy and Granett 1984b) and incorporates measures of the reliability of the mean resistance frequency (Dennedy 1987) to avoid classifying populations as resistant when the suspect pesticide is still efficacious (Dennehy and Granett 1984b).

Thus, the "resistance action threshold" as currently used incorporates two concepts: response to increased resistance frequency and a definition of resistance that emphasizes control failure or near failure. Unfortunately, these two concepts are often incompatible. Resistance management tactics, other than substitution of an alternative control, are often most effective when implemented at very low resistance frequencies, often less than 1% (Chapters 5 and 6), far less than would repeatedly cause control failures.

We suggest it may be more meaningful to refer to the frequency at which resistant phenotypes reliably cause poor control as a "pesticide efficacy threshold" (PET). Further, the resistance frequency at which resistance management tactics should be implemented could be called the resistance "strategy implementation frequency" (SIF). Because control failures depend on density, age structure, and resistance frequency, the pesticide efficacy threshold concept will be most useful only when it is defined in a sliding scale of pest density and the frequency of appropriate life stages.

The relationships between the economic threshold (ET) and economic injury level (EIL) concepts from pest management and these definitions of resistance frequencies are illustrated in Figure 2.6. The ET and EIL are independent of resistance frequency. Similarly, the resistance frequency at which management

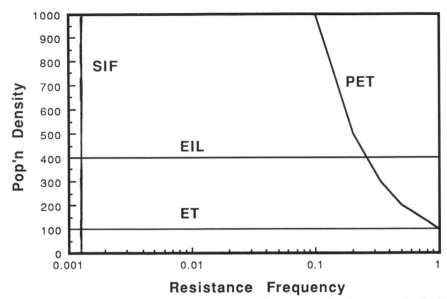

Figure 2.6. Theoretical relationship between the economic threshold (ET) and economic injury level (EIL) concepts of pest management and the strategy implementation frequency (SIF), and pesticide efficacy threshold (PET) concepts proposed for resistance management. See text for discussion.

strategies, such as pesticide mixtures and alternations (Mani 1985, Comins 1986, Chapters 5 and 6), should be implemented (SIF) is independent of population density because these actions must be taken many generations before control failures are likely to occur. On the other hand, very high resistance frequencies will not critically affect control when the population is at a very low density. Thus, the resistance frequency that regularly results in significant loss of control (PET) can be very high at population densities less than the ET, because no applications would be contemplated. Just as with the ET and EIL, the PET can vary with crop stage and relative frequency of less sensitive or more damaging life stages of the pest.

III. Initial Detection of Resistance

A. Pesticide Bioassay

To implement a successful resistance management program to delay the buildup of resistance, it is desirable to detect resistance at low frequencies. Unfortunately, resistance is all too often diagnosed via control failures. The reactive measures taken to manage resistance once it has arisen, such as insecticide rotation (Forrester and Cahill 1987), can only contain resistance. However, baseline levels of susceptibility have now been established for a wide range of insects via a number of standard insecticide bioassay techniques. Thus, if monitored on a regular basis, the appearance of resistance can be detected before control failures occur, at least in some species (see Section IV), which would allow the adoption of more effective "preventive" resistance management tactics (Chapters 5 and 6).

B. Biochemical Assay

As discussed earlier, the initial detection of resistance is often accomplished by insecticide bioassay, almost always due to lack of knowledge of likely biochemical resistance mechanisms that will arise in particular insects. However, the recurring nature of some resistance mechanisms has sometimes allowed early detection in some species on the basis of prior research in others. While screening several mosquito species in Sri Lanka using a microtiter plate assay, *Anopheles nigerrimus* was shown for the first time to possess insecticide-insensitive acetylcholinesterase (Hemingway et al. 1986). Microtiter assays for nonspecific esterase activity also detected phenotypes with elevated esterases, subsequently shown to be correlated with resistance, six months before fenitrothion-resistant *Anopheles albimanus* were documented by susceptibility tests (Brogdon et al. 1988a).

As the biochemistry of resistance mechanisms becomes better understood in a wider range of insects, it may be possible to adopt similar screening tests for more species. However, the fundamental problem for biochemical tests in this role is that without adequate prediction from prior knowledge of resistance mechanisms, a number of potential mechanisms may have to be screened in the place of bioassaying a single relevant insecticide. Despite the potential for multiple screening of resistance mechanisms in the microtiter plate format, a number of assays such as those for mixed function oxidase activity are not presently easily performed in a microtiter plate and therefore remain a major area for research (see Section VI).

IV. Choice of Sample Size and Strategy

A. Sample Size

Roush and Miller (1986) analyzed the relationship between sample size, frequency of resistant individuals, and probability of detection. They showed that very much larger sample sizes are necessary for resistance detection when relying on a dose that kills 95–99% of susceptible individuals rather than a dose that perfectly discriminates between resistant and susceptible individuals. For example, using a "perfectly" discriminating dose (one that causes >99.9% mortality of susceptibles and <0.1% mortality of resistant individuals) or biochemical test, a sample size of approximately only 300 would be necessary to obtain a 95% probability of detecting resistant individuals at a frequency of 1%. In contrast, even when the diagnostic dose kills few resistant individuals and when an LD_{99}

for the susceptible strain is used, a sample of 1,500 individuals would be necessary for the same certainty of detection. This is because statistical tests must be made to determine if the observed number of survivors is greater than would be expected for a susceptible population. When resistance is common, say greater than 10%, sample sizes of 100 individuals will detect resistance quite readily, as long as there is little overlap in dose–mortality regression lines, even if a susceptible LD_{99} must be used. Thus, samples of about 100 individuals per pesticide will often be adequate for making recommendations on pesticide choice that maximize efficacy in areas of multiple resistance problems. Detection of resistance frequencies of 0.1% or less will require samples in the thousands (Roush and Miller 1986).

As noted in Section II.A, a diagnostic dose that kills a large fraction of resistant individuals or allows too many susceptibles to survive (Fig. 2.1) will further increase the necessary sample size. If half of the resistant individuals are killed by a diagnostic dose, the sample size needed to detect a given frequency of resistance will at least double. The problem can be exacerbated if too many susceptible individuals survive the diagnostic test. Under such circumstances, it is necessary to provide even larger sample sizes for statistical testing to determine whether an unusual number of survivors is due to chance or true resistance. If the diagnostic dose is an LD_{95}, more than five times as many individuals must be sampled as would be needed for a perfectly diagnostic test (Roush and Miller 1986).

In cases where the concentration–mortality lines for the susceptible and resistant genotypes unavoidably overlap considerably (see Section II.A)—for example, where the LD_{99} for the susceptible strain kills greater than 70% of the resistant individuals—it may be practically impossible to detect resistance (with greater than 50% certainty) with any reasonable sample size (less than a few thousand per location) when resistance is present at low frequencies. On the other hand, if resistance is present at higher frequencies, say greater than 10%, resistance can generally be detected in sample sizes of a few hundred. Under these circumstances of high resistance frequency and poor discrimination between genotypes (i.e., where dose–response lines overlap more than in Figure 2.1B), the diagnostic dose should be lowered to allow survival of at least 70% of resistant genotypes, which may be as low as a susceptible LD_{90} (R.T. Roush, unpublished).

Biochemical techniques capable of unambiguously identifying resistant genotypes are, in a sense, the ultimate discriminating "dose" and thereby offer a major advantage in reducing sample size. In general, such techniques also eliminate any need to rear field-collected individuals in the laboratory before testing. Even with such techniques, however, the limiting factor for resistance monitoring may be the collection of sufficiently large numbers of individuals from enough sites to provide the desired predictive capabilities. Unless an effective sampling strategy can be devised, it may be necessary to adopt purely preventive resistance management programs (Roush and Miller 1986).

B. Sampling Strategy

The number and spatial distribution of sampling sites should always receive considerable attention, as resistance frequencies in untreated populations may differ considerably not only between regions, as noted earlier, but also between neighboring fields (ffrench-Constant and Devonshire 1988). Following treatment, high frequencies of resistance may be extremely localized. For example, monitoring of nonspecific esterases and insensitive acetylcholinesterase in Guatemalan *Anopheles albimanus* has shown resistance to be highly localized in "foci" associated with insecticide use (Brogdon et al. 1988b). If possible, therefore, small numbers of insects (around 100) should be collected from a number of locations instead of large numbers (thousands) from few locations. Insects should be assayed without rearing in order to maximize sampling efficiency and avoid the distortion of resistance frequencies that might result from poor reproduction in the laboratory.

The ideal sampling strategy will depend on the goals of the monitoring program, pest dispersal, and generation time. A program designed to detect resistance at low frequency will require large samples (and many sites, as discussed earlier), but it might be acceptable to make those collections in any fashion that efficiently provides large numbers, such as pheromone traps, particularly in regions that suffer relatively intense treatment. On the other hand, monitoring designed to avoid ineffective applications may require relatively small samples arranged locally in a very uniform way (e.g., from each field). Resistance can vary significantly across each generation (e.g., Forrester and Cahill 1987, Chapter 9), so samples should be taken at least every few generations.

Sampling strategies may, however, be largely dictated by available resources. For example, one advantage of the adult bioassay for resistance in *Heliothis virescens* was that it facilitated decentralization and a division of labor in monitoring for resistance. Although the production of treated glass vials was centralized, their distribution allowed local participation in resistance monitoring across the cotton belt of the United States (Riley 1989, Chapter 9). This contrasts with the more centralized assay of larvae of *Heliothis armigera* in Australia (Forrester and Cahill 1987), where specialized equipment for topical application was required.

Further improvements in sampling techniques may help to overcome some of the problems inherent in collecting samples from a number of locations. Pheromone traps may be helpful for this, particularly if lures are species specific. For example, topical assay of male moths directly on the adhesive-coated surfaces of the pheromone traps on which they were collected has been used as a sampling method for the codling moth, *Cydia pomonella* (L.) (Riedl et al. 1985). In an extension of this technique for pink bollworm, *Pectinophora gossypiella* (Haynes et al. 1987), and tufted apple bud moth, *Platynota idaeusalis* (Knight and Hull 1989), insecticide was added directly to the adhesive. Another variation on this technique is to add insecticide to yellow sticky cards used for monitoring densities of greenhouse pests, including *Liriomyza trifolii* (Sanderson et al. 1989) and whiteflies (J.P. Sanderson and R.T. Roush, unpublished). However, standardization of this "attracticide" technique is very important, and control mortality is sometimes high (Haynes et al. 1987, Sanderson et al. 1989). Thus, the glass vial technique used on pheromone-trapped moths, similar to that for *Heliothis* (Roush and Luttrell 1987, 1989, Chapter 9), may be more reproducible (Schouest and Miller 1988). Although it only assesses adult susceptibility, the vial technique overcomes many of the limitations of field collecting and may prove useful for species other than *Heliothis*. The use of *M. persicae* and other aphids collected from a nationwide network of suction traps in the United Kingdom has also been suggested for biochemical resistance monitoring (Devonshire et al. 1986), and suitable methods of preservation of aphids in the solution used for collection are currently being investigated (Tatchell et al. 1988).

V. Assessing Factors Relevant to Resistance Management

A. Monitoring Resistance Frequencies

To make successful predictions about resistance management, certain factors—such as the initial frequency of resistance, the extent of cross-resistance, likely rates of selection with different compounds, and the extent of immigration of susceptible insects or reversion of resistance—need to be investigated. Several of these factors are particularly suitable to examination by high-turnover biochemical assays. These techniques can be used to monitor the progress of resistance management programs or trials once in operation.

After detection of resistance, subsequent monitoring allows evaluation of different strategies. Certain biochemical assays allow both good discrimination between genotypes and a rapid processing of insects. The immunoassay of E4 in individual *Myzus persicae* allows the analysis of several thousand insects per day (Devonshire et al. 1986). Thus, the limiting factor in analysis becomes collection of insects. This technique has allowed a clearer definition of the frequency distributions of E4 activity in field populations, their relationship to laboratory strains with known levels of E4 activity, and how distributions shift to higher activities following insecticide selection (ffrench-Constant and Devonshire 1988).

B. Extent of Cross-resistance and Rates of Selection

Following the identification and isolation of a resistance mechanism (or mechanisms) in the target insect, it may be possible to examine the likely extent of cross-resistance conferred by this mechanism at a biochemical level and to test novel compounds in vitro. Biochemical predictions of the efficacy of novel or alternate pesticides can then be tested in the field with the aid of a suitable monitoring technique.

Detailed studies of a range of insensitive acetylcholinesterase variants in house flies using a number of different organophosphorus inhibitors have shown differing cross-resistance spectra (Devonshire and Moores 1984b). As noted earlier, these variants and the corresponding heterozygotes generated by crosses between them can be distinguished by an assay comparing rates of inhibition using a kinetic microplate reader that records rates of reaction over time (Moores et al. 1988a). Thus, it may be possible to analyze the frequencies of different variants in any given population in the field and to formulate a rational choice of insecticide. Modification of this assay for mosquitoes has allowed the detection of homozygous resistant *Anopheles nigerrimus* at low frequencies in field samples (ffrench-Constant and Bonning 1989), whereas detection of this genotype at low frequencies in other mosquito populations is not possible using assays that rely on measurement of the amount of colored end product at a fixed time interval (Brogdon et al. 1988a,b). In contrast to house flies, acetylcholinesterase variants with differing cross-resistance spectra have not yet been found in mosquitoes (ffrench-Constant and Bonning 1989).

Biochemical studies of esterase E4, which degrades organophosphorus, carbamate, and pyrethroid insecticides in *M. persicae,* have shown that the enzyme recovers more slowly following carbamylation than phosphorylation (Devonshire and Moores 1982). These biochemical investigations and the results of leaf residue assays showing lower levels of resistance to carbamates (Sawicki and Rice 1978) suggested that carbamates would select for resistance less rapidly than organophosphorus or pyrethroid insecticides. This hypothesis was confirmed in the field using cages of mixed aphid genotypes monitored by immunoassay for E4 (ffrench-Constant et al. 1987). Resurgence of resistant aphids has also often been shown to occur with pyrethroid use (ffrench-Constant et al. 1988b), Harrington et al. 1989).

Resulting patterns of selection on different aphid genotypes during decay of the insecticide on the crop have also been examined by immunoassay (ffrench-Constant et al. 1988c) in a fashion similar to that used for the Australian sheep blowfly, where larvae were placed into insecticide-treated wounds at varying times after treatment (McKenzie and Whitten 1982). On potatoes, susceptible *M. periscae* survive better than moderately resistant (R_1) aphids following exposure to a deltamethrin–heptenophos mixture, a result not predicted by bioassays, possibly because of stronger avoidance of residues by susceptible aphids. This stresses the strength of biochemical assays in facilitating the monitoring of such processes in large, replicated trials. Consideration of the selection exerted by decaying residues is important when deciding spray dose and frequency in relation to the level of resistance present, and in the latter example confirmed the preferential use of carbamates as the weakest selecting agents.

C. Immigration of Susceptibles, Back-selection, and Reversion

The frequencies of resistance following pesticide selection in a population may decrease either because of the immigration of susceptible individuals or because of "back-selection" (Comins 1986), resulting from lower fitness assumed to be associated with resistance in the absence of pesticide (Georghiou 1983, Mani 1985, Chapter 5). A third possibility, in which individual insects actually genetically "revert" and lose resistance, appears rare (see later). Thus, following the observation that large fitness disadvantages appear to be the exception rather than the rule (Roush and McKenzie 1987, Chapter 5), immigration may be the most important process in decreasing resistance frequencies.

Immigration of susceptible insects does not appear to have been monitored directly. However, consistent declines in resistance frequencies from the end of one season to the beginning of the next have been noted in the Australian *Helicoverpa* (=*Heliothis*) monitoring program using larval insecticide assays. Following studies at individual sites, it seems probable that this decline is due to immigration of susceptibles via the mixing of different crop and noncrop populations (Forrester and Chaill 1987).

Biochemical monitoring of rates of selection following treatment of *M. persicae* populations in the field has incidentally documented susceptible immigration, occurring through higher frequencies of susceptible immigrant alate (winged) aphids diluting the buildup of resistance in the resident selected apterous (wingless) aphids (ffrench-Constant and Devonshire 1986). Resistance assays have enormous potential for investigating manipulated systems both in the laboratory and in the field to quantify the impact of immigration.

Reversion in the strict traditional genetic sense is caused by documented *loss* of resistance in *individuals,* and appears to be uncommon. This is in contrast to examples where resistance is lost by back-selection from laboratory or field populations, often called reversion by resistance workers (e.g., Flexner et al. 1988). However, extremely high levels of resistance (R_3) in *M. persicae* can spontaneously revert to susceptibility, with a corresponding loss of E4 production (Sawicki et al. 1980), while variability of E4 expression in reverted clones still maintains the potential for reselection of resistance in a "susceptible" (lacking elevated E4) population (ffrench-Constant et al. 1988d). This poses problems for an assay designed to measure enzyme quantity; potentially resistant revertants may be overlooked as susceptibles. Following the finding that overproduction of enzyme is caused by duplication of the structural gene for E4 (Field et al. 1988), it has been shown that loss of enzyme production in revertants is associated with the extent of gene methylation. Reversion is not due to the actual loss of genes, as occurs in drug-resistant cell lines (Field et al. 1989a). Thus, susceptible revertants can be detected by the maintenance of high gene copy number in dot-blots of aphid DNA hybridized to the E4 probe (Field et al. 1989b). The combination of immunoassay and dot-blot, which are both performed on the same homogenates from a 96-well microtiter plate, thus provide the ability to monitor observed and *potential* resistance (reselectable E4 expression) frequencies.

VI. Future Advances in Biochemical Assays

Future developments in rapid biochemical assays will probably center on the microtiter plate format. As discussed earlier, the recent development of kinetic microtiter plate readers has facilitated more accurate and faster assays for insensitive acetylcholinesterase (Moores et al. 1988a). The recent addition of an ultraviolet monitoring facility to such plate readers, and the possibility of wavelength scanning, will greatly increase the potential for rapidly performing assays of enzymes such as mixed-function oxidases and glutathione-S-transferases (Brown and Brogdon 1987).

Microtiter plates not only form a suitable receptacle for homogenization of large numbers of insects (ffrench-Constant and Devonshire 1987), but also can be readily integrated with DNA probes using "dot-blotting" procedures for a variety of functions, including resistance confirmation, species identification (Post and Crampton 1988), and detection of plant or animal diseases if the insect is a vector (ole-MoiYoi 1987).

Following the cloning of acetylcholinesterase in *Drosophila melanogaster* (Hall and Spierer 1986), several laboratories are attempting to clone this gene in arthropods possessing insecticide-insensitive variants, including ticks, mosquitoes, and house flies, using conserved segments of the *Drosophila* gene. Recently, acetylcholinesterase has been cloned from *Anopheles stephensi* via this technique (C. Malcolm, personal communication). With the advent of sequencing using the polymerase chain reaction (Mullis and Faloona 1987), it is now possible, having obtained a resistance gene sequence, to determine rapidly the similar sequences in a range of variants. This allows inferences to be drawn about the locations of mutations in relation to the active sites of enzymes and the number of times a given mutation may have arisen in a population. Two related techniques, "oligomer restriction" and "alle-specific oligonucleotides," which rely on hybridization of specifically designed oligonucleotide probes to amplified DNA, have also been used to determine the genotypes of individuals. For example, in sickle-cell β-globins, genotypes having only single base pair differences can be resolved by both methods (Embury et al. 1987, Saiki et al. 1986).

The cloning and sequencing of other resistance genes will allow the development of further diagnostic techniques. If these genes are inaccessible because of the genetic intractability of the species in which they are found, as with the knockdown resistance (*kdr*) gene in house flies, for example, cloning may be achieved via homologous mutants in *Drosophila melanogaster* (Soderlund et al. 1989, Chapter 4) as proposed by Wilson (1988). Further difficulties in gene cloning may arise when resistance mechanisms are polygenic. However, major genes appear to be important in precipitating most control failures (Roush and McKenzie 1987). Even in cases where resistance mechanisms rely on several genes, such as the mixed-function oxidase system, a single (probably regulatory) gene may be responsible for conferring resistance. For example, although a number of genes on chromosome III in *D. melanogaster* appear to be involved with mixed-function oxidase activity (Houpt et al. 1988), resistance in certain field-collected strains is primarily controlled by a single gene on chromosome II (Waters and Nix 1988, Roush unpublished).

VII. Conclusions

Biochemical assays and conventional bioassays are not mutually exclusive; it may be possible to combine some biochemical tests with bioassays to yield more information from the same insects. For example, the linkage between dieldrin resistance (determined by bioassay) and insensitive acetylcholinesterase (microtiter plate assay) in *An. albimanus* has been studied by measuring acetylcholinesterase activities in recently bioassayed, live and dead mosquitoes (Lines et al. 1990). Because of the range of resistance mechanisms found in arthropods, it will probably continue to be desirable to use both traditional and biochemical assays in resistance monitoring. Further research is needed to improve the efficiency of both kinds of assays.

The advantages to resistance monitoring brought about by specific rapid biochemical assays for resistance highlight the need for further research into the basic mechanisms of resistance in different insects. Although this represents a substantial commitment beyond monitoring by conventional bioassays, the increased speed, accuracy, and information supplied will greatly facilitate the assessment of different management options.

References

Anonymous. 1970. Second conference on test methods for resistance in insects of agricultural importance. Bull. Entomol. Soc. Am. 16: 147–153.

Arnold, J. T. A., and M. J. Whitten. 1975. Measurement of resistance in *Lucilia cuprina* larvae and absence of correlation between organophosphorus-resistance levels in larvae and adults. Entomol. Exp. and Appl. 18: 180–186.

Ball, H. J. 1981. Insecticide resistance: a practical assessment. Bull. Entomol. Soc. Am. 27: 261–262.

Beach, R. F., W. G. Brogdon, L. Castanaza, C. Cordon-Rosales, and M. Calderon. 1989. Temperature effect of an enzyme assay for detecting fenitrothion resistance in *Anopheles albimanus*. Bull. WHO 67: 203–208.

Blackman, R. L., and A. J. C. Paterson. 1986. Separation of *Myzus (Nectarosiphon) antirrhinii* (Macchiati) from *Myzus (N.) persicae* (Sulzer) and related species in Europe (Hemiptera: Aphididae). Syst. Ecol. 11: 267–276.

Bloomquist, J. R., and T. A. Miller. 1985. A simple bioassay for detecting and characterizing insecticide resistance. Pestic. Sci. 16: 611–614.

Brent, K. J. 1986. Detection and monitoring of resistant forms: an overview, pp. 298–312. *In* National Academy of Sciences (ed.), Pesticide resistance: strategies and tactics for management. National Academy Press, Washington, D.C.

Brindley, W. A., D. H. Al-Rajhi, and R. L. Rose. 1982. Portable incubator and its use in insecticide bioassays with field populations of lygus bugs, aphids, and other insects. J. Econ. Entomol. 75: 758–760.

Brogdon, W. G., and C. M. Dickinson. 1983. A microassay system for measuring esterase activity and protein concentration in small samples and high pressure liquid chromatography eluate fractions. Anal. Biochem. 131: 499–503.

Brogdon, W. G., J. H. Hobbs, Y. St. Jean, J. R. Jacques, and L. B. Charles. 1988a. Microplate assay analysis of reduced fenitrothion susceptibility in Haitian *Anopheles albimanus*. J. Am. Mosq. Control Assoc. 4: 152–158.

Brogdon, W. G., R. F. Beach, J. T. Stewart, and L. Castanaza. 1988b. Microplate assay analysis of organophosphate and carbamate resistance distribution in Guatemalan *Anopheles albimanus*. Bull. WHO 66: 339–346.

Brown, T. M., and W. G. Brodgon. 1987. Improved detection of insecticide resistance through conventional and molecular techniques. Annu. Rev. Entomol. 32: 145–162.

Busvine, J. R. 1957. A critical review of techniques for testing insecticides. Commonwealth Agricultural Bureau, London.

Comins, H. 1986. Tactics for resistance management using multiple pesticides. Agric. Ecosyst. Environ. 16: 129–148.

Daly, J. C., and P. Gregg. 1985. Genetic variation in *Heliothis* in Australia: species identification and gene flow in the two pest species *H. armigera* (Hubner) and *H. punctigera* (Wallengren) (Lepidoptera: Noctuidae). Bull. Entomol. Res. 75: 169–184.

Daly, J. C., and D. H. Murray. 1988. Evolution of resistance to pyrethroids in *Heliothis armigera* (Hubner) (Lepidoptera: Noctuidae) in Australia. J. Econ. Entomol. 81: 984–988.

Denholm, I., A. W. Farnham, M. W. Rowland, and R. M. Sawicki. 1990. Laboratory evaluation and empirical modelling of resistance-countering strategies, pp. 92–104. *In* W. K. Moberg and H. M. LeBaron (eds.), Managing resistance to agrochemicals. ACS Symposium Series No. 421. American Chemical Scoiety, Washington, D.C.

Denholm, I., R. M. Sawicki, and A. W. Farnham. 1984. The relationship between insecticide resistance and control failure, pp. 527–534. *In* Proceedings, 1984 British Crop Protection Conference, Pests and Diseases, Brighton. British Crop Protection Council, Croydon, England.

Dennehy, T. J. 1987. Decision-making for managing pest resistance to pesticides, pp. 118–126. *In* M. G. Ford, D. W. Holloman, B. P. S. Khambay, and R. M. Sawicki (eds.), Combating resistance to xenobiotics: biological and chemical approaches. Ellis Horwood, Chichester, England.

Dennehy, T. J., and J. Granett. 1984a. Spider mite resistance to dicofol in San Joaquin Valley cotton: inter- and intraspecific variability in susceptibility of three species of Tetranychus (Acari: Tetranychidae). J. Econ. Entomol. 77: 1381–1385.

Dennehy, T. J., and J. Granett. 1984b. Monitoring dicofol-resistant spider mites (Acari: Tetranychidae) in California cotton. J. Econ. Entomol. 77: 1386–1392.

Dennehy, T. J., J. Granett, and T. F. Leigh. 1983. Relevance of slide-dip and residual bioassay comparisons to detection of resistance in spider mites. J. Econ. Entomol. 76: 1225–1230.

Dennehy, T. J., J. Granett, T. F. Leigh, and A. Colvin. 1987a. Laboratory and field investigations of spider mite (Acari: Tetranychidae) resistance to the selective acaricide propargite. J. Econ. Entomol. 80: 565–574.

Dennehy, T. J., E. E. Grafton-Cardwell, J. Granett, and K. Barbour. 1987b. Practitioner-assessable bioassay for detection of dicofol resistance in spider mites (Acari: Tetranychidae). J. Econ. Entomol. 80: 998–1003.

Dennehy, T. J., J. P. Nyrop, W. H. Reissig, and R. W. Weires. 1988. Characterization of resistance to dicofil in spider mites (Acari: Tetranychidae) from New York apple orchards. J. Econ. Entomol. 81: 1551–1561.

Devonshire, A. L., and G. D. Moores. 1982. A carboxylesterase with broad substrate specificity causes organophosphorus, carbamate and pyrethroid resistance in peach-potato aphids *Myzus persicae*. Pestic. Biochem. Physiol. 18: 235–246.

Devonshire, A. L., and G. D. Moores. 1984a. Characterisation of insecticide-insensitive acetylcholin-esterase: Microcomputer-based analysis of enzyme inhibition in homogenates of individual house fly (*Musca domestica*) heads. Pestic. Biochem. Physiol. 21: 341–348.

Devonshire, A. L., and G. D. Moores. 1984b. Different forms of insensitive acetylcholinesterase in insecticide-resistant house flies (*Musca domestica*). Pestic. Biochem. Physiol. 21: 336–340.

Devonshire, A. L., P. H. Needham, A. D. Rice, and R. M. Sawicki. 1975. Monitoring for resistance to organophosphorus insecticides in *Myzus persicae* on sugar beet, pp. 21–25. *In* Proceedings, 1975 British Insecticide and Fungicide Conference, Brighton. British Crop Protection Council, Croydon, England.

Devonshire, A. L., G. D. Moores, and R. H. ffrench-Constant. 1986. Detection of insecticide resistance of immunological estimation of carboxylesterase activity in *Myzus persicae* (Sulzer) and cross reaction of the antiserum with *Phorodon humuli* (Schrank) (Hemiptera: Aphididae). Bull. Entomol. Res. 76: 97–107.

Edge, V. E., and D. G. James. 1986. Organo-tin resistance in *Tetranychus urticae* (Acari: Tetranychidae) in Australia. J. Econ. Entomol. 79: 1477–1483.

Embury, S. H., S. J. Scharf, R. K. Saiki, M. A. Gholson, M. Golbus, N. Arnheim, and H. A. Erlich. 1987. Rapid prenatal diagnosis of sickle cell anemia by a new method of DNA analysis. New Engl. J. Med. 316: 656–661.

Farnham, A. W., K. O'Dell, I. Denholm, and R. M. Sawicki. 1984. Factors affecting resistance to insecticides in house-flies, *Musca domestica* L. (Diptera: Muscidae) 3. Relationship between the level of resistance to pyrethroids, control failure in the field and the frequency of gene *kdr*. Bull. Entomol. Res. 74: 581–589.

Feyereisen, R., J. F. Koener, D. E. Farnsworth, and D. W. Nebert. 1989. Isolation and sequence of cDNA encoding a cytochrome P–450 from an insecticide-resistant strain of the house fly, *Musca domestica*. Proc. Natl. Acad. Sci. USA 86: 1465–1469.

ffrench-Constant, R. J., and B. C. Bonning. 1989. Rapid microtitre plate test distinguishes insecticide resistant acetylcholinesterase genotypes in the mosquitoes *Anopheles albimanus*, *An. nigerrimus* and *Culex pipiens*. Med. Vet. Entomol. 3: 9–16.

ffrench-Constant, R. H., and A. L. Devonshire. 1986. The effect of aphid immigration on the rate of selection of insecticide resistance in *Myzus persicae* by different classes of insecticides, pp. 115–125. *In* Aspects of applied biology 13, Part I, Crop protection of sugar beet and crop protection and quality of potatoes, 1986. AAB, Wellesbourne, UK.

ffrench-Constant, R. H., and A. L. Devonshire. 1987. A multiple homogenizer for rapid sample preparation in immunoassays and electrophoresis. Biochem. Gen. 25: 493–499.

ffrench-Constant, R. H., and A. L. Devonshire. 1988. Monitoring frequencies of insecticide resistance in *Myzus persicae* (Sulzer) (Hemiptera: Aphididae) in England during 1984–1986, by immunoassay. Bull. Entomol. Res. 78: 163–171.

ffrench-Constant, R. H., A. L. Devonshire, and S. J. Clarke. 1987. Differential rate of selection for resistance by carbamate, organophosphorus and combined pyrethroid and organophosphorus insecticide in *Myzus persicae* (Sulzer) Hemiptera: Aphididae). Bull. Entomol. Res. 77: 227–238.

ffrench-Constant, R. H., F. J. Byrne, M. F. Stribley, and A. L. Devonshire. 1988a. Rapid identification of the recently recognized *Myzus antirrhinii* (Machiati) (Hemiptera: Aphididae) by polyacrylamide gel electrophoresis. The Entomologist 107: 20–23.

ffrench-Constant, R. H., R. Harrington, and A. L. Devonshire. 1988b. Effect of repeated applications of insecticides to potatoes on numbes of *Myzus persicae* (Sulzer) Hemiptera: Aphididae) and on the frequencies of insecticide-resistant variants. Crop Protection 7: 55–61.

ffrench-Constant, R. H., S. J. Clark, and A. L. Devonshire. 1988c. Effect of decline of insecticide residues on selection for insecticide resistance in *Myzus persicae* (Sulzer) (Hemiptera: Aphididae). Bull. Entomol. Res. 78: 19–29.

ffrench-Constant, R. H., A. L. Devonshire, and R. P. White. 1988d. Spontaneous loss and reselection of resistance in extremely resistant *Myzus persicae* (Sulzer). Pestic. Biochem. Physiol. 30: 1–10.

Field, L. M., A. L. Devonshire, and B. G. Forde. 1988. Molecular evidence that insecticide resistance in peach-potato aphids (*Myzus persicae* Sulz.) results from amplification of an esterase gene. Biochem. J. 251: 309–312.

Field, L. M., A. L. Devonshire, R. H. ffrench-Constant, and B. G. Forde. 1989a. Positive correlation between methylation and expression of amplified insecticide-resistance genes. FEBS Letters 243: 323–327.

Field, L. M., A. L. Devonshire, and R. H. ffrench-Constant. 1989b. The combined use of immunoassay and a DNA diagnostic technique to identify insecticide-resistant genotypes in the peach-potato aphid *Myzus persicae* (Sulz.). Pestic. Biochem. Physiol. 34: 174–178.

Flexner, J. L., P. H. Westigard, and B. A. Croft. 1988. Field reversion of organotin resistance in two spotted spider mite (Acari: Tetranychidae) following relaxation of selection pressure. J. Econ. Entomol. 81: 1516–1520.

Forrester, N. W., and M. Cahill. 1987. Management of insecticide resistance in *Heliothis armigera* (Hubner) in Australia, pp. 127–137. *In* M. G. Ford, D. W. Holloman, B. P. S. Khambay and R. M. Sawicki (eds.), Combating resistance to xenobiotics; biological and chemical approaches. Ellis Horwood, Chichester, England.

Georghiou, G. P. 1983. Management of resistance in arthropods, pp. 769–792. *In* G. P. Georghiou and T. Saito (eds.), Pest resistance to pesticides. Plenum, New York.

Grafton-Cardwell, E. E., J. Granett, and T. F. Leigh. 1987. Spider mite species (Acari: Tetranychidae) response to propargite: basis for an acaricide resistance management program. J. Econ. Entomol. 80: 579–587.

Grafton-Cardwell, E. E., J. A. Eash and J. Granett. 1988. Isozyme differentiation of *Tetranychus pacificus* from *T. urticae* and *T. turkestani* (Acari: Tetranychidae) in laboratory and field populations. J. Econ. Entomol. 81: 770–775.

Grafton-Cardwell, E. E., J. Granett, T. F. Leigh, and S. M. Normington. 1989. Development and evaluation of a rapid bioassay for monitoring propargite resistance in *Tetranychus* species (Acari: Tetranychidae) on cotton. J. Econ. Entomol. 82: 706–715.

Hall, L. M. C., and P. Spierer. 1986. The Ace locus of *Drosophila melanogaster*: structural gene for acetylcholinesterase with an unusual 5″ leader. EMBO J. 5: 2949–2954.

Halliday, W. R., and G. P. Georghiou. 1985. Cross-resistance and dominance relationships in a permethrin-selected strain of *Culex quinquefasciatus* (Diptera: Culicidae). J. Econ. Entomol. 78: 1227–1232.

Harrington, R., E. Bartlet, D. K. Riley, R. H. ffrench-Constant, and S. J. Clark. 1989. Resurgence of insecticide-resistant *Myzus persicae* on potatoes treated repeatedly with cypermethrin and mineral oil. Crop Protection 8: 340–348.

Hassan, S. A. 1985. Standard methods to test the side-effects of pesticides on natural enemies of insects and mites developed by the IOBC/WPRS Working Group "Pesticides and Beneficial Organisms." EPPO Bulletin 15: 214–255.

Haynes, K. F., T. A. Miller, R. T. Staten, W.-G. Li, and T. C. Baker. 1987. Pheromone trap for monitoring insecticide resistance in the pink bollworm moth (Lepidoptera: Gelechiidae): new tool for resistance management. Environ. Entomol. 16: 84–89.

Hemingway, J., M. Rowland, and K. E. Kisson. 1984. Efficacy of pirimiphos methyl as a larvicide or adulticide against insecticide resistant and susceptible mosquitoes (Diptera: Culicidae). J. Econ. Entomol. 77: 868–871.

Hemingway, J., C. Smith, K. G. I. Jayawardena, and P. R. J. Herath. 1986. Field and laboratory detection of the altered acetylcholinesterase resistance genes which confer organophosphate and carbamate resistance in mosquitoes (Diptera: Culicidae). Bull. Entomol. Res. 76: 559–565.

Hinkle, N. C., D. C. Sheppard, and M. P. Nolan. 1985. Comparing residue exposure and topical application techniques for assessing permethrin resistance in house flies (Diptera: Muscidae). J. Econ. Entomol. 78: 722–724.

Houpt, D. R., J. C. Pursey, and R. A. Morton. 1988. Genes controlling malathion resistance in a laboratory-selected population of *Drosophila melanogaster*. Genome 30: 844–853.

Hoy, M. A., J. Conley, and W. Robinson. 1988. Cyhexatin and fenbutatin-oxide resistance in Pacific spider mite (Acari: Tetranychidae): stability and mode of inheritance. J. Econ. Entomol. 81: 57–64.

Hoyt, S. C., and F. H. Harries. 1961. Laboratory and field studies on orchard-mite resistance to Kelthane. J. Econ. Entomol. 54: 12–16.

Hughes, P. R., N. A. M. van Beek, and H. A. Wood. 1986. A modified droplet feeding method for rapid assay of *Bacillus thuringiensis* and baculoviruses in noctuid larvae. J. Invert. Pathol. 48: 187–192.

Ignoffo, C. M., M. D. Huettel, A. H. McIntosh, C. Garcia, and P. Wilkening. 1985. Genetics of resistance of *Heliothis subflexa* (Lepidoptera: Noctuidae) to *Baculovirus heliothis*. Ann. Entomol. Soc. Am. 78: 468–473.

Keena, M. A., and J. Granett. 1985. Variability in toxicity of propargite to spider mites (Acari: Tetranychidae) from California almonds. J. Econ. Entomol. 78: 1212–1216.

Knight, A. L., and L. A. Hull. 1989. Use of sex pheromone traps to monitor azinphosmethyl resistance in tufted apple bud moth (Lepidoptera: Tortricidae). J. Econ. Entomol. 82: 1019–1026.

Lines, J. D., R. H. ffrench-Constant, and S. H. Kasim. 1990. Testing for genetic linkage of insecticide resistance genes by combining bioassay and biochemical methods. Medical and Veterinary Entomology (in press).

Luttrell, R. G., R. T. Roush, A. Ali, J. S. Mink, M. R. Reid, and G. L. Snodgrass. 1987. Pyrethroid resistance in field populations of *Heliothis virescens* (Lepidoptera: Noctuidae) in Mississippi in 1986. J. Econ. Entomol. 80: 985–989.

Mani, G. S. 1985. Evolution of resistance in the presence of two insecticides. Genetics 109: 761–783.

Martinez-Carrillo, J. L., and H. T. Reynolds. 1983. Dosage-mortality studies with pyrethroids and other insecticides on the tobacco budworm (Lepidoptera: Noctuidae) from the Imperial Valley, California. J. Econ. Entomol. 76: 983–986.

McGaughey, W. H. 1985. Insect resistance to the biological insecticide *Bacillus thuringiensis*. Science 229: 193–195.

McKenzie, J. A., and M. J. Whitten. 1982. Selection for insecticide resistance in the Australian sheep blowfly, *Lucilia cuprina*. Experientia 38: 84–85.

Milio, J. F., P. G. Koehler, and R. S. Patterson. 1987. Evaluation of three methods for detecting chlorpyrifos resistance in German cockroach (Orthoptera: Blattellidae) populations. J. Econ. Entomol. 80: 44–46.

Moores, G. D., A. L. Devonshire, and I. Denholm. 1988a. A microtitre plate assay for characterizing insensitive acetylcholinesterase genotypes of insecticide-resistant insects. Bull. Entomol. Res. 78: 537–544.

Moores, G. D., I. Denholm, F. J. Byrne, A. L. Kennedy, and A. L. Devonshire. 1988b. Characterising acetylcholinesterase genotypes in resistant insect populations, pp. 451–456. *In* Proceedings, 1988 British Crop Protection Conference, Brighton. British Crop Protection Council, Croydon, England.

Mouches, C., M. Magnin, J-B. Berge, M. de Silvestri, V. Beyssat, N. Pasteur, and G. P. Georghiou. 1987. Overproduction of detoxifying esterase in organophosphate-resistant *Culex* mosquitoes and their presence in other insects. Proc. Natl. Acad. Sci. 84: 2113–2116.

Mullins, W., and E. P. Pieters. 1982. Weight versus toxicity: a need for revision of the standard method of testing for resistance of the tobacco budworm to insecticides. J. Econ. Entomol. 75: 40–42.

Mullis, K. B., and F. A. Faloona. 1987. Specific synthesis of DNA *in vitro* via a polymerase catalysed chain reaction. Meth. Enzymol. 155: 335–350.

ole-MoiYoi, O. K. 1987. Trypanosome species specific DNA probes to detect infection in tsetse flies. Parasitol. Today 3: 371–374.

Onstad, D. W. 1987. Calculation of economic-injury levels and economic thresholds for pest management. J. Econ. Entomol. 80: 297–303.

Oppenoorth, F. J. 1985. Biochemistry and genetics of insecticide resistance, pp. 731–773. *In* G. A. Kerkut and L. I. Gilbert (eds.), Comprehensive insect physiology, biochemistry, and pharmacology, Vol. 12. Pergamon, Oxford.

Pasteur, N., and G. P. Georghiou. 1981. Filter paper test for rapid determination of phenotypes with high esterase activity in organophosphate resistant mosquitoes. Mosq. News 41: 181–183.

Pasteur, N., and G. P. Georghiou. 1989. Improved filter paper test for detecting and quantifying increased esterase activity in organophosphate-resistant mosquitoes (Diptera: Culicidae). J. Econ. Entomol. 82: 347–353.

Post, R. J., and J. M. Crampton. 1988. The taxonomic use of variation in repetitive DNA sequences in the *Simulium damnosum* complex, pp. 245–256. *In* M. W. Service (ed.), biosystematics of haematophagous insects. The Systematics Association Special, Vol. 37. Clarendon Press, Oxford.

Riedl, H., A. Seeman, and F. Henrie. 1985. Montoring susceptibility to azinphosmethyl in field populations of the codling moth (Lepidoptera: Tortricidae) with pheromone traps. J. Econ. Entomol. 78: 692–699.

Riley, S. L. 1989. Pyrethroid resistance in *Heliothis virescens:* Current U.S. management programs. Pestic. Sci. 26: 411–421.

Robertson, J. L., K. C. Smith, N. E. Savin, and R. J. Lavigne. 1984. Effects of dose selection and sample size on precision of lethal dose estimates in dose-mortality regression. J. Econ. Entomol. 77: 833–837.

Roush, R. T. 1989. Designing pesticide management programs: how can you choose? Pestic. Sci. 26: 423–441.

Roush, R. T., and M. A. Hoy. 1978. Relative toxicity of permethrin to a predator, *Metaseiulus occidentalis* and its prey, *Tetranychus urticae.* Environ. Entomol. 7: 287–288.

Roush, R. T., and R. G. Luttrell. 1987. The phenotypic expression of pyrethroid resistance in *Heliothis* and implications for resistance management, pp. 220–224. *In* Proceedings, 1987 Beltwide Cotton Production Research Conference. National Cotton Council of America, Memphis.

Roush, R. T., and R. G. Luttrell. 1989. Expression of resistance to pyrethroid insecticides in adults and larvae of tobacco budworm (Lepidoptera: Noctuidae): implications for resistance monitoring. J. Econ. Entomol. 82: 1305–1310.

Roush, R. T., and G. L. Miller. 1986. Considerations for design of insecticide resistance monitoring programs. J. Econ. Entomol. 79: 293–298.

Roush, R. T., and J. A. McKenzie. 1987. Ecological genetics of insecticide and acaricide resistance. Annu. Rev. Entomol. 32: 361–380.

Roush, R. T., R. L. Combs, T. C. Randolph, J. MacDonald, and J. A. Hawkins. 1986. Inheritance and effective dominance of pyrethrooid resistance in the horn fly (Diptera: Muscidae). J. Econ. Entomol. 79: 1178–1182.

Rowland, M., and J. Hemingway. 1987. Changes in malathion resistance with age in *Anopheles stephensi* from Pakistan. Pestic. Biochem. Physiol. 27: 239–247.

Saiki, R. K., T. L. Bugawan, G. T. Horn, K. B. Mullis, and H. A. Erlich. 1986. Analysis of enzymatically amplified B-globin and HLA-DQo DNA with allele-specific oligonucleotide probes. Nature 324: 163–166.

Sanderson, J. P., M. P. Parrella, and J. T. Trumble. 1989. Monitoring insecticide resistance in *Liriomyza trifolii* (Diptera: Agromyzidae) with yellow sticky cards. J. Econ. Entomol. 82: 1011–1018.

Sawicki, R. M. 1978. Unusual response of DDT-resistant houseflies to carbinol analogues of DDT. Nature 275: 443–444.

Sawicki, R. M. 1987. Definition, detection and documentation of insecticide resistance, pp. 105–117. *In* M. G. Ford, D. W. Holloman, B. P. S. Khambay, and R. M. Sawicki (eds.), Combating resistance to xenobiotics; biological and chemical approaches. Ellis Horwood, Chichester, England.

Sawicki, R. M., and A. D. Rice. 1978. Response of susceptible and resistant peach-potato aphid *Myzus persicae* (Sulzer) to insecticides in leaf-dip bioassays. Pestic. Sci. 9: 513–516.

Sawicki, R. M., A. L. Devonshire, R. W. Payne, and S. M. Petzing. 1980. Stability of insecticide resistance in the peach-potato aphid *Myzus persicae* (Sulzer). Pestic. Sci. 11: 33–42.

Schouest, L. P., Jr., and T. A. Miller. 1988. Factors influencing pyrethroid toxicity in pink bollworm (Lepidoptera: Gelechiidae): implications for resistance management. J. Econ. Entomol. 81: 431–436.

Schouest, L. P., Jr., N. Umetsu, and T. A. Miller. 1983. Solvent modified desposition of insecticide on house fly (Diptera: Muscidae) cuticle. J. Econ. Entomol. 76: 973–982.

Scott, J. G., S. B. Ramaswamy, F. Matsumara, and K. Tanaka. 1986. Effect of method of application on resistance to pyrethroid insecticides in *Blattella germanica* (Orthoptera: Blattellidae). J. Econ. Entomol. 79: 571–575.

Scott, J. G., R. T. Roush, and D. A. Rutz. 1989. Insecticide resistance of house flies from New York dairies (Diptera: Muscidae). J. Agric. Entomol. 6: 53–64.

Soderlund, D. M., J. R. Bloomquist, F. Wong, L. L. Payne, and D. C. Knipple. 1989. Molecular neurobiology: insights for insecticide action and resistance. Pestic. Sci. 26: 359–374.

Sparks, T. C., J. A. Lockwood, R. L. Byford, J. B. Graves, and B. R. Leonard. 1989. The role of behaviour in insecticide resistance. Pestic. Sci. 26: 383–399.

Staetz, C. A. 1985. Susceptibility of *Heliothis virescens* (F.) (Lepidoptera: Noctuidae) to permethrin from across the cotton belt; a five year study. J. Econ. Entomol. 78: 505–510.

Tabashnik, B. E., and N. L. Cushing. 1987. Leaf residue vs. topical bioassays for assessing insecticide resistance in the diamond-back moth, *Plutella xylostella* L. FAO Plant Prot. Bull. 35: 11–14.

Tabashnik, B. E., N. L. Cushing, and M. W. Johnson. 1987. Diamondback moth (Lepidoptera: Plutellidae) resistance to insecticides in Hawaii: intra-island variation and cross-resistance. J. Econ. Entomol. 80: 1091–1099.

Tatchell, G. M., M. Thorn, H. D. Loxdale, and A. L. Devonshire. 1988. Monitoring for insecticide resistance in migrant populations of *Myzus persicae*, pp. 559–564. *In* Proceedings, 1988 British Crop Protection Conference Brighton. British Crop Protection Council, Croydon, England.

Walker, W. F., A. L. Boswell, and F. F. Smith. 1973. Resistance of spider mites to acaricides: comparison of slide dip and leaf dip methods. J. Econ. Entomol. 66: 549–550.

Waters, L. C., and C. E. Nix. 1988. Regulation of insecticide resistance-related cytochrome P-450 expression in *Drosophila melanogaster*. Pestic. Biochem. Physiol. 30: 214–227.

Watkinson, I. A., J. Wiseman, and J. Robinson. 1984. A simple test kit for field evaluation of the susceptibility of insect pests to insecticides, pp. 559–564. *In* Proceedings, 1984 British Crop Protection Conference. Brighton. British Crop Protection Council, Croydon, England.

Welty, C., W. H. Reissig, T. J. Dennehy, and R. W. Weires. 1987. Cyhexatin resistance in New York populations of European red mite (Acari: Tetranychidae). J. Econ. Entomol. 80: 230–236.

Welty, C., W. H. Reissig, T. J. Dennehy, and R. W. Weires. 1988. Comparison of residual bioassay methods and criteria for assessing mortality of cyhexatin-resistant European red mite (Acari: Tetranychidae). J. Econ. Entomol. 81: 442–448.

Welty, C., W. H. Reissig, T. J. Dennehy, and R. W. Weires. 1989. Relationship between field efficacy and laboratory estimates of susceptibility to cyhexatin in populations of European red mite (Acari: Tetranychidae). J. Econ. Entomol. 82: 354–364.

W.H.O. 1970. Insecticide resistance and vector control. Seventeenth report of the WHO expert committee on insecticides. WHO Technical Report Series, No. 433.

W.H.O. 1976. Resistance of vectors and reservoirs of disease to pesticides. Twenty-second report of the WHO expert committee on insecticides. WHO Technical Report Series, No. 585.

W.H.O. 1980. Resistance of vectors of disease to pesticides. Fifth report of the WHO expert committee on vector biology and control. WHO Technical Report Series, No. 655.

Wilson, T. G. 1988. *Drosophila melanogaster* (Diptera: Drosophilidae): A model insect for insecticide resistance studies. J. Econ. Entomol. 81: 22–27.

Winks, R. G. 1986a. The significance of response time in detection and measurement of fumigant resistance in insects with special reference to phosphine. Pestic. Sci. 17: 165–174.

Winks, R. G. 1986b. The biological efficacy of fumigants: time/dose response phenomena, pp. 211–221. *In* Pesticides and humid tropical grain storage systems, ACIAR Proceedings No. 14. ACIAR, Canberra, Australia.

Winks, R. G., and C. J. Waterford. 1986. The relationship between concentration and time in the toxicity of phosphine to adults of a resistant strain of *Tribolium castaneum* (Herbst). J. Stored Prod. Res. 22: 85–92.

Wolfenbarger, D. A., J. R. Raulston, A. C. Bartlett, G. E. Donaldson, and P. P. Lopez. 1982. Tobacco budworm: selection for resistance to methyl parathion from a field-collected strain. J. Econ. Entomol. 75: 211–215.

3

Investigating Mechanisms of Insecticide Resistance: Methods, Strategies, and Pitfalls

Jeffrey G. Scott

I. Introduction

Pesticide resistance is a severe and important problem in situations where chemicals are used to kill pests. However, apart from the economic, social, and environmental costs associated with this problem, resistant insects are a physiological marvel. Some strains have become so resistant to a given insecticide that they can survive exposure to virtually any dose. Pesticide resistance is truly one of the most amazing cases of evolutionary adaptation to environmental change, especially when we consider that it has occurred relatively quickly in terms of evolutionary time.

There are numerous reasons for studying the mechanisms by which pests become resistant to pesticides. Such studies are important for both the applied and basic aspects of insecticide resistance, as well as providing valuable information for workers in allied fields, as exemplified later. First, if the biochemical basis of the resistance can be determined, it may be possible to design a highly sensitive monitoring technique, which is one of the key factors in developing a successful resistance management program (Chapter 2). This has recently been demonstrated in *Myzus persicae*, where the resistance is due to increased amounts of a single esterase (Esterase–4, or E4) (Devonshire and Moores 1982). By purifying this esterase, raising antibodies against it, and using appropriate staining techniques, researchers can use immunoassays to determine the phenotype and level of resistance in individual aphids (ffrench-Constant and Devonshire 1988, Chapter 2).

Second, a common response by pesticide applicators facing a resistant pest is to try one or more different pesticides. This haphazard approach may or may not work, depending on whether or not there is cross-resistance to the new pesticide. However, if the mechanism of resistance is known, a replacement pesticide that would not be affected by the mechanism could be rationally chosen. Additionally, many of the resistance management strategies, such as alternations or mixtures, can only be successful if there is no cross-resistance between the insecticides used (Georghiou 1983). Clearly, identification of actual or potential resistance mechanisms is needed to maximize the likelihood of success in a resistance management program.

A third reason for studying the mechanisms of resistance is that by understanding the mechanism we may be able to devise methods for overcoming it. One of the best examples is the use of synergists to control resistant pests (Georghiou 1983, Wilkinson 1983). This strategy has recently been implemented in Australia for the control of *Heliothis armigera*. In cases where pyrethroid resistance is severe, piperonyl butoxide is being recommended as a synergist to block the primary resistance mechanism (oxidative metabolism) and thus maintain the usefulness of these insecticides (Forrester 1988, Chapter 5).

Apart from the more practical reasons for studying insecticide resistance, resistant pests offer valuable tools for research in many areas: basic biochemical and physiological processes, mode of action of insecticides, evolution, metabolism, pharmacokinetics, and molecular genetics. For example, resistant strains were a key factor used to elucidate the mode of action of cyclodiene (Matsumura 1985a) and organophosphate (Oppenoorth 1985) insecticides. Additionally, resistant insects were a key to the first successful purification of an arthropod cytochrome P–450 (Wheelock and Scott 1989), the cloning of an insect P–450 gene (Feyereisen et al. 1989), and possibly an improved understanding of the biochemical regulation of the cytochrome P–450 microsomal monooxygenases (Scott and Georghiou 1986c).

This chapter outlines common and easily used methods for determining physiological mechanisms of resistance to *neurotoxic* insecticides, how to interpret the results, and possible pitfalls to avoid. I assume that the reader is not trained in toxicology but has some background in chemistry, physiology, and statistics. Because it would be impossible to detail every method used to identify insecticide resistance mechanisms, I will present detailed descriptions for methods that are most often used. Other, more technical methods, for which good reviews are available, are presented only briefly and the reader may consult the references for more details. Recent reviews of the behavioral basis of resistance (Lockwood et al. 1984, Sparks et al. 1989) and the mechanisms of insecticide resistance (Georghiou 1965, Oppenoorth 1985, Oppenoorth and Welling 1979, Wilkinson 1983, Chapter 4) are available and will not be covered in detail here. The goal of this chapter is to help researchers identify the resistance mechanisms specific to their situation, in sufficient detail for practical application to insect biology and resistance management.

II. General Considerations for Resistance Studies

A. Relatedness of the Strains Used

Genetics is a useful tool in evaluating and confirming resistance mechanisms. Although genetics is covered in detail in Chapter 5, two points are highlighted here. First is the importance of using genetically similar strains. Unless the resistance level (intensity) is very large, it may be difficult to distinguish if different responses (e.g., to a pesticide ± synergist) are actually due to a particular physiological resistance mechanism or simply due to the differences that can occur between strains of dissimilar origins, even when reared under similar conditions. This is especially critical in attempting to identify possible minor mechanisms of resistance.

Genetics also allows one to test if a putative mechanism is really the one that determines resistance. For example, if a putative resistance mechanism has been identified in a resistant versus a susceptible insect, the expression of the mechanism in F_1 (resistant × susceptible) progeny should be correlated with the level of resistance seen in the bioassay. This technique was recently used by Ahn et al. (1986) to identify nervous system insensitivity (*kdr*, see later) as a mechanism of pyrethroid resistance in Japanese strains of house flies.

B. Bioassay Techniques

The choice of the bioassay technique can be of major importance in studies on resistance mechanisms. Different application methods may change the relative importance of each of the physiological processes involved (Welling and Patterson 1985) in the intoxication process (as well as resistance), radically alter the level of resistance observed (Dennehy et al. 1983, Scott et al. 1986b, Chapter 2), and limit one's ability to identify the resistance mechanism. The lower the level of resistance, the more difficult it becomes to show statistically significant differences between strains and unequivocally establish the mechanism(s) of resistance. The slope of the log-dose probit line is particularly important when synergists are used to help determine the mechanism of resistance. A shallow (flat) slope generally gives wide confidence intervals (Finney 1971, Chapter 2), which makes it very difficult to determine accurately the effect of a synergist (see later). The ideal bioassay reveals high levels of resistance, gives steep slopes of the log-dose probit lines, is representative of the method of exposure in the field, and is cheap and fast. Since all these characteristics are not always available, it is up to the individual researcher to choose the most important criterion for each situation.

Insecticide bioassays can use either time or dose (or both) as the variable. Although the use of time as the variable may be useful as a resistance-monitoring method (e.g., Cochran 1989), this technique is more limited for investigating mechanisms of resistance. In particular, the resistance ratio can vary with the dose chosen for the time-variable bioassay (J. Scott unpublished). Additionally, by this method, synergism data can be difficult or impossible to interpret. The dose-variable type of bioassay is best for studying resistance mechanisms. This includes several routes of application, including topical, residual, sprays, and so on. In this chapter I will limit my discussion of insecticide bioassays to the use of dose-variable bioassays.

III. Methods for Identifying Resistance Mechanisms

Insects have three basic resistance mechanisms: decreased cuticular penetration, increased metabolic detoxication and target site insensitivity. Sequestration is another mechanism of resistance (Devonshire and Moores 1982), although it appears to be rare. Identification of the mechanism(s) involved in resistance is one of the major challenges in insect toxicology and resistance management. Strategies for identifying each possible mechanism follow and are outlined in Table 3.1. Many of the methods described are relatively easy and could be carried out in many laboratories without much expensive equipment.

A. Penetration Studies

Decreased cuticular penetration has been known to be a resistance mechanism since first described in the early 1960s (Fine et al. 1963, Forgash et al. 1962, Matsumura and Brown 1963a). By itself this mechanism usually confers only low levels (<3-fold) of resistance (Plapp and Hoyer 1968); however, it does seem to provide protection to a wide variety of insecticides (Oppenoorth 1985, Plapp and Hoyer 1968). The simplest test for this mechanism involves exposing the insect to the insecticide and then comparing the amount absorbed over time in the resistant and susceptible strains. At fixed time periods after exposure, insecticide remaining on the surface of the insect is rinsed off with an appropriate solvent. The amount of insecticide in the external rinse, in the holding container, and inside the insect is then quantified by gas–liquid chromatography, high-performance liquid chromatography, or scintillation counting (if a radioisotope is used). Penetration studies are usually followed as a time course over a relatively short timespan (less than 8 h) (e.g., DeVries and Georghiou 1981, Fine et al. 1963, Forgash et al. 1962, Matsumura and Brown 1963a, Plapp and Hoyer 1968, Scott and Georghiou 1986a). Results at later time points are often difficult to interpret and sometimes inconclusive because of the increasing effects of metabolism, redistribution, and excretion. Decreased penetration has been documented as a resistance mechanism only at the level of the insect cuticle. However, any biological membrane has the potential to serve as a barrier (Welling and Patterson 1985) and thus confer resistance.

B. Observations of Test Insects

Observations of test animals can be helpful in identifying possible resistance mechanisms. The best examples of this are the *kdr* type (described later) and decreased cuticular penetration (described earlier) mechanisms. In addition to conferring protection to kill by pyrethroids, *kdr* also protects against the rapid onset of ataxia (knockdown) produced by DDT (Milani 1956) and many pyrethroids (Oppenoorth and Welling 1979). The decreased penetration mechanism

Table 3.1. A General Listing of Resistance Mechanisms and Criteria Useful in their Identification

Mechanism	Cross-Resistance Pattern	General Observations	Partial Listing of Potential Synergists[a]	Preferred Method for Identifying
Penetration	Organophosphates Carbamates (?) Pyrethroids Pyrethrins Cyclodienes Abamectin DDT Organotins Others ?	Generally low levels of protection May delay onset of symptoms and/or knockdown	—	Monitor rate of insecticide penetration over time
Metabolism:		Insects *may* show symptoms but then recover		
MFO	Insecticides with similar functional groups		PBO, Sesamex, others	*In vivo* and *in vitro* metabolism studies
Hydrolase	Insecticides with similar functional groups		DEF, TPP and others	*In vivo* and *in vitro* metabolism studies
GSH S-transferase	Preference for methoxy versus ethoxy substituted organophosphates		Diethylmaleate (DEM) and others (Note: DEM is water soluble and may not penetrate the cuticle of some insects)	*In vivo* and *in vitro* metabolism studies
DDTase	DDT and trichlorethane analogues of DDT		DMC	*In vivo* and *in vitro* metabolism studies

Table 3.1. *Continued*

Mechanism	Cross-Resistance Pattern	General Observations	Partial Listing of Potential Synergists[a]	Preferred Method for Identifying
Insensitivity of the nervous system				
kdr-type	DDT Pyrethroids Pyrethrins	Delays knockdown May be selected for by either DDT, pyrethroid or pyrethrins	—	Electrophysiological study of isolated nervous system
cyclodiene-type	cyclodienes		—	Electrophysiological study of isolated nervous system
altered AChE	Certain organophosphates and/or carbamates Pattern of cross-resistance depends upon the AChE isozyme	Rate of acetylthiocholine hydrolysis may be increased, decreased or unchanged	—	Determine inhibition rates (or I_{50}) for AChE Electrophysiological study of isolated nervous system

[a]See Casida (1970) or Raffa and Priester (1985) for a more complete listing.

also produces a similar delay in knockdown (Oppenoorth and Welling 1979). However, the decreased cuticular penetration mechanism would not cause a delay in knockdown if the insecticide were injected, whereas the *kdr*-type mechanism would still delay knockdown. Although today the utility of observing insects may be limited to clarification of the involvement of *kdr* and penetration, generalizations may be drawn for other mechanisms in the future.

C. Cross-resistance Patterns

The use of cross-resistance (i.e., protection from more than one insecticide through the action of a single mechanism) patterns is important in resistance studies for two reasons. First, cross-resistance patterns are a valuable tool for identifying known resistance mechanisms. For example, *kdr*-type resistance is characterized by cross-resistance to DDT and its analogs and to pyrethroids, although the exact resistance ratios will vary. Second, whenever a new resistance mechanism is identified, an important part of its characterization is to determine the cross-resistance spectrum.

One critical factor for cross-resistance studies is that the resistant strain should not exhibit multiple resistance (i.e., coexistence of different resistance mechanisms in one strain) (Georghiou 1965). It may be possible to interpret cross-resistance studies if other resistance mechanisms are present, but *only* if the other mechanisms are clearly limited to insecticides other than those being investigated. Isolation of specific resistance mechanisms in isogenic strains is an optimal approach for these types of studies (Chapter 5).

Three major types of target site insensitivity resistance mechanisms are known, and the insecticides to which they confer protection have been characterized. The nervous system insensitivity mechanism commonly known as *kdr* confers resistance to DDT, DDT analogs, and all pyrethroids tested to date (DeVries and Georghiou 1981, Farnham 1973, Farnham et al. 1987, Milani 1956, Plapp and Hoyer 1968, Scott and Georghiou 1986a, Chapter 4), independent of the physicochemical properties of the pyrethroid (Scott et al. 1986a). Insensitivity of the nervous system to cyclodienes was first demonstrated by Yamasaki and Narahashi (1958). In German cockroaches this mechanism appears to be due to an altered GABA receptor–ionophore complex, probably caused by a change in the picrotoxinin binding site where cyclodiene insecticides are thought to act (Ghiasuddin and Matsumura 1982, Kadous et al. 1983, Lawrence and Casida 1984, Matsumura 1985, Tanaka et al. 1984, Chapter 4). Perhaps the best-characterized of the target site insensitivity mechanisms is the altered acetylcholinesterase responsible for resistance to organophosphate and carbamate insecticides (Chapter 4). However, this mechanism is more specific than the first two described, because it confers resistance only to other structurally similar carbamates or organophosphates (Devonshire and Moores 1984, Yamamoto et al. 1983). In fact, in some insects such as house flies, there seem to be several isozymes of acetylcholinesterase, each conferring its own characteristic cross-resistance spectrum (Devonshire and Moores 1984). This may not appear to be the case for all insects, as only one isozyme was found in studies on insecticide-resistant mosquitoes (ffrench-Constant and Bonning 1989, Chapter 2).

Because of the large number of enzymes that may be involved, several cross-resistance spectra conferred by known types of metabolic resistance have been characterized. DDT dehydrochlorinase, for example, is an important mechanism of resistance to DDT, but it confers resistance only to dehydrochlorinatable analogs (Lipke and Kearns 1960). Another relatively specific metabolism-mediated resistance is malathion carboxylesterase, which confers protection against malathion and a close structural analogue (phenthoate), but not to other organophosphates (Beeman and Schmidt 1982, Bigley and Plapp 1962, Dyte and Rowlands 1968, Hughes et al. 1984, Matsumura and Brown 1963b, Matsumura and Hogendijk 1964). Recently, MFO-mediated pyrethroid resistance was shown to be specific for pyrethroids having a phenoxybenzyl or similar functional group (Scott and Georghiou 1986a). It appears to be a general rule of thumb that metabolic resistance usually does not confer resistance to all insecticides within a given class.

D. Insecticide Synergists

One of the easiest and fastest ways to gain preliminary information about possible mechanisms of resistance is by the use of insecticide synergists. Synergists are "compounds that greatly enhance the toxicity of an insecticide, although they are usually practically nontoxic by themselves" (Matsumura 1985b). Insecticide synergists are thought to act primarily by inhibiting a given type of metabolism (Casida 1970, Georghiou 1983, Sun and Johnson 1972). Therefore, insecticides that are readily metabolized become much more toxic when this process is blocked. One of the best-known examples is pyrethrins, which are potent knockdown agents but are only moderately toxic to most insects without the addition of a synergist (usually piperonyl butoxide). Three common insecticide synergists are piperonyl butoxide (PBO), DEF (S,S,S-tributylphosphorotrithioate), and DMC (4,4'-dichloro-α-methylbenzhydrol), which inhibit cytochrome P–450 microsomal monooxygenases (MFOs or PSMOs), hydrolases, and DDT-dehydrochlorinase ("DDT-ase"), respectively. A listing of many insecticide synergists that have been used, or are potentially useful, and the detoxication pathway(s) they are proposed to inhibit has been prepared by Raffa and Priester (1985). Although this listing is a particularly helpful starting point, the authors did not attempt to evaluate critically the supporting data for each proposed insecticide. Therefore, care must be taken when choosing a synergist, and a thorough literature search is needed to confirm the reported action and to check the specificity of any synergist. Even a widely used synergist, such as DEF, is not completely specific for hydrolases, because it can also block oxidases at high concentrations (in vitro at $>10^{-4}$ M, Scott unpublished). Because of their widespread availability it is desirable to test *both* oxidase and hydrolase inhibitors (at least one of each) before attempting to reach a conclusion.

By using synergists in a comparison between susceptible and resistant insects, it is sometimes possible to show that a given metabolic pathway is involved in the resistance. For example, if resistance is due to increased metabolism via the cytochrome P–450 microsomal mixed-function oxidases (MFOs), using the MFO inhibitor PBO would likely overcome the resistance. A simple comparison of the resistance ratio (e.g., LD_{50} of the resistant strain/LD_{50} of the susceptible strain) in the presence and absence of synergist can be used to test the effectiveness of a synergist at overcoming the resistance. Obviously toxicity must be determined with and without synergist for both the susceptible and resistant strains. The synergistic ratio (SR; e.g., LD_{50} without synergist/LD_{50} with synergist) is an intrastrain calculation and, therefore, of less importance for resistance studies.

To understand better the use of synergists in resistance studies, consider the example presented in Table 3.2. Resistance to the hypothetical insecticide Knocksemdead is 30-fold for strains A, B, C, and D. Resistance to Knocksemdead + PBO is 1.2-, 11-, 30-, and 40-fold for strains A, B, C, and D, respectively. Thus, resistance in strain A is completely PBO suppressible, and in strain B it is partially PBO suppressible. These results suggest the cytochrome P–450 monooxygenases are responsible for resistance in strain A and are one of the mechanisms responsible for resistance in strain B. Resistance in strains C and D is not suppressed by PBO, suggesting resistance is not monooxygenase mediated. The fact that resistance actually increased in strain D with the addition of PBO likely has no bearing on the resistance. Instead, the results imply that the susceptible strain was better able to detoxify Knocksemdead oxidatively than strain D. This is probably due to the lack of genetic relatedness of the strains, as mentioned earlier. Although other calculations have been proposed to examine the effect of synergists on resistance (Brindley and Selim 1984), the method described above presents a much simpler and more widely accepted analysis of the data.

Historically, two different methods have been used to apply synergists in

Table 3.2. Hypothetical Bioassay Results of Knocksemdead ± PBO to Five House Fly Strains

	Treatment			
	Knocksemdead		Knocksemdead + PBO	
Strain	LD_{50}	RR^a	LD_{50}	RR^a
Susceptible	4	—	1	—
A	120	30	1.2	1.2
B	120	30	11	11
C	120	30	30	30
D	120	30	40	40

[a]Resistance Ratio = LD_{50} resistant strain/LD_{50} susceptible strain.

studies of resistance: (1) a constant ratio of synergist to insecticide or (2) a fixed dose of synergist. In earlier reports, a constant ratio of synergist to insecticide was used. However, this method has several potential problems. When a ratio is used, inhibition of the metabolic system may be unequal across the range of doses used. When comparisons are made between strains having significantly different dose–response ranges, this effect may become severe. Consider, for example, two strains of house fly; one is susceptible (LD_{50} = 2 ng per fly) and the other has a 10,000-fold resistance to deltamethrin. If the synergist is applied at a constant ratio (e.g., 1:10, insecticide:synergist), then the susceptible strain will receive between 5 and 80 ng per fly of the synergist, whereas the resistant strain may receive up to 16,000 ng per fly. For some synergists this would mean that the metabolism in the susceptible strain is inhibited very little, whereas the synergist itself may be toxic at the levels tested against the resistant strain. Even if the synergist is not used at such high doses, this example shows how ratios would tend to show greater effect on the resistant strain, and thus overestimate the effect of the synergist. Because the goal of using synergists is to compare the toxicity of an insecticide in the presence or absence of a given metabolic system, the synergist should be applied at an equal dose to both susceptible and resistant strains. This dose should be high enough to cause maximal inhibition of the metabolic system without causing mortality from the synergist alone. This dose can be established for any species and test method by evaluating the toxicity of the synergist over a range of doses and determining the maximum dose that causes no observable toxicity. Many researchers will often use one third or one quarter of this dose to assure that the synergist is not causing mortality.

Another factor that becomes important when examining the effect of synergists is the amount of time between application of the synergist and insecticide. Sawicki (1962) found that pretreatment of insects with synergist increases the amount of synergism observed. Because maximum inhibition of the metabolic system in question is desired, it is best to pretreat insects with the synergist one or two hours prior to insecticide application. In cases where this may not be possible (i.e., insects are sensitive to handling), synergists can still be reliably used in simultaneous application with the insecticide, but the degree of synergism will likely be reduced.

Although synergists are useful in the study of resistance mechanisms, there are a few pitfalls to avoid. The lack of synergism by a known enzymatic inhibitor on a given resistance cannot be taken as proof that the metabolic pathway is not involved in the resistance. The metabolic systems discussed here are composed of a number of different enzymes and/or isozymes. Although synergists such as PBO or DEF inhibit general classes of enzymes, the possibility exists that they may not inhibit the specific isozyme responsible for the resistance (Chapter 4). For example, in house flies, Pimprikar and Georghiou (1979) found that sesamex reduced resistance to diflubenzuron, whereas PBO did not, suggesting that only sesamex inhibited the cytochrome P–450 isozyme(s) responsible for the resistance. For GSH-transferases other than DDT-ase, no highly reliable synergists exist, although several have been suggested (Raffa and Priester 1985); thus, currently this important mechanism cannot usually be diagnosed by this method.

Another potential problem is that even synergists whose effects are reasonably well defined, such as the compounds mentioned earlier, need to be tested for their effectiveness when a new species or method of application is investigated. For example, a synergist may not be capable of passing through the cuticle of all insects or acarines. The lack of reduction in the resistance ratio by a synergist will not be meaningful if the synergist does not reach the target tissue. One possible way to test the effectiveness of a synergist would be to evaluate its effectiveness at enhancing the toxicity of an insecticide known to be metabolized by that given pathway. For example, if PBO can enhance the toxicity of pyrethrins against the insect in question, this implies that the synergist is in fact penetrating, blocking metabolism, and suitable for use in resistance studies. By a similar rationale, DEF could be used with malathion to check for synergism, or a proposed GSH-S-transferase inhibitor could be tested with azinphosmethyl.

A third problem with the use of insecticide synergists in determining resistance mechanisms is that there is no widely accepted statistical method for determining if one resistance ratio (i.e., without synergist) is significantly different from another (i.e., with synergist). For example, in Table 3.2 is the resistance ratio of strain D significantly changed by the addition of PBO? Such a statistical analysis is sorely needed, given the increasing use of synergists in resistance studies and many other research areas (Hedin et al. 1988, Raffa and Priester 1985). In lieu of formal methods, most researchers only claim that a synergist has had an effect on resistance if the effect was large. For example, the reduction of resistance from 6,000-fold to 33-fold in the presence of the oxidative inhibitor PBO provided conclusive evidence that oxidative metabolism played a major role in resistance (Scott and Georghiou 1986a). However, a reduction in resistance from 10-fold to six-fold makes the results less than certain. Furthermore, it has been suggested that synergists will generally have a greater effect on resistant strains due solely to the pharmacodynamic interaction between poisoning processes (Oppenoorth 1985), although this certainly *does not* appear to be a common result (Scott et al. 1985, J. G. Scott unpublished).

A fourth possible problem with synergists is that they may enhance toxicity by some means other than blocking a specified type of metabolism. Sun and Johnson (1960) presented indirect evidence that, for 3 of 18 compounds they studied, synergism was due not only to decreased metabolism but also to increased penetration. However, a recent study detected no differences in penetration of radiolabeled permethrin between PBO-treated and nontreated susceptible house flies (Scott and Georghiou 1986a). Overall, it appears that for a *few* insecticides, especially those with low toxicity (Sun and Johnson 1960), synergists may block metabolism and also facilitate cuticular transport.

Occasionally, the resistance ratio is greater in the presence of synergist than in its absence. Although this is an unusual result, it probably reflects the lack of relatedness (i.e., differences in origin) between the strains rather than an effect of the resistance gene(s). For example, in a comparison between susceptible and pyrethroid-resistant house flies, the permethrin resistance ratio increased from 5,900-fold to 19,000-fold in the presence of DEF. This occurred because of greater hydrolysis (nonoxidative metabolism) of permethrin in the susceptible compared with the resistant strain (Scott and Georghiou 1986a).

It is obvious that synergists can be very helpful in identification of possible mechanisms of resistance; however, when evaluating synergism results, "it is clear that data relating to resistance mechanisms should be interpreted with caution and should be confirmed by *in vivo* and *in vitro* metabolic studies" (Wilkinson 1983).

E. Enzyme Assays

Enzyme assays (excluding the acetylcholinesterase assay that is described in Section III.G) are generally of minor importance in identifying resistance mechanisms. Insects contain hundreds of different enzymes, and running an assay that measures the sum of several undefined enzymes, from different and often unrelated strains, will frequently show a difference that may not be in any way related to the resistance. A common assay used in conjunction with insecticide resistance studies has been the general esterase assay using one of a few standard substrates (most notably α-napthyl acetate). In one species of mosquito, *Culex quinquefasciatus*, this enzyme assay has been associated with resistance to temephos and a few other organophosphates (Georghiou and Pasteur 1978, Georghiou et al. 1980). However, use of the α-napthyl acetate assay in a different mosquito species as a means to identify the mechanism of organophosphate resistance led to equivocal results (Hemingway 1982), probably because this assay is not correlated with insecticide resistance or metabolism in many species (Beeman 1983, Dowd and Sparks 1984, Scott and Georghiou 1986b, Chapter 4). The true utility of enzyme assays comes after the mechanism has been identified, when an appropriate enzyme assay can be used to monitor resistance (Chapter 2).

F. Metabolic Detoxication

Metabolism studies are important when investigating resistance for three reasons: (1) Such studies can conclusively determine if resistance is due to increased detoxication (Chapter 4), (2) the enzyme system responsible for the resistance can be accurately identified, and (3) a better understanding of metabolism in insects may be achieved. It is important to note that metabolism does not equal detoxication for all insecticides. This is especially important for the organophosphates where metabolic activation (e.g., parathion $\rightarrow$ paraoxon) is common (Eto 1974). In this case resistance may arise from a lower rate of metabolic activation (Konno et al. 1988).

In general, both in vivo and in vitro metabolism studies are needed to determine resistance mechanisms conclusively. For in vivo studies susceptible and resistant insects are treated (usually with radiolabeled insecticide) and the *internal, external,* and *excreted* levels of parent compound and metabolites are determined over time. Such studies are useful because they are the most conclusive in implicating increased detoxication as a resistance mechanism. Additionally, with the use of synergists, such studies are useful for identifying the possible enzyme system(s) involved. However, in vivo studies may be difficult to interpret since penetration and excretion differences (if they exist) may obscure the metabolic differences.

For in vitro studies, a given enzyme may be purified or a specified centrifugal fraction used as the enzyme source (Hodgson 1985, Lee and Scott 1989, Terriere 1979, Wilkinson 1979). The presence or absence of inhibitors and/or cofactors in the incubation will clarify the enzymes involved, their cellular location, and the rate of metabolism (Scott and Georghiou 1986a, Scott et al. 1987, Terriere 1979, Wilkinson 1979). Comparisons between susceptible and resistant strains will then help to identify the resistance mechanisms. It is necessary to remember that in vitro studies need not accurately mimic the conditions found in vivo. This is especially true when trying to relate different rates of metabolism observed in vitro to their relative effects in vivo (i.e., in resistance). A more detailed discussion of in vitro studies is presented in two excellent reviews (Terriere 1979, Wilkinson 1979).

G. Target Site Sensitivity

1. Electrophysiology

Electrophysiological techniques are undoubtedly among the most powerful tools for studying the sensitivity of the nervous system to neurotoxic insecticides. These methods are the most reliable for obtaining clear evidence of an altered target site as a mechanism of resistance to neurotoxic pesticides (Chapter 4), with the exception of the acetylcholinesterase assay described later. Useful texts that list the common equipment and preparations used to study insects (Miller 1979) and information on the setup of electrophysiological equipment (Oakley and Schafer 1978) should be consulted for specific details and methods.

The two main types of electrophysiological studies used to identify resistance mechanisms are in vivo and in situ tests. In vivo tests using electrodes implanted into insects are probably less useful because the effects of nonneural mechanisms can come into play. Electrophysiological monitoring of poisoned insects in vivo is little better than the close observation of test animals in a bioassay when identification of the resistance mechanism is the goal. In vivo assays can be modified so that detoxication mechanisms are blocked by removal of body regions and/or the use of synergists; however, such preparations are still subject to interference by decreased penetration mechanisms (Scott and Georghiou

1986a,c). The more useful in situ studies are carried out using two types of preparations: A specific part of the nervous system, such as the abdominal nerve cord of the German cockroach, is entirely removed from the body cavity (Scott and Matsumura 1983), or the body cavity can be opened and nonneural tissue (gastroinestinal tract, fat bodies, etc.) removed (Salgado et al. 1983, Scott and Matsumura 1981). By these types of studies, the effect of other resistance mechanisms can be minimized.

Electrophysiological analysis of resistance can be carried out using either time or dose as the variable. If time is the variable, identical doses are applied to neural preparations from susceptible and resistant animals, and the time to onset of neural poisoning symptoms is compared between the strains (e.g., Scott and Matsumura 1981). If dose is the variable, then strains are compared using an identical observation period (e.g., Scott and Georghiou 1986a,c). Care must be taken to know exactly what type of symptom represents neural poisoning. DDT produces typical repetitive discharges (Welsh and Gordon 1947), whereas cyclodienes or organophosphates produce bursts of activity (Lund 1985, Shankland 1979). Pyrethroid insecticides have been classed into Type I or II based on a number of differences, including their causing repetitive discharges (Type I) or neural block (Type II) (Gammon et al. 1981, Lund and Narahashi 1983, Scott 1988, Scott and Matsumura 1983). Whereas pyrethroids having primarily Type I action provide a readily identifiable symptom, pyrethroids with primarily Type II action will require more sophisticated equipment (i.e., stimulator) to monitor for nerve conduction block (Pap et al. 1986).

2. Acetylcholinesterase

Although the electrophysiological techniques outlined previously provide a powerful tool for studying the target site sensitivity between strains, a much more widely used technique for investigating organophosphate or carbamate resistance involves in vitro measurement of acetylcholinesterase activity in the presence or absence of different doses of insecticide. Dramatic differences have been noted where acetylcholinesterase can be inhibited completely in susceptible strains, but only partially (<20%) in resistant strains. Assays are typically carried out by the method of Ellman et al. (1961) using acetylthiocholine as the substrate. Although this is not the true substrate, it appears to be an acceptable substitute for most species studied to date (Oppenoorth 1985). The acetylcholinesterase is typically prepared by homogenizing the heads of insects in a buffer solution. Occasionally, the addition of a small amount of detergent, such as 1% Triton X–100, is necessary to obtain optimum activity. Descriptions of the enzyme kinetics, methods for determination of the rate constants, calculations used and potential problems to be avoided have been presented by several authors (Devonshire 1987, Devonshire and Moores 1982, Eto 1974, Matsumura 1985b, Oppenoorth and Welling 1979).

IV. Conclusions

From the preceding information it is clear that identification of the mechanisms of resistance necessitates the use of several different techniques that require careful experimental design and interpretation. Although by themselves not all of these techniques provide irrefutable evidence, the use of *multiple methods* (i.e., synergists + cross-resistance patterns + metabolism + others) can allow conclusive identification of the resistance mechanism(s) involved (e.g., Scott and Georghiou 1985, 1986a,c). Unfortunately, it is difficult to devise a flow chart for researchers to follow in order to identify resistance mechanisms. Usually the best strategy is to begin with synergism and cross-resistance data. From this information it is usually possible to formulate some idea about what the mechanism may be. The hypothesis that is generated should then be critically evaluated by the other relevant methods described in the preceding sections.

Advances in understanding the mechanisms of resistance have been substantial in the last 25 years. However, as will be pointed out in the next chapter, our understanding of the molecular basis of many of these mechanisms is still largely incomplete. Identification of new mechanisms in new pests and the molecular basis for resistance mechanisms will continue to be one of the most challenging areas of research in insect toxicology and resistance management.

References

Ahn, Y.-J., T. Shono, and J. Fukami. 1986. Inheritance of pyrethroid resistance in a housefly strain from Denmark. J. Pestic. Sci. 11: 591–596.

Beeman, R. W. 1983. Inheritance and linkage of malathion resistance in the red flour beetle. J. Econ. Entomol. 76: 737–740.

Beeman, R. W., and B. A. Schmidt. 1982. Biochemical and genetic aspects of malathion-specific resistance in the Indianmeal moth (Lepidoptera: Pyralidae). J. Econ. Entomol. 75: 945–949.

Bigley, W. S., and F. W. Plapp, Jr. 1962. Metabolism of malathion and malaoxon by the mosquito *Culex tarsalis*. J. Insect Physiol. 8:545–557.

Brindley, W. A., and A. A. Selim. 1984. Synergism and antagonism in the analysis of insecticide resistance. Environ. Entomol. 13: 348–353.

Casida, J. E. 1970. Mixed-function oxidase involvement in the biochemistry of insecticide synergists. Agric. Food Chem. 18: 753–772.

Cochran, D. G. 1989. Monitoring for insecticide resistance in field-collected strains of the German cockroach (Dictyoptera: Blattellidae). J. Econ. Entomol. 82: 336–341.

Dennehy, T. J., J. Grannett, and T. F. Leigh. 1983. Relevance of slide dip and residual bioassay comparisons to detection of resistance in spider mites. J. Econ. Entomol. 76: 1225–1230.

Devonshire, A. L. 1987. Biochemical studies of organophosphorus and carbamate resistance in houseflies and aphids, pp. 239–255. *In* M. G. Ford, D. W. Holloman, B. P. S. Khambay, and R. M. Sawicki (eds.), Combating Resistance to Xenobiotics. Ellis Horwood Ltd., London.

Devonshire, A. L. and G. D. Moores. 1982. A carboxylesterase with broad substrate specificity causes organophosphorus, carbamate and pyrethroid resistance in peach-potato aphids (*Myzus persicae*). Pestic. Biochem. Physiol. 18: 235–246.

Devonshire, A. L., and G. D. Moores. 1984. Different forms of insensitive acetylcholinesterase in insecticide-resistant house flies (*Musca domestica*). Pestic. Biochem. Physiol. 21: 336–340.

DeVries, D. H., and G. P. Georghiou. 1981. Decreased nerve sensitivity and decreased cuticular penetration as mechanisms of resistance to pyrethroids in a (1 R)-*trans*-permethrin-selected strain of the house fly. Pestic. Biochem. Physiol. 15: 234–241.

Dowd, P. F., and T. C. Sparks. 1984. Developmental changes in *trans*-permethrin an α-Napthyl acetate ester hydrolysis during the last larval instar of *Pseudoplusia includens*. Pestic. Biochem. Physiol. 21: 275–282.

Dyte, C. E., and D. G. Rowlands. 1968. The metabolism and synergism of malathion in resistant and susceptible strains of *Tribolium castaneum*. J. Stored Prod. Res. 4: 157–173.

Ellman, G. L., K. D. Courtney, V. Andres, Jr., and R. M. Featherstone. 1961. A new and rapid colorimetric determination of acetylcholinesterase activity. Biochem. Pharmacol. 7: 88–94.

Eto, M. 1974. Organophosphorus Pesticides: Organic and Biological Chemistry. CRC Press, Cleveland. 387 pp.

Farnham, A. W. 1973. Genetics of resistance of pyrethroid-selected houseflies, *Musca domestica* L. Pestic. Sci. 4: 513–520.

Farnham, A. W., A. W. A. Murray, R. M. Sawicki, I. Denholm, and J. C. White. 1987. Characterization of the structure-activity relationship of *kdr* and two variants of *super-kdr* to pyrethroids in the house fly (*Musca domestica* L.). Pestic. Sci. 19: 209–220.

Feyereisen, R., J. F. Koener, D. E. Farnsworth, and D. W. Nebert. 1989. Isolation and sequence of cDNA encoding a cytochrome P–450 from an insecticide-resistant strain of house fly, *Musca domestica*. Proc. Natl. Acad. Sci. USA 86: 1465–1469.

ffrench-Constant, R. H., and A. L. Devonshire. 1988. Monitoring frequencies of insecticide resistance in *Myzus persicae* (Sulzer) (Hemiptera: Aphididae) in England during 1985–86 by immunoassay. Bull. Entomol. Res. 78: 163–171.

ffrench-Constant, R. H., and B. C. Bonning. 1989. Rapid microtitre plate test distinguishes insecticide resistant acetylcholinesterase genotypes in the mosquitoes *Anopheles albimanus*, *An. nigerrimus* and *Culex pipiens*. Med. Vet. Entomol. 3: 9–16.

Fine, B. C., P. J. Godin, and E. M. Thain. 1963. Penetration of pyrethrin I labelled with carbon–14 into susceptible and pyrethroid resistant houseflies. Nature 199: 927–928.

Finney, D. J. 1971. Probit Analysis, 3rd ed. Cambridge Univ., London. 333 pp.

Forrester, N. W. 1988. Field selection for pyrethroid resistance genes, Australian Cottongr. 9: 48–51.

Forgash, A. J., B. J. Cook, and R. C. Riley. 1962. Mechanisms of resistance in diazinon-selected multi-resistant *Musca domestica*. J. Econ. Entomol. 55: 544–551.

Gammon, D. W., M. A. Brown, and J. E. Casida. 1981. Two classes of pyrethroid action in the cockroach. Pestic. Biochem. Physiol. 15:181–191.

Georghiou, G. P. 1965. Genetic studies on insecticide resistance, pp. 171–230. *In* R. L. Metcalf (ed.), Advances in Pest Control Research, Vol. VI. John Wiley and Sons, New York.

Georghiou, G. P. 1983. Management of resistance in arthropods, pp. 769–792. *In* G. P. Georghiou and T. Saito (eds.), Pest Resistance to Pesticides. Plenum Press, New York.

Georghiou, G. P., and N. Pasteur. 1978. Electrophoretic esterase patterns in insecticide-resistant and susceptible mosquitoes. J. Econ. Entomol. 71: 201–205.

Georghiou, G. P., N. Pasteur, and M. K. Hawley. 1980. Linkage relationships between organophosphate resistance and a highly active esterase-B in *Culex quinquefasciatus* from California. J. Econ. Entomol. 73: 301–305.

Ghiasuddin, S. M., and F. Matsumura. 1982. Inhibition of gamma-aminobutyric acid (GABA)-induced chloride uptake by gamma-BHC and heptachlor epoxide. Comp. Biochem. Physiol. 73C, 141–144.

Hedin, P. A., W. L. Parrott, J. N. Jenkins, J. E. Mulrooney, and J. J. Mean. 1988. Eludicating mechanisms of tobacco budworm resistance to allelochemicals by dietary tests with insecticide synergists. Pestic. Biochem. Physiol. 32: 55–61.

Hemingway, J. 1982. The biochemical nature of malathion resistance in *Anopheles stephensi* from Pakistan. Pestic. Biochem. Physiol. 17:149–155.

Hodgson, E. 1985. Microsomal mono-oxygenases, pp. 225–321. *In* G. A. Kerkut and L. I. Gilbert (eds.), Comprehensive Insect Physiology, Biochemistry and Pharmacology, Vol. 11. Pergamon, Oxford.

Huges, P. B., P. E. Green, and K. G. Reichmann. 1984. Specific resistance to malathion in laboratory and field populations of the Australian sheep blowfly, *Lucilia cuprina* (Diptera: Calliphoridae). J. Econ. Entomol. 77: 1400–1404.

Kadous, A. A., S. M. Ghiasuddin, F. Matsumura, J. G. Scott, and K. Tanaka. 1983. Difference in the picrotoxinin receptor between the cylodiene-resistant and susceptible strains of the German cockroach. Pestic. Biochem. Physiol. 19: 157–166.

Konno, T., E. Hodgson, and W. C. Dauterman, 1989. Studies on methyl parathion resistance in *Heliothis virescens*. Pestic. Biochem. Physiol. 33: 189–199.

Lawrence, L. J., and J. E. Casida. 1984. Interactions of lindane, toxaphene and cyclodienes with brain-specific *t*-butylbicyclophosphorothionate receptor. Life Sci. 35: 171–178.

Lee, S. S. T., and J. G. Scott. 1989. An improved method for the preparation, stabilization and storage of house fly (Diptera: Muscidae) microsomes. J. Econ. Entomol. 82: 1559–1563.

Lipke, H., and C. W. Kearns. 1960. DDT-dehydrochlorinase, pp. 253–287. *In* R. L. Metcalf (ed.), Advances in Pest Control Research, Vol. III. Interscience, New York.

Lockwood, J. A., T. C. Sparks, and R. N. Story. 1984. Evolution of insect resistance to insecticides: a reevaluation of the roles of physiology and behavior. Bull. ESA 30(4): 41–51.

Lund, A. E. 1985. Insecticides: effects on the nervous system, pp. 9–56. *In* G. A. Kerkut and L. I. Gilbert (eds.), Comprehensive Insect Physiology, Biochemistry and Pharmacology, Vol. 12. Pergamon, Oxford.

Lund, A. E., and T. Narahashi. 1983. Kinetics of sodium channel modification as the basis for variation in the nerve membrane effects of pyrethroids and DDT analogs. Pestic. Biochem. Physiol. 20: 203–216.

Matsumura, F. 1985a. Involvement of picrotoxinin receptor in the action of cyclodiene insecticides. Neurotoxicol. 6: 139–164.

Matsumura, F. 1985b. Toxicology of Insecticides, 2nd ed. Plenum Press, New York 598 pp.

Matsumura, F., and A. W. A. Brown. 1963a. Studies on organophosphorus tolerance in *Aedes aegypti*. Mosq. News 23: 26–31.

Matsumura, F., and A. W. A. Brown. 1963b. Studies on the carboxylesterase in malathion-resistant *Culex tarsalis*. J. Econ. Entomol. 56: 381–388.

Matsumura, F., and C. J. Hogendijk. 1964. The enzymatic degradation of malathion in organophosphate resistant and susceptible strains of *Musca domestica*. Entomol. exp. Appl. 7: 179–193.

Milani, R. 1956. Mendellian inheritance of knock-down resistance to DDT and correlation between knockdown and mortality in *Musca domestica* L. Selected Sci. Papers Instit. Super. Sanita. I, Part 1: 176–182.

Miller, T. A. 1979. Insect Neurophysiological Techniques. Springer-Verlag. 308 pp.

Oakley, B., and R. Schafer. 1978. Experimental Neurobiology. Univ. of Michigan Press, Ann Arbor, Michigan. 367 pp.

Oppenoorth, F. J. 1985. Biochemistry and genetics of insecticide resistance, pp. 731–773. *In* G. A. Kerkut and L. I. Gilbert (eds.), Comprehensive Insect Physiology, Biochemistry and Pharmacology, Vol. 12. Pergamon, Oxford.

Oppenoorth, F. J., and W. Welling. 1979. Biochemistry and physiology of resistance, pp. 507–551. *In* C. F. Wilkinson (ed.), Insecticide Biochemistry and Physiology. Plenum Press, New York.

Pap, L., E. R. Hegedus, K. Bauer, I. Ujvary, and G. Matolesy. 1986. A rapid method for evaluation of nerve conduction blocking compounds. Comp. Biochem. Physiol. 85C: 347–352.

Pimprikar, G. D., and G. P. Georghiou. 1979. Mechanisms of resistance to diflubenzuron in the house fly *Musca domestica* (L.). Pestic. Biochem. Physiol. 12: 10–22.

Plapp, F. W., Jr., and R. F. Hoyer. 1968. Insecticide resistance in the house fly: decreased rate of absorption as the mechanism of action of a gene that acts as an intensifier of resistance. J. Econ. Entomol. 61: 1298–1303.

Raffa, K. F., and T. M. Priester. 1985. Synergists as research tools and control agents in agriculture. J. Agric. Entomol. 2: 27–45.

Salgado, V. L., S. N. Irving, and T. A. Miller. 1983. Depolarization of motor nerve terminals by pyrethroids in susceptible and kdr-resistant house flies. Pestic. Biochem. Physiol. 20: 100–114.

Sawicki, R. M. 1962. Insecticidal activity of pyrethrum extract and its four insecticidal constituents against house flies, III.—knock-down and recovery of flies treated with pyrethrin extract with and without piperonyl butoxide. J. Sci. Food Agric. 13: 283–292.

Scott, J. G. 1988. Pyrethroid insecticides. ISI Atlas Sci. Pharmacol. 2: 125–128.

Scott, J. G., and G. P. Georghiou. 1985. Rapid development of high-level permethrin resistance in a field-collected strain of the house fly (Diptera: Muscidae) under laboratory selection. J. Econ. Entomol. 78: 316–319.

Scott, J. G., and G. P. Georghiou. 1986a. Mechanisms responsible for high levels of permethrin resistance in the house fly. Pestic. Sci. 17: 195–206.

Scott, J. G., and G. P. Georghiou. 1986b. Malathion-specific resistance in Anopheles stephensi from Pakistan. J. Am. Mosq. Cont. Assoc. 2: 29–32.

Scott, J. G., and G. P. Georghhiou. 1986c. The biochemical genetics of permethrin resistance in the Learn-PyR strain of house fly. Biochem. Genet. 24: 25–37.

Scott, J. G., and F. Matsumura. 1981. Characteristics of a DDT-induced case of cross-resistance to permethrin in Blattella germanica. Pestic. Biochem. Physiol. 16: 21–27.

Scott, J. G., and F. Matsumura. 1983. Evidence for two types of toxic actions of pyrethroids on susceptible and DDT-resistant German cockroaches. Pestic. Biochem. Physiol. 19: 141–150.

Scott, J. G., C. J. Palmer, and J. E. Casida. 1987. Oxidative metabolism of the GABA$_A$ receptor antagonist t-butylbicycloortho[^{3}H]benzoate. Xenobiotica 17: 1085–1093.

Scott, J. G., R. B. Mellon, O. Kirino, and G. P. Georghiou. 1986a. Insecticidal activity of substituted benzyl dichlorovinylcyclopropanecarboxylates on susceptible and kdr-resistant strains of the southern house mosquito, Culex quinquefasciatus. J. Pestic. Sci. 11: 475–477.

Scott, J. G., S. B. Ramaswamy, F. Matsumura, and K. Tanaka. 1986b. Effect of method of application on resistance to pyrethroid insecticides in Blattella germanica (Orthoptera: Blattellidae). J. Econ. Entomol. 79: 571–575.

Shankland, D. L. 1979. Action of dieldrin and related compounds on synaptic transmission, pp. 139–153. In T. Narahashi (ed.), Neurotoxicology of Insecticides and Pheromones. Plenum, New York.

Sparks, T. C., J. A. Lockwood, R. L. Byford, J. B. Graves, and B. R. Leonard. 1989. The role of behavior in insecticide resistance. Pestic. Sci. (in press).

Sun, Y.-P., and E. R. Johnson. 1960. Synergistic and antagonistic actions of insecticide-synergist combinations and their mode of action. J. Agric. Food Chem. 8: 261–266.

Sun, Y.-P., and Johnson, E. R. 1972. Quasi-synergism and penetration of insecticides. J. Econ. Entomol. 65: 349–353.

Tanaka, K., J. G. Scott, and F. Matsumura. 1984. Picrotoxinin receptor in the central nervous system of the American cockroach: its role in the action of cyclodiene insecticides. Pestic. Biochem. Physiol. 22: 117–124.

Terriere, L. C. 1979. The use of in vitro techniques to study the comparative metabolism of xenobiotics, pp. 285–320. In G. D. Paulson, D. S. Frear, and E. P. Marks (eds.), Xenobiotic Metabolism: In Vitro Methods. American Chemical Society, Washington, DC.

Welling, W., and G. D. Paterson. 1985. Toxicodynamics of insecticides, pp. 603–645. In G. A. Kerkut and L. I. Gilbert (eds.), Comprehensive Insect Physiology, Biochemistry and Pharmacology, Vol. 12. Pergamon, Oxford.

Welsh, J. H., and H. T. Gordon. 1947. The mode of action of certain insecticides on the arthropod nerve axon. J. Cell. Comp. Physiol. 30: 147–171.

Wheelock, G. D., and J. G. Scott. 1989. Simultaneous purification of a cytochrome P-450 and cytochrome b$_5$ from the house fly, Musca domestica L. Insect Biochem. 19: 481–488.

Wilkinson, C. F. 1979. The use of insect subcellular components for studying the metabolism of xenobiotics, pp. 249–284. In G. D. Paulson, D. S. Frear, and E. P. Marks (eds.), Xenobiotic Metabolism: In Vitro Methods. American Chemical Society, Washington, DC.

Wilkinson, C. F. 1983. Role of mixed-function oxidases in insecticide resistance, pp. 175–205. In G. P. Georghiou and T. Saito (eds.), Pest Resistance to Pesticides. Plenum, New York.

Yamamoto, I., Y. Takahashi, and N. Kyomura. 1983. Suppression of altered acetylcholinesterase of the green rice leafhopper *N*-propyl and *N*-methyl carbamate combinations, pp. 579–594. *In* G. P. Georghiou and T. Saito (eds.), Pest Resistance to Pesticides. Plenum Press, New York.

Yamasaki, T., and T. Narahashi. 1958. Resistance of house flies to insecticides and susceptibility of nerve to insecticides: studies on the mechanism of action of insecticides (XVII). Botyu Kagaku 23: 146–157.

4

Molecular Mechanisms of Insecticide Resistance

David M. Soderlund and *Jeffrey R. Bloomquist*

I. Introduction

A. Context and Scope

Four decades of intensive use of synthetic organic insecticides to control arthropod pests and disease vectors have led to the selection of insecticide or acaricide resistance in approximately 450 arthropod species (Georghiou 1986). In the most extreme cases, such as the Colorado potato beetle (*Leptinotarsa decemlineata*) in parts of the eastern United States, populations are resistant to virtually all chemicals available for control (Forgash 1984). The deleterious consequences of pesticide resistance in arthropods include increased levels of environmental contamination and risks of applicator and agricultural worker exposure from higher rates of pesticide application; increases in pest control costs; disruption of ecologically sound pest control strategies; increased incidence of human, animal, and plant diseases in which transmission depends on insect vectors; and, in the most extreme case, the complete destruction of agricultural production systems on a local or regional basis.

In the past, resistant populations were controlled by increasing the amounts of pesticide applied or replacing older chemicals with new, more effective compounds. Both of these strategies are of limited value today. Increased application rates increase risks of environmental contamination and in many cases are not cost effective for newer, more expensive compounds. Moreover, the rate of discovery and development of new insecticides and acaricides has declined, so that alternatives are not available for populations that are resistant to organophosphorus (OP), carbamate, and pyrethroid insecticides (Hammock and Soderlund 1986). These considerations lend urgency to the need for other measures to prevent or reduce the impact of resistance. These measures, which collectively have been called *resistance management*, involve rational and informed strategies of pesticide use in the context of integrated pest management to prevent, delay, or reverse the development of resistance. Knowledge of the mechanisms of pesticide resistance is essential to the development of such strategies.

The scientific literature on all aspects of arthropod resistance to pesticides is extensive, and numerous review articles published in the past five years have covered that portion of the resistance literature relevant to the definition of resistance mechanisms (Booth et al. 1983, Dauterman 1983, Devonshire 1987, Hama 1983, Hodgson and Kulkarni 1983, Matsumura 1983, Miller et al. 1983, Narahashi 1983, Oppenoorth 1985, Plapp 1986, Saito et al. 1983, Terriere 1983, Yasutomi 1983). Other reviews on insecticide-metabolizing enzymes of insects also address the roles of these enzymes in resistance (Agosin 1985, Dauterman 1985, Hodgson 1983, Hodgson 1985). In this chapter we will not attempt to duplicate or update the broad coverage of these reviews. Instead, we will focus specifically on recent progress in defining the molecular mechanisms of pesticide resistance in arthropods through the application of contemporary methods and approaches in biochemistry, pharmacology and molecular biology.

B. Overview of Resistance Mechanisms

The intoxication of an insect by an insecticide encompasses three levels of pharmacokinetic interactions (Welling and Paterson 1985): penetration of barrier tissues; distribution, storage, and metabolism in internal tissues; and molecular interaction with the ultimate target site. The measured toxicity of an insecticide therefore depends not only on its intrinsic potency at its site of action but also on the pharmacokinetic parameters that describe its uptake following contact, fumigation or dietary exposure, its distribution and partitioning among insect tissues, and the extent of its metabolic activation or detoxication in these tissues. Mutations that affect the rate constants of any of these processes may reduce either the delivery of the ultimate toxicant to its target or the affinity of the toxicant–target interaction, thereby conferring resistance at the level of the whole organism. Despite the diversity of both chemical structures and arthropod species in the instances of resistance compiled by Georghiou (1986), the number of identified resistance mechanisms is small. These include reduced cuticular penetration of toxicants; enhanced metabolism by cytochrome P_{450}-dependent monooxygenases, hydrolases, or glutathione-S-transferases; and reduced sensitivity of mutant acetylcholinesterases to OP and carbamate insecticides and of other neuronal targets to pyrethroids, DDT and analogs, and chlorinated cyclodienes (Oppenoorth 1985). All these mechanisms are nonspecific in that they generally confer cross-resistance to other structurally related toxicants and, in several cases, to chemically unrelated compounds as well. Since these mechanisms span all three pharmacokinetic levels of insect–insecticide interactions, the occurrence of multiple resistance mechanisms acting at different levels has the potential to produce highly synergistic interactions that can result in virtual immunity in some populations.

Of these mechanisms, reduced penetration is both the least significant as a single resistance mechanism and the least understood. Since no new information on the molecular basis of this mechanism has emerged in the literature since Oppenoorth's (1985) review, we omit discussion of the mechanisms underlying reduced penetration in this chapter. In the following sections we summarize the current level of understanding of the molecular alterations that confer the major metabolic and reduced neuronal sensitivity mechanisms.

II. Resistance Conferred by Enhanced Detoxication

A. Cytochrome P_{450}-Dependent Monooxygenases

1. Properties of Cytochrome P_{450}-Dependent Monooxygenases

Our understanding of the biochemistry of cytochrome P_{450}-dependent monooxygenases is based primarily on the extensive studies of these enzymes in mammalian liver. An excellent overview of the structure and function of this system, derived primarily from studies with mammalian hepatic enzymes, is given by Hodgson (1985). The following summary based on this review is provided as a brief orientation and is not intended to be exhaustive.

The monooxygenase system has a very broad substrate specificity, catalyzing hydroxylations at both aromatic and aliphatic carbon atoms, expoxidation of double bonds, desulfuration of phosphorothionates, and oxidation of sulfides. Cytochrome P_{450}-dependent monooxygenases preferentially metabolize lipophilic substrates to products with increased water solubility or with functional groups that enable conjugation reactions, thereby promoting excretion. This breadth of substrate recognition by the monooxygenase system is achieved through the coexistence in mammalian liver of multiple forms of cytochrome P_{450}, each having a broad but unique pattern of substrate specificity. Chronic exposure of organisms to certain lipophilic compounds can cause the induction of high levels of cytochrome P_{450}, through either general proliferation of the cell endoplasmic reticulum, the subcellular location of the monooxygenase system, or increased production of a specific form of cytochrome P_{450}.

The oxidized cytochromes P_{450} bind both substrates and oxygen and undergo a reduction–oxidation cycle, releasing oxidized substrate and water as the products. The reduction of oxidized cytochrome P_{450} forms by NADPH is catalyzed by the enzyme NADPH-cytochrome P_{450} reductase. Another cytochrome, cytochrome b_5, is a possible second source of electrons in the reduction of oxidized cytochrome P_{450}, but it has not been shown to be an essential component of monooxygenase-dependent oxidation in purification and reconstitution experiments.

Detailed studies of the biochemistry of cytochrome P_{450}-dependent monooxygenases in insects have lagged far behind those using mammalian preparations. Spectroscopic characterization of the cytochromes P_{450} and their binding of substrates and inhibitors, together with assays of model substrate and insecticide oxidation, suggest that the principal characteristics of this system evident in mammalian preparations are also found in insects. The literature on biochemical studies of reactions catalyzed by insect monooxygenases is so extensive that a summary table of these citations covered 27 pages in a recent review (Hodgson 1985). Almost half of this literature involves studies of a single species, the house fly (*musca domestica*), and many of the studies include comparisons of susceptible and resistant strains. Despite this extensive documentation of monooxygenase reactions, further progress with insect preparations has been limited by the inability to achieve sufficient yields of cytochrome P_{450} upon solubilization and purification and by the instability of the insect cytochromes P_{450} obtained upon purification by methods optimized for mammalian preparations.

2. Evidence for Enhanced Oxidation in Resistant Insects

Enhanced oxidative metabolism has been implicated as a major mechanism of resistance for all insecticide classes except the chlorinated cyclodienes. Much of this evidence is based solely on the ability of piperonyl butoxide (PBO) and related compounds, which are inhibitors of cytochrome P_{450}-dependent monooxygenases, to reduce the magnitude of resistance observed when they are used as synergists with insecticides in bioassays. The routine use of PBO and the routine assumption that synergism in resistant insects is diagnostic for an oxidative resistance mechanism are so pervasive that a comprehensive summary of such data in the context of this chapter is impossible.

The reliance on synergists in bioassays to detect oxidative metabolism in resistant insects can be misleading in the absence of subsequent biochemical confirmation (Chapter 3). For example, methylenedioxyphenyl compounds coapplied with carbamate insecticides have been shown to increase the cuticular penetration of the insecticide, thereby producing a component of "synergism" completely unrelated to the inhibition of oxidative metabolism (Sun and Johnson 1972). The common practice of formulating synergist and insecticide at a constant ratio for bioassays may also be unreliable if the potency of the insecticide necessitates dilution of treatment solutions to a level where the concentration of synergist is too low to be effective. This problem, which is particularly prevalent in bioassays of pyrethroids and other extremely potent compounds, can be overcome by holding the synergist dose constant. Finally, all forms of cytochrome P_{450} are not equally susceptible to inhibition by piperonyl butoxide or any other single enzyme inhibitor or chemical class of inhibitors, so that lack of synergism may not reflect lack of oxidative metabolism (Chapter 3). An extreme recent example of this phenomenon involves a pyrethroid-resistant strain of *Heliothis virescens* in which no synergism by PBO was observed, yet pyrethroid toxicity was synergized by propynyl ethers, another class of monooxygenase inhibitors (Brown and Payne 1986).

Despite these potential pitfalls, synergism of toxicity in resistant strains usually provides the first clue that enhanced oxidative metabolism contributes to resistance (Chapter 3). Confirmation of this finding requires biochemical studies to compare resistant and susceptible strains. Such studies involve comparisons of the rates of metabolism by enzyme preparations of model monooxygenase substrates or, preferably, the insecticide for which resistance has been observed. Biochemical documentation of enhanced oxidative metabolism in resistant strains is now available for numerous combinations of insect species and insecticides (Agosin 1985, Hodgson 1985, Oppenoorth 1985).

3. Mechanisms Underlying Enhanced Oxidation

Several molecular mechanisms might account for the enhanced rates of insecticide oxidation observed in enzyme preparations from resistant insect strains. First, resistant insects might possess generally higher levels of all components of the monooxygenase system. Second, rapid oxidation may result from overexpression of a single constitutive form of cytochrome P_{450} having high catalytic activity toward the insecticide substrate. Third, resistant insects may have alterations in cytochrome P_{450} gene regulation so that a form of cytochrome P_{450} that is observed only upon induction in susceptible insects is instead constitutively expressed in resistant insects. Finally, resistant insects may possess a mutant form of cytochrome P_{450} with unique properties and high catalytic activity toward the insecticide substrate.

Studies directed at exploring these mechanisms have been limited by the technical difficulties involved in the solubilization, purification, and characterization of insect cytochromes P_{450}. A number of early studies showed that total cytochrome P_{450}, measured by the extinction coefficient of the reduced cytochrome P_{450}–carbon monoxide complex at 450 nm, was increased in several insecticide-resistant strains of the house fly. These findings have been reviewed in detail elsewhere (Agosin 1985, Hodgson 1983, Hodgson 1985). In these cases, as well as in a more recent study of pyrethroid-resistant house flies (Scott and Georghiou 1986a), the increase in cytochrome P_{450} content is less than four-fold. In the Learn-PyR strain of house fly, increased levels of cytochrome P_{450} are accompanied by a threefold increase in NADPH–cytochrome P_{450} reductase activity, which is under separate genetic control (Scott and Georghiou 1986b). Enhanced levels of the reductase component of the monooxygenase system have not previously been reported in resistant insects. Scott and Georghiou (1986a) hypothesize that higher reductase levels can contribute to enhanced substrate oxidation by increasing the rate of electron transport. Direct evidence to support this hypothesis is currently not available.

Qualitative differences in the cytochromes P_{450} of resistant and susceptible insects have been inferred from the unique spectral properties of cytochrome P_{450}–carbon monoxide and other cytochrome P_{450}–ligand complexes in microsomes prepared from resistant flies. These studies, which are reviewed comprehensively elsewhere (Hodgson 1985), suggest that resistant flies may contain one or more unique cytochrome P_{450} forms expressed at a high level, but they are unable to address the several mechanisms by which these forms might arise. However, a recent study of cytochrome P_{450} induction in resistant and susceptible strains suggests that induction produces only quantitative changes in cytochrome P_{450} in a resistant strain but produces both quantitative and qualitative changes in a susceptible strain (Vincent et al. 1985). These findings also provide indirect evidence for qualitative differences between the constitutive cytochromes P_{450} of resistant and susceptible insects and suggest that the relationship between "resistant" and specifically inducible forms of cytochrome P_{450} bears further investigation.

The purification and characterization of individual cytochrome P_{450} isozymes from resistant and susceptible flies is the method of choice to define the molecular basis of enhanced oxidative metabolism. Several laboratories have worked to purify insect cytochrome P_{450} since the mid-1970s, but progress has been modest. Multiple forms of solubilized house fly (Capdevila et al. 1975, Fisher and Mayer 1984, Moldenke et al. 1984, Yu and Terriere 1979) and fruit fly (*Drosophila melanogaster*) (Waters et al. 1984) cytochrome P_{450} have been resolved by column chromatography and electrophoresis. In some of these studies, different forms are recognizable in resistant and susceptible insects (Waters et al. 1984, Yu and Terriere 1979). Application of newer methods for the purification of insect cytochrome P_{450} (Wheelock and Scott 1989) may allow recovery of cytochrome P_{450} proteins in greater yields and at greater purities. Recent progress in the efficient purification of insect NADPH–cytochrome P_{450} reductase (Vincent et al. 1983) and in the reconstitution of cytochrome P_{450}-dependent monooxygenase activity from its purified components (Moldenke et al. 1984, Vincent et al. 1983) may allow functional comparisons of different cytochrome P_{450} forms from resistant and susceptible insects, but no studies of this type have been reported to date.

The ability to detect electrophoretically distinct forms of cytochrome P_{450} in the fruit fly has permitted an analysis of the separate expression and regulation of those forms in susceptible and resistant strains of this species. Cytochrome P_{450}-B, which is the electrophoretic variant found only in resistant strains (Waters et al. 1984), was mapped to a region of chromosome II that was previously shown to contain a gene conferring insecticide resistance (Waters and Nix 1988). Moreover, high levels of expression of this form also required the presence of one or more regulatory elements on chromosome III (Waters and Nix 1988). These findings suggest that the expression of high levels of cytochrome P_{450} in resistant insects may involve a complex interaction between structural and regulatory elements.

Ultimately, the nature, properties, and genetic regulation of cytochrome P_{450} in resistant insects must be addressed using molecular genetic techniques. Although the molecular biology and genetics of mammalian cytochrome P_{450} are well advanced (Nebert and Gonzales 1987), the powerful tools available from these studies are only beginning to be applied to the insect system. Recently, Feyereisen and coworkers (1989) reported the isolation and characterization of a cDNA comprising the principal cytochrome P_{450} from insecticide-resistant, phenobarbital-induced house flies. The deduced amino acid sequence contains key structural elements that are consistent with mammalian cytochrome P_{450} proteins but differs sufficiently from mammalian sequences to implicate the existence of a separate cytochrome P_{450} gene family in insects. The availability of this sequence for an insect cytochrome P_{450} should rapidly accelerate studies of the structure, multiplicity, and regulation of cytochrome P_{450} proteins in many insect species and permit molecular analyses of cytochrome P_{450}-mediated resistance.

B. Hydrolases

1. Hydrolases in Insecticide Metabolism

Hydrolases capable of cleaving carboxylester and phosphorotriester bonds play a significant role in the metabolism of OP and pyrethroid insecticides (Dauterman 1985). Among OP compounds two distinct hydrolytic pathways are found: (1) cleavage of carboxylester groups in malathion, phenthoate, and related compounds and (2) cleavage of the phosphate ester bond of a wide variety of compounds. The latter reactions were originally ascribed to "phosphatase" or "phosphorotriesterase' activity (Dauterman 1985), but it now appears that these classifications are artificial and do not represent distinctive families of enzymes in insects. Among pyrethroids, hydrolysis of the central carboxyl ester bond is an important detoxication reaction (Soderlund et al. 1983). Most of these enzymes are also inhibited by some phosphate esters, but the substrate and inhibitor specificities vary considerably between enzymes. The irreversible or slowly reversible inhibition of hydrolases by some relatively nontoxic OP esters has led to the use of these compounds as synergists to probe the significance of hydrolytic detoxication in limiting insecticide toxicity (Soderlund et al. 1983).

Insect hydrolases exist in multiple forms, which can be resolved by electrophoresis and detected using simple spectrophotometric assays with substrates such as the acetate and butyrate esters of p-nitrophenol and the acetate esters of α- and β-naphthol (αNA and βNA). Using these methods, Maa and Terriere (1983) resolved up to 15 distinct hydrolases in adult susceptible and resistant (Rutgers) house flies. It is likely that only a few of these enzymes participate in insecticide hydrolysis, so that overall comparisons of general esterase activity using simple substrates are not likely to reveal differences in the levels or activities of insecticide-hydrolyzing esterases. The biochemical basis of hydrolytic detoxication may be even more complex in the case of pyrethroids, since insects possess multiple pyrethroid-hydrolyzing esterases that exhibit different specificities for pyrethroid isomers as substrates and OP esters as inhibitors (Soderlund et al. 1983).

2. Evidence for Enhanced Hydrolysis in Resistant Insects

The involvement of hydrolytic mechanisms in resistance has been recognized or inferred in three ways: detection of high levels of insecticide hydrolysis products in metabolism studies with resistant insects; synergism of insecticide toxicity in resistant insects by nontoxic OP esterase inhibitors such as TPP (*O,O,O*-triphenyl phosphate), DEF (*S,S,S*-tributyl phosphorotrithioate, a cotton defoliant), or IBP (*O,O*-bis[1-methylethyl] *S*-phenylmethyl phosphorothioate, a fungicide) (Chapter 3); or detection of high levels of general esterase activity using simple substrates and either spectrophotometric assays with insect or tissue homogenates or staining of electrophoresis gels (Chapter 2). Of these, evidence from metabolism studies is most convincing, but requires the most time and effort. Positive results with synergists also provide strong evidence for a hydrolytic mechanism, but the multiplicity and varied inhibitor specificities of insect esterases render a lack of synergism ambiguous, reflecting either lack of hydrolytic metabolism or the failure of the synergist to inhibit relevant esterases (Chapter 3). Evidence based on studies with simple substrates is least reliable, since the hydrolysis of these compounds may be accomplished by enzymes other than those hydrolyzing insecticides. The involvement of hydrolases in resistance recognized by these methods is summarized in Oppenoorth's (1985) comprehensive review.

As with enhanced oxidation, the role of enhanced hydrolysis in resistance must be confirmed by characterization of the enhanced esterase activity in resistant strains. For most insecticides, these studies require assays of enzyme activity using radiolabeled substrates. However, studies of malathion hydrolysis are aided by the availability of a coupled colorimetric assay for this substrate (Talcott 1979). Recent studies of malathion resistance in mosquitoes exemplify the use of this assay with insect enzymes (Hemingway 1985, Scott and Georghiou 1986c). Studies on insecticide-hydrolyzing esterases in resistant insects published prior to 1983 are reviewed elsewhere (Oppenoorth 1985, Dauterman 1985). In the following sections, we summarize recent progress to define hydrolytic mechanisms in several species at the molecular level.

3. Malathion Carboxylesterases in the House Fly and Other Species

Hydrolysis of either the α or β ethyl ester group of malathion is an important detoxication reaction for this compound. Early studies of malathion-resistant house fly strains (reviewed by Oppenoorth 1985) showed that increased malathion hydrolysis was correlated with decreased αNA hydrolysis. These findings led to the "mutant ali-esterase" theory of hydrolytic detoxication in resistance, in which mutations in the structural gene for a nonspecific esterase produced an altered enzyme that rapidly cleaved malathion (as well as phosphotriester insecticides) with a concomitant reduction in esterase activity toward αNA. More recently, the Hirokawa strain of house fly from Japan has been shown to possess high malathion carboxylesterase activity without loss of activity toward simple substrates (Kao et al. 1984, Motoyama et al. 1980, Picollo de Villar et al. 1983). Esterases in this strain hydrolyze malathion much more rapidly than those in susceptible flies and also hydrolyze the carboxylester moiety of malaoxon. Purification of carboxylesterases from the Hirokawa strain and a susceptible strain by chromatofocusing provided evidence for enhanced hydrolysis of malathion by a single esterase (pl 5.1 esterase) in the Hirokawa strain (Kao et al. 1985). Differences in substrate affinity, inhibitor sensitivity, and efficacy of paraoxon hydrolysis between the pl 5.1 esterases of the Hirokawa and susceptible strains suggested that this esterase is modified in the resistant strain.

Biochemical studies of hydrolytic mechanisms of malathion resistance in mosquitoes illustrate the pitfalls of using general esterase substrates as an index of insecticide-hydrolyzing activity (Chapter 3). In both *Anopheles stephensi* (Hemingway 1982, 1984) and *A. arabiensis* (Hemingway 1983), the toxicity of malathion to resistant strains was greatly increased by the use of esterase-inhibiting synergists, thus implicating a hydrolytic resistance mechanism. Although susceptible and resistant strains did not differ in their ability to hydrolyze αNA and βNA (Hemingway 1982, 1983), subsequent studies demonstrated increases in malathion-specific hydrolysis in the resistant strains (Hemingway 1985, Scott and Georghiou 1986c). A similar situation was also found in malathion-resistant *Culex tarsalis*, which exhibited extensive malathion hydrolysis in vivo and in vitro but negligible increases in the in vitro hydrolysis of αNA (Ziegler et al. 1987). Electrophoretic separation of hydrolases from susceptible and resistant insects followed by assay of gel regions for the hydrolysis of αNA and malathion revealed a single peak of intense malathion-hydrolyzing activity in the resistant strain that did not correspond to any of the esterases detected by αNA staining.

Although enhanced malathion carboxylesterase activity has also been inferred in other species, the esterases involved have been characterized in very few cases. In one of the earliest studies of malathion resistance, Matsumura and Voss (1965) partially purified and characterized the malathion-hydrolyzing activity of malathion resistant and susceptible two-spotted spider mites (*Tetranychus urticae*). The enzyme from resistant mites had a 20-fold higher apparent affinity for malathion than the corresponding activity from susceptible mites and also exhibited unique physical properties. In malathion-resistant Indianmeal moth larvae, enhanced malathion hydrolysis is also correlated with reduced esterase activity toward αNA (Beeman and Schmidt 1982). These authors also cite unpublished data showing that the kinetics of malathion hydrolysis differ in enzyme preparations from resistant and susceptible insects.

In many of the cases where the properties of malathion-hydrolyzing esterases from resistant and susceptible insects have been compared, the enzymes from resistant insects exhibit altered kinetic and physical properties and, in some cases, altered substrate and inhibitor specificities. These observations are consistent with the "mutant ali-esterase" hypothesis for the origin of enhanced malathion hydrolysis, but this hypothesis has yet to be tested at the genetic level. The most rigorous test would involve cloning and sequencing the gene for malathion carboxylesterase from a resistant strain and the corresponding carboxylesterase locus from a susceptible strain and pursing a complete structural and functional analysis of both genes and their products.

4. Overproduction of an Insecticide-Hydrolyzing Esterase in Myzus persicae

High levels of αNA hydrolysis were found in organophosphate-resistant strains of the peach-potato aphid (*Myzus persicae*) (Devonshire 1975). The increased αNA-hydrolyzing activity was localized in the soluble fraction of aphid homogenates and associated with exceptionally high levels of activity of a single electrophoretically distinct esterase, E4 (Devonshire 1977). Purification of E4 revealed that the enzyme from susceptible and resistant aphids had identical kinetic properties, thereby implicating overproduction of an unaltered enzyme as the mechanism of resistance (Devonshire 1977). Differences in dimethoate resistance between 10 aphid strains were closely correlated with differences in αNA-hydrolyzing activity (Devonshire 1975). In highly resistant aphid clones, the amount of this enzyme reaches 10–12 pmol/aphid, or approximately 3–4% of the total aphid protein (Oppenoorth 1985). Resistant aphid populations obtained from Ferrarra, Italy, contain a mutant form of E4 (called FE4), which exhibits similar activity toward insecticides but hydrolyzes αNA more rapidly than E4 and has slightly different electrophoretic mobility (Devonshire et al. 1983). For a series of seven aphid clones with increasing levels of resistance, the amounts of E4 per aphid also increased in a relationship that closely approximated a geometric progression with a factor of 2 (Devonshire and Sawicki 1979). This finding suggested that the increased amounts of E4 resulted from progressive duplications of the structural gene for this enzyme. The subsequent cloning of the E4 gene has permitted the direct demonstration of duplication of the E4 locus in resistant aphids (Field et al. 1988).

E4 is a typical carboxylesterase, which rapidly hydrolyzes αNA and other simple substrates and is inhibited by OP and carbamate insecticides, which are cleaved to form an inactivated acylated enzyme (Devonshire and Moores 1982). The rate of reactivation of the acylated E4 depends on the nature of the acyl moiety: The dimethylphosphorylated enzyme recovers most rapidly, followed by the diethylphosphorylated and monomethyl- or dimethylcarbamylated enzymes, whereas the presence of larger alkyl substituents generally results in a permanently inactivated enzyme. The slow deacylation rates for these reactions make most

OP and carbamate insecticides very poor substrates for E4. Nevertheless the large amounts of E4 present in resistant aphids are able to react with and sequester sufficient amounts of penetrating insecticide to cause the observed resistance.

5. Overproduction of Esterases Associated with Resistance in Culex Species

Studies of esterase polymorphism in *Culex quinquefasciatus* from California and *C. pipiens* from France have correlated broad cross-resistance to OP insecticides with the appearance of new, strongly staining esterase bands following electrophoretic separations (Georghiou and Pasteur 1978, Pasteur and Sinegre 1975). In *C. quinquefasciatus*, the highly active enzyme (designated esterase B2) preferentially hydrolyzes βNA, whereas in *C. pipiens* the active enzyme (designated esterase A' or esterase–3) is observed only in the presence of EDTA and preferentially hydrolyzes αNA. Genetic analyses of esterase polymorphisms in both species have shown that the esterases B2 and A' are products of loci that are located on the same chromosomes and tightly linked with the genes conferring insecticide resistance (Georghiou et al. 1980, Pasteur et al. 1981a,b). These analyses, together with the efficacy of DEF as a synergist in resistant *C. quinquefasciatus* (Georghiou et al. 1985), have led to the hypothesis that the observed esterases are involved in insecticide hydrolysis or sequestration, as has been found for resistant *Myzus persicae*. Moreover, in *C. quinquefasciatus,* three different staining intensities of esterase B were observed, suggesting that gene duplication may play a role in the different levels of resistance observed with this species (Georghiou and Pasteur 1980).

Recent efforts have been directed at testing the gene duplication hypothesis using the tools of molecular genetics. Mouches et al. (1985) detected amplified sequences of DNA in restriction enzyme digests of total genomic DNA from resistant *C. quinquefasciatus* and in an OP resistant house fly strain, but did not detect unique amplified sequences in resistant *C. pipiens*. In subsequent studies (Fournier et al. 1987), esterase B1 from resistant *C. quinquefasciatus* was purified to homogeneity and used to prepare polyclonal antibodies that specifically immunoprecipitate this enzyme. Immunoassays using these antibodies have shown that highly resistant *C. quinquefasciatus* contain at least 500-fold higher levels of esterase B1 than susceptible strains (Mouches et al. 1987). A portion of the gene encoding esterase B1 has been cloned, and this clone has been used as a probe to detect esterase B1–encoding sequences in restriction enzyme digests of genomic DNA (Mouches et al. 1986). These studies identified an amplified 2.1-kilobase fragment that hybridized with the labeled probe and presumably contains a portion of the esterase B1 gene. This fragment is identical in size to one of the several amplified fragments found in earlier studies. Taken together, these results provide strong evidence for multiple duplication of the esterase B1 gene in resistant *C. quinquefasciatus*. The same mechanism may be involved in the high levels of esterase A' in *C. pipiens*, but in this case the levels of enzyme detected by immunoassay are considerably lower than those in *C. quinquefasciatus* (Mouches et al. 1987), and preliminary evidence for amplified gene sequences has not been confirmed using a specific probe.

Although the extent of molecular genetic information on esterases B1 and A' is greater than that for most other enzymes implicated in insecticide resistance, the actual roles of these esterases in insecticide detoxication have not been shown directly. In neither case has the insecticide-hydrolyzing capability of the enzyme been characterized, nor have effective synergists been shown to block the enzyme and prevent labeling of the active site by insecticides. Thus, the role of these enzymes in resistance is inferred only from the tight linkage of the esterase and resistance genes and from the expectation that such an enzyme ought to hydrolyze or sequester OP insecticides. Until interactions between the esterases and insecticides are documented in susceptible and resistant insects, these enzymes should be viewed as reliable biochemical markers for resistance rather than as "detoxifying esterases."

6. Esterases in Pyrethroid Resistance

Despite the central role of esterases in pyrethroid metabolism in arthropods, very few instances are known in which resistance is clearly linked with enhanced esterase activity. In three strains of the Egyptian cotton leafworm (*Spodoptera littoralis*), low levels of pyrethroid resistance were correlated with enhanced hydrolysis of αNA, and both resistance and the extent of αNA hydrolysis were reduced by DEF (Riskallah 1983). In field-collected strains of the house fly, strong resistance to trichlorfon and malathion and slight resistance to pyrethroids was correlated with the high αNA-hydrolyzing activity of an electrophoretically distinct esterase, $E_{0.39}$ (Sawicki et al. 1984). As with OP-resistant *Culex* mosquitoes, genetic studies demonstrated very close linkage between the esterase and resistance genes. These findings implicate hydrolytic involvement in pyrethroid resistance in these two species, but further biochemical studies using pyrethroids as esterase substrates are required.

Esterase-dependent cross-resistance between OPs and pyrethroids has also been shown in the aphid, *Myzus persicae* (Devonshire and Moores 1982). In addition to its role in OP resistance, the esterase E4 also hydrolyzes the 1*S,trans* isomer of permethrin but not the 1*R,trans* isomer or either *cis* isomer. Since the 1*S,trans* isomer of permethrin does not possess significant insecticidal activity (Burt et al. 1974), the observed cross-resistance of these aphids to permethrin and other pyrethroids is not explained by metabolic detoxication. However, it is possible that the large amounts of E4 bind but do not hydrolyze the toxic pyrethroid isomers in amounts sufficient to cause the observed resistance.

Metabolism studies and bioassays with synergists have implicated enhanced hydrolysis of pyrethroids as one of two resistance mechanisms in the cattle tick,

Boophilus microplus (Schnitzerling et al. 1983). Biochemical studies (de Jersey et al. 1985, Riddles et al. 1983) on the esterases of susceptible (Y) and resistant (M) strains resolved four esterase fractions from each strain (designated Y1–Y4 and M1–M4), corresponding to a total of at least 15 distinct isozymes, using preparative isoelectric focusing and detection by αNA staining. Esterase fractions Y1, M1, and M2 exhibited unusual substrate specificity, rapidly hydrolyzing *trans*-cypermethrin with little or no activity toward *trans*-permethrin. Esterase fractions Y3 and Y4 preferentially hydrolyzed *trans*-permethrin, whereas fractions Y2, M3, and M4 hydrolyzed both pyrethroids at approximately equal rates. In the M strain, hydrolysis of *trans*-cypermethrin was increased almost two-fold in fraction 1, the cypermethrin-specific activity, whereas the hydrolysis of both pyrethroid substrates by fraction 4 enzymes was increased more than five-fold. In contrast, the hydrolysis of both substrates by fraction 3 esterases was decreased in the M strain. These results suggest that the high activity of fraction 4 enzymes in the M strain is the principal cause of the esterase-dependent resistance. However, since each isoelectric focusing fraction contains more than one enzyme, the results are unable to show whether the activity and specificity difference between strains involves a mutant enzyme mechanism or an overproduction of one or more constitutively expressed esterases.

Recently, a hydrolytic mechanism of pyrethroid resistance has been described in the tobacco budworm, *Heliothis virescens* (Dowd et al. 1987). Profenofos, an OP insecticide, increased the toxicity of *trans*-permethrin to the resistant strain to a level similar to that found for susceptible insects, thus implicating a major role for esterases in resistance. Homogenates of resistant larvae hydrolyzed *trans*-permethrin more rapidly than equivalent preparations from susceptible larvae. Isoelectric focusing of esterases revealed both higher levels of esterase activity and the presence of an additional peak of activity in the resistant strain. These results suggest that both qualitative and quantitative changes are involved in the enhanced hydrolytic capabilities of the resistant strain.

Hydrolysis is also implicated as one of two mechanisms of resistance to deltamethrin in *Spodoptera exigua* populations in Guatemala (Delorme et al. 1988). Resistant larvae exhibited higher hemolymph esterase activity toward αNA than susceptible larvae, and whole homogenates of resistant larvae converted radiolabeled deltamethrin to ester cleavage products more extensively than equivalent preparations from susceptible larvae. The extent of hydrolysis of both αNA and deltamethrin by preparations of resistant larvae was decreased by the esterase inhibitors DEF and paraoxon. In this species the mechanism underlying increased hydrolysis remains to be investigated.

In the predatory mite, *Amblyseius fallacis,* pyrethroid resistance is also correlated with enhanced esterase activity (Chang and Whalon 1986). Resistant mites exhibit enhanced activity in several electrophoretic fractions toward both αNA and *trans*-permethrin as substrates. This study was not sufficiently detailed to determine whether the high esterase activities in resistant mites resulted from mutant esterases or from duplication of genes for enzymes present at low levels in the susceptible strain.

C. Glutathione-*S*-Transferases

1. Glutathione-S-Transferases in Insecticide Metabolism

Glutathione-*S*-transferases (GSH-transferases) catalyze the nucleophilic attack of the endogenous tripeptide glutathione (GSH) on a variety of reactive substrates. These enzymes are of particular importance in the metabolism of OP insecticides, in which either the alkyl or aryl substituents on phosphorus can be cleaved (Dauterman 1985). The role of GSH-transferases in OP insecticide metabolism cannot be directly inferred from either bioassays or metabolism studies. No specific enzyme inhibitors are available for use as synergists in bioassays, and the phosphorus ester metabolites resulting from GSH-transferase attack are often identical to those produced by the action of monooxygenases or hydrolases on the same insecticide substrate. Thus, evidence for GSH-transferase involvement in OP insecticide metabolism *in vivo* requires demonstration of the formation of GSH conjugates of the cleaved alkyl or aryl moieties using appropriately radiolabeled insecticides. Subsequent biochemical confirmation requires the demonstration of GSH-dependent stimulation of insecticide metabolism by tissue homogenates and subfractions that is not sensitive to inhibition by monooxygenase and hydrolase inhibitors.

Recent studies have shown that insect GSH-transferases, like monooxygenases and hydrolases, exist in multiple molecular forms with distinct physical and catalytic properties (Clark and Dauterman 1982, Clark et al. 1984, Motoyama et al. 1983). Consequently, correlations of enzyme activity with insecticide resistance are complicated by the specificities of the individual enzymes for both insecticides and model substrates used in spectrophotometric assays of GSH-transferase activity.

2. Evidence for Enhanced GSH-Transferase Activity in Resistant Insects

Nearly all the studies of the role of GSH-transferases in insecticide resistance involve various OP-resistant strains of the house fly. Biochemical studies using whole insect or insect tissue homogenates and subfractions provide the most convincing evidence for enhanced GSH-transferase activity in resistant strains. Resistant house fly strains show increases ranging from two-fold to 30-fold in the rates of model substrate conjugation and increases up to 100-fold in the conjugation of insecticide substrates when compared to insecticide susceptible strains (Oppenoorth 1985). Large variations in the ratios of the rates of conjugation of two model substrates, CDNB (1-chloro-2,4-dinitrobenzene) and DCNB (1,2-dichloro-4-nitrobenzene), and in the ratios of O-alkyl and O-aryl conjugation of diazinon in partially purified preparations from several house fly strains suggested that these reactions involved separate GSH-transferases under separate genetic control (Motoyama et al. 1983).

The existence in house flies of multiple GSH-transferases having unique physical properties is evident in recent enzyme purification studies. Purification by affinity chromatography and preparative isoelectric focusing revealed two clearly defined groups of enzymes: one with relatively low isoelectric points and high specificity for CDNB over DCNB as a substrate and a second with relatively high isoelectric points and less substrate specificity (Clark et al. 1984). Insecticide degradation appears to be specifically associated with the second group of enzymes (Clark et al. 1986). Two OP-resistant strains differed both from a susceptible strain and from each other in the electrophoretic mobility and substrate specificity of the major GSH-transferases obtained upon purification (Clark et al. 1984). Electrophoresis of purified enzymes under denaturing conditions revealed dimeric structures with three size classes of subunits represented. These workers concluded that at least three types of GSH-transferase activity are present in each strain and that the differing specificities and levels of activity observed reflect the relative proportions of individual enzymes in each strain. It is particularly noteworthy that affinity chromatography yields a much greater factor of purification of GSH-transferase activity from the susceptible strain than from either resistant strain (Clark et al. 1984). These data suggest that the high levels of enzyme activity in resistant insects result from overproduction of one or more GSH-transferases. In this context, it is interesting that a resistant house fly strain with high GSH-transferase activity was found to contain amplified DNA sequences not found in a susceptible strain (Mouches et al. 1985).

GSH-transferases from susceptible and resistant house flies also differ in their inducibility by phenobarbital and certain pesticides. Phenobarbital induction was greatest with strains having low basal levels of GSH-transferase and conferred some protection from OP insecticide intoxication (Hayaoka and Dauterman 1982, Ottea and Plapp 1981). Subsequent studies provided evidence for specific induction of one form of GSH-transferase (Hayaoka and Dauterman 1983, Ottea and Plapp 1984). In one case the induced enzyme preferentially conjugated DCNB and the O-alkyl groups of diazinon (Hayaoka and Dauterman 1983).

3. DDT-Dehydrochlorinase as a Specialized GSH-Transferase

The enzymatic, glutathione-dependent dehydrochlorination of DDT to DDE is an important detoxication reaction and was selected as a major mechanism of DDT resistance, particularly in house flies (Oppenoorth 1985). This enzyme activity, named "DDT-dehydrochlorinase" or "DDT-ase," has been studied for more than 30 years, but its relationship to known enzyme activities remained obscure until recently. Using the affinity chromatography–isoelectric focusing strategy developed for house fly GSH-transferases, Clark and Shamaan (1984) found that two peaks of DDT-ase activity from DDT-resistant flies copurified with a neutral GSH-transferase activity exhibiting specificity toward DCNB over CDNB. Isoelectric focusing revealed additional acidic GSH-transferases that attacked CDNB but not DCNB or DDT. The highly purified neutral enzyme having the majority of DDT-ase activity was found to be a heterodimer, a subunit structure found in the analyses of GSH-transferases from OP-resistant house flies. These findings strongly suggest that DDT-ase is one of several GSH-transferases present in house flies.

III. Resistance Conferred by Reduced Neuronal Sensitivity

A. Altered Acetylcholinesterase

1. Occurrence of Altered Acetylcholinesterase in Arthropods

OP and carbamate insecticides exert their neurotoxic effects by inhibiting the enzyme acetylcholinesterase (AChE), thereby prolonging the residence time of acetylcholine at cholinergic synapses and producing hyperexcitation of cholinergic pathways. AChE that is less sensitive to inhibition by OP and carbamate insecticides has been documented in resistant strains of several insect, tick, and mite species. Most of the biochemical work describing the activity and inhibitor sensitivity of mutant enzymes was published prior to 1982 and is reviewed elsewhere (Hama 1983, Oppenoorth 1985). More recent studies have reported assay methods capable of determining the biochemical phenotype of individual insects (Devonshire and Moores 1984a, Hemingway et al. 1986, Raymond et al. 1985a,b) and have employed these methods to detect differing mutant enzymes in the house fly (Devonshire and Moores 1984b) and identify altered AChE as a resistance mechanism in mosquito populations (Hemingway et al. 1986, Raymond et al. 1985a,b; Chapter 2).

2. Progress in Defining the Molecular Basis of Altered Inhibitor Sensitivity

Information on the molecular basis of the reduced sensitivity of AChE is limited. In resistant house flies, cattle ticks, and green rice leafhopper (*Nephotettix cincticeps*) the reduced sensitivity of the resistant AChE to inhibitors appears to result from reduced affinity of the enzyme for the inhibitor molecule rather than from alterations in the rate constant for acylation (Oppenoorth 1985). In the case of the green rice leafhopper, steric hindrance by bulky substituents on phenylmethyl carbamates plays a significant role in the reduced affinity of the resistant enzyme for these inhibitors (Hama 1983). In one house fly strain, the mutant enzyme displays a somewhat higher affinity for the colorimetric substrate, acetyl thiocholine (Devonshire and Moores 1984b). More detailed or comprehensive analyses of altered inhibitor interactions with mutant AChE are lacking.

The existence of multiple forms of mutant AChE in the house fly and the cattle tick is inferred from the unique resistance spectra of enzymes from different strains or from assays of individual insects (Devonshire and Moores 1984b, Oppenoorth 1982, 1985). It is of obvious interest to correlate these biochemical phenotypes with structural changes at the genetic level. Recent progress in cloning and sequencing the *Ace* (AChE-encoding) locus from *Drosophila melanogaster* (Hall and Spierer 1986) provides an entry point for the application of the techniques of molecular biology to this problem. The use of probes derived from *Drosophila* (Berge et al. 1986, Chapter 5) may offer the opportunity to define the specific mutations that confer reduced sensitivity of AChE to inhibitors.

B. Reduced Neuronal Sensitivity to DDT and Pyrethroids

1. *Neurotoxic Actions of DDT and Pyrethroids*

Elucidation of the mechanisms that might result in reduced neuronal sensitivity to DDT and pyrethroids depends on knowledge of the mode of action of these compounds. DDT and pyrethroids are capable of disrupting the normal function of many enzymes, neuroreceptors, and ion channels in vitro, but the toxic actions of these compounds in insects and at the level of isolated nerves are best explained by their action on the voltage-sensitive sodium channel of nerve membranes (Soderlund and Bloomquist 1989). Consequently, we consider the voltage-sensitive sodium channel to be the principal site of action of DDT and pyrethroids and have synthesized available information on possible mechanisms of reduced neuronal sensitivity to these compounds from this perspective.

2. *Knockdown Resistance (kdr) in the House Fly*

a. Properties and Resistance Spectrum. A factor conferring resistance to both the rapid paralytic ("knockdown") and lethal actions of DDT and pyrethrins was first observed in adult house flies in 1951 (Busvine 1951) and was isolated genetically in 1954 (Milani 1954). Several alleles of this recessive factor (called *kdr*) have been described and mapped to autosome 3, including alleles (called *super-kdr*) that confer much greater resistance than that found in *kdr* strains (Farnham 1977, Farnham et al. 1987, Sawicki 1978). Recently, the development of a bioassay to measure rapid paralysis of house fly larvae has shown that the *kdr* and *super-kdr* phenotypes are expressed in immature stages as well (Bloomquist and Miller 1985). The possible involvement of reduced neuronal sensitivity in *kdr* house flies was suggested by the failure of synergists to increase the toxicity of DDT and pyrethroids and by the location of the *kdr* gene on autosome 3, a locus not known to be involved in metabolic mechanisms of resistance to these and other insecticides (Farnham 1977). Subsequent studies of pyrethroid metabolism in *kdr* strains have confirmed the absence of enhanced detoxication mechanisms (Nicholson et al. 1980a). Overviews of *kdr* as a mechanism of pyrethroid resistance are given by Sawicki (1985) and Shono (1985).

The *kdr* mechanism is recognized in part by the unique spectrum of cross-resistance it affords (Chapter 3). In adult house flies, the *kdr* mechanism confers resistance to DDT and to all known pyrethroid and pyrethroid-derived insecticides (Farnham 1977, Farnham et al. 1987, Pedersen 1986, Sawicki 1978) but not to dicofol (Sawicki 1978) (a carbinol metabolite of DDT that is a potent miticide) or to cyclodiene, OP or carbamate insecticides (Farnham 1977, Sawicki 1978). Although all house fly strains with the *kdr* factor show broad cross-resistance to pyrethroids and DDT, allelic variants at the *kdr* locus exhibit unique pyrethroid resistance spectra (Farnham et al. 1987). The central role of the voltage-sensitive sodium channel of nerve membranes in the action of DDT and pyrethroids has prompted studies of the possible cross-resistance of *kdr* insects to other sodium channel–directed neurotoxins. Larvae of *kdr* and *super-kdr* house flies were at least 16-fold resistant to the plant alkaloid aconitine, a sodium channel activator that binds to a site that is thought to be different from but coupled to the insecticide recognition site (Bloomquist and Miller 1986). Bioassays with adult house flies confirm the resistance of the same *kdr* strain to aconitine but this resistance does not extend to veratridine, another alkaloid activator thought to act at the same binding site as aconitine (A. E. Lund personal communication). In contrast, larvae of both strains were not resistant to tetrodotoxin, procaine, or the α-toxin present in the venom of the scorpion *Leiurus quinquestriatus*, all of which are known to act at other binding sites on the sodium channel (Bloomquist and Miller 1986). These results suggest that the *kdr* mechanism involves selective modification of only some of the neurotoxin recognition properties of the sodium channel.

It must be emphasized that *kdr* in strictest terms refers only to resistance-conferring alleles in the house fly that map to a common locus on autosome 3. Other pyrethroid-resistant house fly strains that exhibit unaltered pyrethroid metabolism (De Vries and Georghiou 1981a) or reduced nerve sensitivity in physiological assays (Ahn et al. 1987, De Vries and Georghiou 1981b) also appear to possess a *kdr*-like mechanism, but in the absence of genetic mapping studies it is not clear whether these factors result from mutations at the *kdr* locus or from mutations at other loci that also confer reduced neuronal sensitivity.

b. Physiological Characterization. Intrinsic reduced sensitivity of the adult house fly central nervous system to the neuroexcitatory effects of DDT was first shown in 1965 (Tsukamoto 1965). It is likely that the strain used in these studies, which carried a nonmetabolic recessive resistance factor on autosome 3, was in fact a *kdr* strain. Several subsequent studies have shown that the following nerve preparations from both adults and larvae of *kdr* and *super-kdr* strains are resistant to both DDT and pyrethroids: adult thoracic ganglia (Miller et al. 1979), indirect flight muscle motor neurons (Scott and Georghiou 1986a), and leg muscle motor neurons (Ahn et al. 1986); larval neuromuscular junction (Salgado et al. 1983a); and larval sensory neurons (Osborne and Hart 1979, Osborne and Smallcombe 1983). In each case, production of abnormal nerve function in preparations from *kdr* and *super-kdr* insects required longer periods of insecticide exposure or higher insecticide concentrations than in equivalent assays with preparations from susceptible insects. These data show that reduced sensitivity to insecticides is broadly distributed within the nervous system of *kdr* insects.

Physiological assays have also been used to explore the cross-resistance of *kdr* insects to other sodium channel–directed neurotoxins. Larval sensory nerves from susceptible and *kdr* insects did not differ in their sensitivity to tetrodotoxin, a sodium channel blocker, or *Leiurus quinquestriatus* venom, which contains a polypeptide toxin that prolongs the open state of the sodium channel (Osborne and Smallcombe 1983). However, these preparations in *kdr* larvae were approximately 100-fold resistant to the actions of veratridine, an alkaloid that opens sodium channels. Surprisingly, no resistance was noted to aconitine, another alkaloid activator thought to act at the same binding site as veratridine. In contrast, Salgado et al. (1983a) found that motor nerve terminals of *kdr* larvae were clearly resistant to aconitine. These findings stand in sharp contradiction to each other, and the sensory nerve data (Osborne and Smallcombe 1983) also contradict available bioassay data for aconitine and veratridine (Bloomquist and Miller 1986, A. E. Lund personal communication). Further studies are needed to resolve these inconsistencies and determine the extent of cross-resistance to alkaloid sodium channel activators afforded by the *kdr* mechanism.

c. Biochemical Characterization. Despite the wealth of physiological data implicating the voltage-sensitive sodium channel as the principal target site for DDT and pyrethroids (Soderlund and Bloomquist 1989), the molecular basis of the interaction of insecticides with this target remains obscure. The lack of biochemical and pharmacological methods to explore and define these interactions in insect nervous tissue has necessarily hampered a more detailed examination of the molecular basis of the *kdr* mechanism. The limited number of studies published to date illustrate both the difficulties inherent in these approaches and the insights that might be gained by further studies.

One approach to defining insecticide–target site interactions is to use the radiolabeled insecticide itself as a ligand to describe the insecticide target site. Chang and Plapp (1983a,b) used [^{14}C]DDT and [^{14}C] [1RS,cis]-permethrin in an effort to define insecticide binding to house fly head membranes in relation to insecticide action and the mechanism of *kdr*. The binding of DDT and [1RS,cis]-permethrin measured in these studies was interpreted to reflect the toxicologically relevant target. Moreover, the 40% reduction in the apparent density of insecticide binding sites inferred from the results obtained with preparations from resistant insects was proposed as a mechanism of reduced neuronal sensitivity (Osborne and Smallcombe 1983). Unfortunately, the value of these studies in defining the DDT/pyrethroid target site and its possible modification in resistant insects is compromised by significant methodological flaws. First, the resistant strain used by these authors exhibited only threefold resistance to permethrin and required continued selection with DDT to maintain constant levels of resistance. In contrast, the well-characterized *kdr* strain 538ge typically exhibits at least 10-fold resistance to pyrethroids that is stable in continuous culture without selection (Farnham 1977). These observations suggest that the strain used by Chang and Plapp was not homozygous for the *kdr* trait (Oppenoorth 1985). Second, the low specific activities of the radioligands used in these studies restricted all observations to those describing low-affinity binding. It is unlikely that the measurable interactions include toxicologically relevant high-affinity interactions between these potent neurotoxicants and their target site. Finally, the studies with [1*RS,cis*]-permethrin were not controlled for the stereospecific recognition of only the toxic 1*R,cis* isomer by the pyrethroid site of action. In light of these difficulties, the conclusions drawn from these data regarding a reduction in insecticide target site density in *kdr* insects provide limited insight into the mechanism of this resistance factor.

A second, related approach is based on the assumption that the pyrethroid–DDT target site is intimately associated with the voltage-sensitive sodium channel. In this case, variations in sodium channel density or neurotoxin recognition properties can be determined by the characterization of binding sites for known sodium channel radioligands in subcellular preparations from susceptible and *kdr* insects. Rossignol (1988) described the binding of [^{3}H]saxitoxin to sodium channels in house fly head membranes. This ligand, a potent and selective blocker of sodium channels, is particularly suitable for this approach. These studies documented a 40–60% reduction in the number of saxitoxin binding sites in membranes from *kdr* flies with no significant difference in affinity for this ligand between strains. Since the number of binding sites for α-bungarotoxin, which specifically labels acetylcholine receptors, was identical in preparations from both strains, the apparent reduction in sodium channel density in *kdr* membranes was not due to a lower overall yield of nerve membranes in head preparations from this strain. In contrast to these findings, recent studies in our laboratory (Grubs et al. 1988) using well-defined susceptible and *kdr* house fly strains failed to document any differences between susceptible and resistant insects in affinity or binding site density for [^{3}H]saxitoxin at assay temperatures ranging from 4°C to 37°C. Similar comparisons of the binding capacity for [^{3}H]saxitoxin in susceptible and *super-kdr* flies also failed to detect any strain-specific differences (Pauron et al. 1989, Sattelle et al. 1988). Thus, it appears that a reduction in sodium channel density is not specifically associated with the *kdr* mechanism.

[^{3}H]Batrachotoxinin A 20-α-benzoate (BTX-B), an analog of batrachotoxin that labels the alkaloid activator site of the sodium channel, can be used to assess indirectly the interactions of pyrethroids and DDT with their binding domain on the sodium channel. The precedent for this approach has been established in recent studies (Brown et al. 1988, Lombet et al. 1988, Payne and Soderlund 1989) that demonstrate that DDT and neurotoxic pyrethroids enhance the binding of [^{3}H]BTX-B to rat and mouse brain sodium channels. A recent extension of this approach to house fly head membrane preparations documented a similar effect of deltamethrin on the binding of [^{3}H]BTX-B to membranes prepared from two susceptible house fly strains but failed to detect the enhancement of binding in equivalent preparations from strains carrying the *super-kdr* resistance mechanism (Pauron et al. 1989). These results provide evidence for alterations in the binding domain for DDT and pyrethroids on the voltage-sensitive sodium channel as the molecular basis of the *kdr* mechanism.

A third approach has considered the possible role of alterations in the nerve membrane lipid environment of the sodium channel in conferring reduced neuronal sensitivity. Chiang and Devonshire (1982) measured the temperature-dependent catalytic activity of membrane-bound acetylcholinesterases in house fly head preparations to determine indirectly the transition temperatures of the nerve membranes of susceptible, *kdr*, and *super-kdr* insects. Both resistant strains exhibited higher transition temperatures than the susceptible strain, with a higher value for *super-kdr* preparations than that from *kdr* insects. Moreover, the inheritance of abnormal transition temperatures was recessive, as would be expected for a trait involved in the *kdr* mechanism. These results implicate a role for alterations in the lipid composition of *kdr* nerve membranes in the mechanism of resistance.

3. Reduced Neuronal Sensitivity Mechanisms in Other Species

a. Criteria for Recognizing Reduced Neuronal Sensitivity Mechanisms. The possibility that the widespread use of DDT may have previously selected *kdr*-like resistance mechanisms conferring broad cross-resistance to pyrethroids in a variety of agricultural pests and disease vectors has intensified efforts to characterize these mechanisms. The involvement of reduced nerve sensitivity mechanisms in species other than the house fly has often been invoked when diagnostic tests for other mechanisms have given negative results. Although the failure of synergists to increase the toxicity of DDT or pyrethroids to resistant strains and the lack of enhanced esterase or monooxygenase activity toward model substrates in biochemical assays may suggest the involvement of such a mechanism, two additional important types of experiments should be done to confirm that reduced nerve sensitivity is involved. First, comparative metabolism studies using radiolabeled insecticide are required to rule out completely the involvement of metabolic mechanisms. Second, appropriate physiological assays of insecticide actions on nerve preparations must be established, and these preparations must exhibit intrinsic reduced sensitivity in resistant strains. Finally, it must be noted that cross-resistance between DDT and pyrethroids or even broad cross-resistance to all pyrethroids may not be a feature of all reduced nerve sensitivity mechanisms. Reliance on the details of the house fly system as a paradigm for all *kdr*-like phenomena may hinder the elucidation of the unique details of such mechanisms in other species.

By these criteria, there appear to be a limited number of other species or species groups where substantial evidence supports the existence of reduced nerve sensitivity resistance to DDT and pyrethroids. In other cases where such a mechanism has been proposed, critical evidence is lacking.

b. Mosquitoes. Evidence for a *kdr*-like mechanism of resistance in mosquitoes was first reported by Plapp and Hoyer (1968), who identified a recessive gene unrelated to DDT-ase in DDT-resistant *Culex tarsalis* that conferred cross-resistance to a pyrethrins/PBO combination. Subsequent studies have identified similar mechanisms in *Aedes aegypti* (Chadwick et al. 1977, 1984), *C. quinquefasciatus* (Halliday and Georghiou 1985, Priester and Georghiou 1978, 1980), and *Anopheles stephensi* (Omer et al. 1980). Although no subsequent studies with the *C. tarsalis* strain of Plapp and Hoyer have been reported, additional characterization of the latter three species has provided substantial evidence supporting a *kdr*-like nerve insensitivity mechanism.

A failure of bioresmethrin to control *A. aegypti* in Bangkok, Thailand, in 1975 led to the characterization of DDT and pyrethroid resistance in this species (Chadwick et al. 1977). The field-collected (BKK) strain was highly resistant to DDT and moderately resistant to several pyrethroids, and neither PBO nor a DDT-ase inhibitor was able to reduce the level of resistance observed. Further mass selection of this strain with permethrin produced a substrain (BKPM) having greater resistance to several pyrethroids and DDT (Chadwick et al. 1984), and single-family sib selection of the BKPM strain with permethrin for three generations produced a strain (BKPM3) exhibiting apparent homogeneity for permethrin resistance (Malcolm and Wood 1982a). Comparisons of the penetration and metabolism of [^{3}H] [1*R*,*trans*]-permethrin in adult susceptible and BKPM mosquitoes failed to identify differences in rates of penetration or metabolism between strains (Brealey et al. 1984). Moreover, these studies showed that BKPM insects required a much higher internal concentration of insecticide to produce toxic effects, thus indirectly providing evidence for a nerve insensitivity mechanism. Genetic studies with the BKPM3 strain identified a single pyrethroid resistance factor (R^{py}) on chromosome III (Malcolm and Wood 1982b) and DDT resistance factors (R^{DDT} and R^{DDT2}) on chromosomes II and III (Malcolm 1983a). Detailed mapping studies showed that R^{DDT} conferred a DDT-ase-dependent resistance mechanism, whereas R^{DDT2} and R^{py} were allelic and conferred the observed non-metabolic DDT–pyrethroid cross-resistance (Malcolm 1983b). When isolated in a susceptible genetic background, R^{DDT2}/R^{py} conferred approximately 10-fold resistance to DDT and 20-fold resistance to permethrin. Unlike the *kdr* factor in house flies, R^{py} was incompletely dominant rather than recessive in inheritance (Malcolm and Wood 1982b). Although direct demonstration of reduced neuronal sensitivity is missing for the BKK, BKPM, and BKPM3 strains, the weight of evidence from bioassays and metabolic and genetic experiments strongly implicates a *kdr*-like mechanism in this species.

Selection of field-collected *C. quinquefasciatus* with [1*R,trans*]-permethrin produced a strain having >4,000-fold resistance to the selection compound and high levels of resistance to a variety of other pyrethroids and to DDT (Priester and Georghiou 1978, 1980). Substantial resistance remained in the presence of synergists capable of inhibiting monooxygenases, hydrolases, or DDT-ases, thereby implicating a nonmetabolic mechanism as a major resistance factor in this strain. Genetic studies suggest that nonmetabolic resistance to both permethrin and DDT is inherited similarly, but allelism has not been shown in mapping studies (Halliday and Georghiou 1985). A larval neuromuscular preparation from this strain was >1,000-fold less sensitive to [1*R,trans*]-permethrin than an equivalent preparation from susceptible insects, thus providing physiological confirmation of reduced neuronal sensitivity in this species (Salgado et al. 1983b).

Cross-resistance between DDT and pyrethroids was also documented in a field-collected colony of *An. stephensi* that was selected first with DDT and then with DDT plus PBO and chlorfenethol (a DDT-ase inhibitor) (Omer et al. 1980). Larval neuromuscular preparations from this strain were resistant to the actions of [1*R,cis*]-permethrin, but the magnitude of resistance was much lower than that found for *C. quinquefasciatus* (Omer et al. 1980, Salgado et al. 1983b).

c. Lepidoptera. The possible existence of *kdr*-like resistance mechanisms has been explored in larvae of *Spodoptera littoralis* and *Heliothis virescens*. A field-derived strain of *S. littoralis* exhibiting fourfold resistance to permethrin but no detectable resistance to cypermethrin was assayed for possible differences in permethrin penetration, metabolism, and neuronal sensitivity. Susceptible and resistant larvae did not differ in their accumulation or metabolism of [^{14}C]permethrin (Holden 1979), but physiological assays using ventral nerve cord preparations showed increased latency of permethrin-dependent neuroexcitatory activity (Gammon 1980). This assay was unable to document possible differences in neuronal sensitivity to cypermethrin, since this compound did not produce repetitive firing in nerve preparations from either strain. There was no correlation between nerve block induced by either compound and resistance.

Resistant strains of *H. virescens* established from field collections made in the Imperial Valley of California were evaluated for differences in the rates of penetration and metabolism of [^{3}H] [1*R,trans*]-permethrin and for differences in neuronal sensitivity using physiological assays (Nicholson and Miller 1985). Metabolism studies revealed more extensive permethrin metabolism in the resistant strains than in a susceptible reference strain but were unable to determine whether enhanced metabolism was the result of more rapid oxidation or hydrolysis of the parent molecule. Assays of nerve sensitivity using a larval neuromuscular preparation showed that threshold concentrations for the production of pyrethroid-dependent neuroexcitatory effects were 10- to 50-fold higher in resistant insects. These findings provide evidence for resistance to pyrethroids in field-derived strains of *H. virescens* resulting from a combination of mechanisms in which reduced neuronal sensitivity plays an important role.

Recent studies have described the isolation of two strains of *H. virescens* that are homozygous for a high level (>500-fold) of permethrin resistance through an intensive program of outcrossing followed by selection of highly resistant inbred lines (Payne et al. 1988). In both strains, two mechanisms have been identified: enhanced oxidation, which can be suppressed by propynyl ether synergists (Brown and Payne 1986), and reduced neuronal sensitivity, which is detected in physiological assays using larval neuromuscular preparations (G. T. Payne and T. M. Brown, personal communication). In both strains, resistance is stable without selection and extends to a variety of pyrethroids (Payne et al. 1988). The availability of these strains provides an excellent opportunity to characterize a *kdr*-like mechanism in this species.

   d. Cockroaches.   The possible existence of a reduced nerve sensitivity resistance mechanism in the German cockroach (*Blattella germanica*) analogous to *kdr* was examined using a DDT-resistant strain (VPIDLS) having no known metabolic mechanisms of resistance. Initial studies demonstrated very low levels of cross-resistance to several pyrethroids in this strain (Scott and Matsumura 1981), but selection of the VPIDLS strains for three generations with DDT increased the levels of resistance to DDT as well as to a series of six structurally diverse pyrethroids (Scott and Matsumura 1983). In DDT-selected insects, resistance measured as acute toxicity following treatment by surface contact was 50-fold for DDT and at least 10-fold for pyrethrins, allethrin, permethrin, and fenvalerate, but less than three-fold for cypermethrin and deltamethrin. However, administration of the latter two compounds by topical application revealed levels of resistance similar to those observed for other pyrethroids administered by this route (Scott et al. 1986, Chapter 2). These data demonstrate a DDT-selectable resistance factor in the German cockroach that confers broad cross-resistance to pyrethroids in the absence of any pyrethroid selection, a finding consistent with a *kdr*-like nerve insensitivity mechanism.

Biochemical and physiological experiments were undertaken to explore the possible mechanisms involved in this resistance. The in vitro degradation of permethrin by microsomal fractions prepared from whole body homogenates of susceptible and resistant cockroaches failed to reveal any differences in rates of degradation (Scott and Matsumura 1981). In the ventral nerve cord of unselected VPIDLS insects, a modest increase was noted in the latency of spontaneous burst discharges following application of either DDT or permethrin at high concentrations (Scott and Matsumura 1981). Selection for one generation with DDT produced a further increase in the latency of spontaneous burst activity (Scott and Matsumura 1983).

e. **Other Species.** In other arthropod species, evidence for putative nerve insensitivity mechanisms is much less complete. For example, the possible existence of a *kdr*-like mechanism in the cattle tick is based principally on cross-resistance to pyrethroids in DDT-resistant strains (Nolan et al. 1977). Although there is some evidence for reduced neuronal sensitivity in adult ticks (Nicholson et al. 1980b), the interpretation of this finding is complicated by the existence of a strong hydrolytic detoxication mechanism in the strain employed (Schnitzerling et al. 1983).

More recently, conditional neurological mutants of *Drosophila melanogaster* that are thought to have altered sodium channel function have drawn attention as possible models of mechanisms capable of conferring resistance by virtue of reduced neuronal sensitivity. The *nap^{ts}* (no action potential, temperature-sensitive) strain exhibits nerve conduction failure at nonpermissive temperatures (Wu et al. 1978) and has a sodium channel density approximately half that of wild-type flies (Jackson et al. 1984). Thus, the *nap^{ts}* strain offers an opportunity to test the significance of reduced sodium channel density as a mechanism of resistance. Kasbekar and Hall (1988) found that *nap^{ts}* flies exhibited resistance to fenvalerate and flucythrinate and that this resistance was genetically inseparable from the *nap^{ts}* locus. Studies in our laboratory have confirmed that the *nap^{ts}* trait confers low levels of resistance to the lethal effects of DDT and a variety of pyrethroids but confers more substantial resistance to the rapid paralytic effects of rapidly acting compounds. The resistance to the rapid paralytic effects of fenfluthrin is correlated with the reduced sensitivity of the nervous system of resistant flies to this compound in physiological assays (Bloomquist et al. 1989). These findings suggest that reduced sodium channel density can confer resistance to DDT and pyrethroids by reducing the sensitivity of the nervous system to these compounds but that the magnitude of resistance caused by this mechanism is smaller than that found in *kdr* and *super-kdr* house flies (Farnham 1988, Farnham et al. 1987, Sawicki 1978).

In addition to the *nap^{ts}* strain, three other temperature-sensitive paralytic mutants of *D. melanogaster* have been described that also exhibit altered sodium channel properties: *para^{ts}* (paralysis); *sei^{ts}* (seizure); and *tip-E^{ts}* (temperature-induced paralysis, locus E) (Hall 1986). Of these mutant strains, *para^{ts}* is of particular interest because the cloning and sequencing of this locus has revealed significant structural homology with vertebrate sodium channel structural genes (Loughney et al. 1989). Moreover, a preliminary report (Hall and Kasbekar 1989) suggests that different alleles of *para^{ts}* may exhibit either resistance or hypersensitivity to pyrethroids. These findings suggest that the continued genetic and molecular analysis of the many alleles available at the *para^{ts}* locus may provide unique insight into the elements of sodium channel structure that comprise the binding domain for DDT and pyrethroids.

4. Mechanisms of Reduced Neuronal Sensitivity

Although reduced neuronal sensitivity to DDT and pyrethroids is well character-ized in the house fly and strongly implicated in other species, the molecular mechanisms that confer reduced sensitivity are poorly understood. Knowledge of such mechanisms requires knowledge of toxicant–target interactions; in the case of DDT and pyrethroids, the molecular details of the interactions between these compounds and their presumptive target, the voltage-dependent sodium channel, remain to be elucidated. Nevertheless, three hypotheses have been advanced to explain the *kdr* trait in the house fly.

The first hypothesis postulates that reduced neuronal sensitivity results from a lower density of insecticide target sites in the nerves of resistant insects (Chang and Plapp 1983b, Plapp 1986). Technical difficulties have limited the direct assessment of this mechanism using insecticides as radioligands. However, if the target site for DDT and pyrethroids is assumed to be one of the neurotoxin-binding domains on the voltage-dependent sodium channel (Soderlund and Bloomquist 1989), then [^{3}H]saxitoxin binding can be used to determine target site density indirectly. Results of such studies (Grubs et al. 1988, Pauron et al. 1989, Rossignol 1988, Sattelle et al. 1988) are not in complete agreement, but the weight of evidence suggests that reduced sodium channel density is not specifically associated with the *kdr* trait of the house fly. However, the results of experiments with the *nap^ts* strain of *D. melanogaster* (Bloomquist et al. 1989, Kasbekar and Hall 1988) show that a reduction in sodium channel density, as measured by [^{3}H]saxitoxin binding, is able to produce modest levels of resistance.

A second hypothesis, advanced by Chiang and Devonshire (1982), relates reduced neuronal sensitivity to alterations in the fluidity of nerve membranes. Although their data provide evidence for a strong correlation between membrane fluidity and resistance, they do not define a causal relationship. Thus, the altered membrane environment may confer altered neurotoxin sensitivity of structurally unaltered sodium channels, or alterations in the structure of the target may require altered membrane composition, achieved through feedback regulatory mechanisms, to preserve normal nerve function. The third hypothesis, formally stated by Salgado et al. (1983a), ascribes altered insecticide and neurotoxin sensitivity in resistant house flies to structural changes in the voltage-sensitive sodium channel that selectively alter the binding domains for insecticides and at least some of the alkaloid activators. Although this hypothesis is also consistent with available data, there is no direct evidence of altered sodium channel structure in resistant insects.

At present, there is no basis for preferring either of the latter two hypotheses as the basis for the *kdr* mechanism in the house fly, and there is no specific evidence implicating either of these mechanisms in other species. Studies de-signed to discriminate between sodium channel environment and sodium channel structure as a cause of reduced neuronal sensitivity in the house fly must take into account the differences in membrane fluidity that may be found in resistant insects and must devise methods to define the structure and neurotoxin sensitivity of sodium channels from susceptible and resistant insects apart from their native membrane environments. One approach is to solubilize, purify, and characterize sodium channels from resistant and susceptible house flies and reconstitute them functionally in model membranes of known composition. These methods are well established for vertebrate sodium channels (Catterall 1986), but purification methods developed for vertebrate systems have not yet been applied successfully to insect preparations.

A second approach takes advantage of recent progress in the molecular biology of the sodium channel. Genes for sodium channels of electric eel electroplax (Noda et al. 1984) and rat brain (Noda et al. 1986a) have been cloned and sequenced, and sodium channel mRNA from a variety of vertebrate sources has been functionally expressed following injection into the developing oocytes of the frog, *Xenopus laevis* (Lester 1988, Noda et al. 1986b, Sumikawa et al. 1984). Two putative sodium channel genes from *D. melanogaster* have also been cloned and characterized. The first of these is a gene isolated on the basis of its structural homology with the electric eel sodium channel gene (Salkoff et al. 1987), whereas the second is encoded by the *para^{ts}* locus (Loughney et al. 1989). However, the functional identity of both of these putative insect sodium channel genes remains to be established in appropriate expression assays. The complete characterization of sodium channel genes from the fruit fly should permit the isolation and characterization of sodium channel genes from susceptible and resistant house flies and other insects. These studies, together with the functional characterization of the gene products in the controlled membrane environment of the *Xenopus* oocyte expression assay, offer an opportunity to determine whether alterations in the sodium channel itself or in its local membrane environment are the primary mechanism of reduced insecticide sensitivity.

C. Reduced Neuronal Sensitivity to Chlorinated Cyclodienes

1. Evidence for an Altered Target Site

Reduced neuronal sensitivity has been proposed as a mechanism of broad cyclodiene cross-resistance in several insects. Much of the original work in this area was published prior to 1970 and is extensively reviewed by Brooks (1974). In two of these instances, bioassays and genetic studies appear to support the hypothesis that an altered target site is involved. In house flies, a single gene conferring broad cross-resistance to many structurally diverse cyclodienes and lindane, but not to other insecticide classes, has been mapped to autosome 4, a linkage group not known to bear genes for any major metabolic mechanisms. A similar single-gene mechanism has also been isolated in the German cockroach. In each case, enhanced detoxication has not been identified as an important resistance mechanism.

2. *Pharmacological Studies with German Cockroaches*

Although reduced neuronal sensitivity to cyclodienes has been suspected for many years, progress in defining the molecular basis of this resistance has been hampered by lack of knowledge of the mechanism of action of these compounds. Recently, several studies in insect (Matsumura and Ghiasuddin 1983, Tanaka et al. 1984) and mammalian (Abalis et al. 1985, 1986, Bloomquist and Soderlund 1985, Bloomquist et al. 1986, Gant et al. 1987, Lawrence and Casida 1984) nerve preparations have provided persuasive evidence that the cyclodienes exert their neurotoxic effects by blocking γ-aminobutyric acid (GABA)-dependent chloride flux at the GABA$_A$ receptor–chloride ionophore complex. There is some evidence that lindane also acts at this target (Abalis et al. 1985, Lawrence and Casida 1984), but other studies suggest that at least in mammals other targets are more important for this compound (Bloomquist and Soderlund 1985, Bloomquist et al. 1986). Thus, reduced neuronal sensitivity in resistance to the cyclodienes and lindane may result from alterations in the recognition site for these compounds on the chloride ionophore component of this complex.

To date, these possibilities have been investigated only in studies with cyclodiene-resistant German cockroaches. Several cyclodiene-resistant strains of this species were also found to be resistant to picrotoxinin, which binds to the chloride ionophore component of the GABA receptor complex and blocks GABA-dependent chloride fluxes (Kadous et al. 1983). In the LPP strain, the neuroexcitatory actions of both dieldrin and picrotoxinin on ventral nerve cord preparations were delayed relative to those measured in equivalent preparations from susceptible insects. Although cyclodienes and lindane displace the specific binding of a tritiated picrotoxinin derivative to rat brain and American cockroach nerve membranes (Matsumura and Ghiasuddin 1983, Tanaka et al. 1984), the extension of this approach to equivalent preparations from susceptible and resistant German cockroaches is limited by the much smaller size of the latter insect and the high levels of nonspecific binding observed with this lipophilic ligand. Results of such experiments comparing purified membrane preparations from susceptible and resistant German cockroaches show very small differences between strains (Kadous et al. 1983) and are therefore inconclusive. However, more recent studies using crude mitochondrial–synaptosomal fraction of head homogenates demonstrated a 90% reduction in the density of [^{3}H]dihydropicrotoxinin binding sites in the LPP strain together with a 10-fold increase in the affinity for this ligand (Tanaka 1987). Although these findings support the hypothesis of an altered picrotoxinin receptor in the LPP strain, the significance of the properties of this altered binding in conferring resistance to the chloride channel–blocking actions of these insecticides remains to be demonstrated in functional assays.

A number of technical difficulties have prevented a more detailed analysis of the interactions between cyclodienes and the GABA receptor complex of resistant insects. For example, the convulsant TBPS (*t*-butylbicyclophosphorothionate) is a potent ligand for the picrotoxinin binding site on the mammalian GABA receptor–chloride ionophore complex and [^{35}S]TBPS has proved to be an excellent probe for defining the interactions of cyclodienes with this site (Squires et al. 1983). However, binding studies in insects with [^{35}S]TBPS (Cohen and Casida 1986) and [^{3}H]n-propylbicyclophosphate (Ozoe et al. 1986), a related compound, have not demonstrated that these ligands label an equivalent site in insect nervous tissue and have therefore been of little use in probing altered cyclodiene binding in resistant insects. Similarly, radioisotopic assays of GABA-dependent chloride uptake have proven useful in studies of cyclodienes and lindane in mammalian brain preparations (Abalis et al. 1986, Bloomquist and Soderlund 1985, Bloomquist et al. 1986), but extension of these studies to insect nervous tissue produces marginal results even in species, such as the American cockroach, in which large amounts of tissue are available (Wafford et al. 1987). Recently, the genes encoding the α and β subunits of the GABA receptor complex from bovine brain have been cloned and sequenced, and functional receptor–ion channel complexes have been expressed in *Xenopus* oocytes upon injection with mRNAs synthesized in vitro from these genes (Levitan et al. 1988, Schofield et al. 1987). If components of the GABA receptor complex exhibit a high degree of structural conservation between mammals and insects, these findings may allow the molecular characterization of the insect receptor and its possible alteration in insecticide-resistant strains.

IV. Conclusions

Despite a heightened awareness of the practical problems caused by insecticide resistance, knowledge of the molecular mechanisms of resistance has progressed slowly. The ultimate goals of mechanistic studies are to identify and characterize the products of genes that confer resistance and to correlate altered function of these products in mutant strains with the resistant phenotype at the organismal level. Substantial progress in achieving these goals has been realized only in two areas: the overproduction of carboxylesterases associated with resistance to OP and carbamate insecticides in aphids and mosquitoes and the occurrence of mutant forms of acetylcholinesterase in a small number of insect species. Even in many of these cases, molecular characterization remains incomplete. In other areas, progress has been limited to the development of physiological or biochemical assays to identify specific mechanisms in resistant strains.

The failure to employ successfully the tools of classical biochemistry to characterize the limited number of known resistance mechanisms is due in large part to substantial technical and conceptual barriers encountered in pursuing this research in insects. For example, the purification and characterization of mammalian cytochromes P$_{450}$ represent years of effort by many laboratories. The extension of these investigations to insect systems has been confounded by both the inherent difficulties posed by the small biomass of insects and the unique properties of the insect cytochromes that have rendered them less amenable to purification. Conceptual barriers to the molecular definition of resistance mechanisms are evident in the cases of reduced neuronal sensitivity to DDT, pyrethroids, and cyclodienes. For example, the molecular interactions between DDT or pyrethroids and the voltage-sensitive sodium channel are only poorly understood, and the identification of the GABA receptor–ionophore complex as the target of cyclodiene action is a recent development that has only just begun to influence studies of mechanisms involving reduced neuronal sensitivity to these compounds.

The continued development of methods for the application of molecular biology to a wide variety of biological systems offers new strategies for elucidating resistance mechanisms at both the functional and genetic levels. The application of these approaches to problems of insecticide resistance is a very recent phenomenon, but a growing arsenal of recombinant DNA probes for enzymes involved in insecticide detoxication and for neuronal target macromolecules, most of which are the product of prior biomedical research with mammalian species, is now available for use in isolating and characterizing genes and gene products of interest from resistant insects. The strategy of gene isolation using recombinant DNA probes from other species, coupled with transformation and expression assays, offers new approaches to defining the molecular mechanisms of resistance that are able to overcome the technical limitations inherent in doing classical biochemistry in insect systems.

Recent advances in the molecular characterization of resistance mechanisms have already made substantial contributions to the detection and monitoring of resistance and the determination of the genotypes of individual insects in heterogeneous populations (Brown and Brogden 1987, Chapter 2). Such techniques are most easily developed when the product of the resistance-conferring gene is an enzyme that can be detected using simple spectrophotometric assays or by immunoassay. Much additional work will be required to develop rapid diagnostic assays to detect other mechanisms, such as *kdr*-like reduced neuronal sensitivity. Recombinant DNA probes for resistance-conferring genes may permit the determination of the genotype of resistant insects, but the inherent specificity of this approach, which depends on the detection of particular altered nucleotide sequences in hybridization assays, may limit its general applicability even for detecting a single resistance mechanism in a single species having a broad geographical distribution.

Molecular characterization of resistance mechanisms has also been viewed as a prerequisite to the use of resistance-conferring genes as a resource for the genetic improvement of beneficial species (Beckendorf and Hoy 1985). At present, however, limitations in knowledge of resistance mechanisms are overshadowed by the technical difficulties inherent in achieving stable transformation with foreign genes of any insect species other than *D. melanogaster*. The ultimate use of resistance-conferring genes for the genetic engineering of beneficial species will require the discovery and description of unique transposable elements in other insect species, if they in fact exist, or will depend on random integration events as a source of transformants. Recent studies with mosquitoes (Miller et al. 1987) illustrates the difficulties inherent in attempting to achieve transformation in other species and suggests that advances in this area will depend as much on progress in insect molecular biology as on the molecular definition of resistance mechanisms.

References

Abalis, I. M., M. E. Eldefrawi, and A. T. Eldefrawi. 1985. High-affinity binding of cyclodiene insecticides and γ-hexachlorocyclohexane to γ-aminobutyric acid receptors of rat brain. Pestic. Biochem. Physiol. 24: 95–102.

Abalis, I. M., M. E. Eldefrawi, and A. T. Eldefrawi. 1986. Effects of insecticides on GABA-induced chloride influx into rat brain microsacs. J. Toxicol. Environ. Health 18: 13–23.

Agosin, M. 1985. Role of microsomal oxidations in insecticide degradation, pp. 647–712. In G. A. Kerkut and L. I. Gilbert (eds.), Comprehensive insect physiology biochemistry and pharmacology, Vol. 12. Pergamon, Oxford.

Ahn, Y.-J., T. Shono, and J. Fukami. 1986. Inheritance of pyrethroid resistance in a housefly strain from Denmark. J. Pestic. Sci. 11: 591–596.

Ahn, Y.-J., E. Funaki, N. Motoyama, T. Shono, and J. Fukami. 1987. Nerve insensitivity as a mechanism of resistance to pyrethroids in a Japanese colony of house flies. J. Pestic. Sci. 12: 69–75.

Beckendorf, S. K., and M. A. Hoy. 1985. Genetic improvement of arthropod natural enemies through selection, hybridization, or genetic engineering techniques, pp. 167–187. In M. A. Hoy and D. C. Herzog (eds.), Biological control in agricultural IPM systems. Academic, Orlando.

Beeman, R. W., and B. A. Schmidt. 1982. Biochemical and genetic aspects of malathion-specific resistance in the Indian mealmoth (Lepidoptera: Pyralidae). J. Econ. Entomol. 75: 945–949.

Berge, J. B., C. Mouches, and D. Fournier. 1986. Molecular biology of some insecticide resistant genes suitable to improve resistance in beneficial arthropods. Abstract 3E–14, VIth Int. Cong. Pestic. Chem. Ottawa.

Bloomquist, J. R., and T. A. Miller. 1985. A simple bioassay for detecting and characterizing insecticide resistance. Pestic. Sci. 16: 611–614.

Bloomquist, J. R., and T. A. Miller. 1986. Sodium channel neurotoxins as probes of the knockdown resistance mechanism. Neurotoxicology 7: 217–224.

Bloomquist, J. R., and D. M. Soderlund. 1985. Neurotoxic insecticides inhibit GABA-dependent chloride uptake into mouse brain vesicles. Biochem. Biophys. Res. Commun. 133: 37–43.

Bloomquist, J. R., P. M. Adams, and D. M. Soderlund. 1986. Inhibition of γ-aminobutyric acid-stimulated chloride flux in mouse brain vesicles by polychlorocycloalkane and pyrethroid insecticides. Neurotoxicology 7: 11–20.

Bloomquist, J. R., D. M. Soderlund, and D. C. Knipple. 1989. Knockdown resistance to dichlorodiphenyltrichloroethane and pyrethroid insecticides in the *nap*[ts] mutant of *Drosophila melanogaster* is correlated with reduced neuronal sensitivity. Arch. Insect Biochem. Physiol. 10: 293–302.

Booth, G. M., D. J. Weber, L. M. Ross, S. D. Burton, W. S. Bradshaw, W. M. Hess, and J. R. Larsen. 1983. Mechanisms of pesticide resistance in non-target organisms, pp. 387–409. In G. P. Georghiou and T. Saito (eds), Pest resistance to pesticides. Plenum, New York.

Brealey, C. J., P. L. Crampton, P. R. Chadwick, and F. E. Rickett. 1984. Resistance mechanisms to DDT and transpermethrin in *Aedes aegypti*. Pestic. Sci. 15: 121–132.

Brooks, G. T. 1974. Chlorinated insecticides, Vol. II, Biological and environmental aspects, pp. 3–62. CRC Press, Cleveland.

Brown, G. B., J. E. Gaupp, and R. W. Olsen. 1988. Pyrethroid insecticides: stereospecific allosteric interaction with the batrachotoxinin-A benzoate binding site of mammalian voltage-sensitive sodium channels. Mol. Pharmacol. 34: 54–59.

Brown, T. M., and W. G. Brogdon. 1987. Improved detection of insecticide resistance through conventional and molecular techniques. Ann. Rev. Entomol. 32: 145–162.

Brown, T. M., and G. T. Payne. 1986. Synergists for permethrin in *Heliothis virescens*. Abstract 3D–17, VIth Int. Congr. Pestic. Chem. Ottawa.

Burt, P. E., M. Elliott, A. W. Farnham, N. F. Janes, P. H. Needham, and D. A. Pulman. 1974. The pyrethrins and related compounds. Part XIX, Geometrical and optical isomers of 2,2-dimethyl-3-(2,2-dichlorovinyl)-cyclopropanecarboxylic acid and insecticidal esters with 5-benzyl-3-furylmethyl and 3-pehnoxybenzyl alcohols. Pestic. Sci. 5: 791–799.

Busvine, J. R. 1951. Mechanism of resistance to insecticide in house flies. Nature 168: 193–195.

Capdevila, J., N. Ahmad, and M. Agosin. 1975. Soluble cytochrome P–450 from house fly microsomes. Partial purification and characterization of two hemoprotein forms. J. Biol. Chem. 250: 1048–1060.

Catterall, W. A. 1986. Molecular properties of voltage-sensitive sodium channels. Ann. Rev. Biochem. 55: 953–985.

Chadwick, P. R., J. F. Invest, and M. J. Bowron. 1977. An example of cross-resistance to pyrethroids in DDT-resistant *Aedes aegypti*. Pestic. Sci. 8: 618–624.

Chadwick, P. R., R. Slatter, and M. J. Bowron. 1984. Cross-resistance to pyrethroids and other insecticides in *Aedes aegypti*. Pestic. Sci. 15: 112–120.

Chang, C. P., and F. W. Plapp, Jr. 1983a. DDT and pyrethroids: receptor binding and mode of action in the house fly. Pestic. Biochem. Physiol. 20: 76–85.

Chang, C. P., and F. W. Plapp, Jr. 1983b. DDT and pyrethroids: receptor binding in relation to knockdown resistance (*kdr*) in the house fly. Pestic. Biochem. Physiol. 20: 86–91.

Chang, C. K., and M. E. Whalon. 1986. Hydrolysis of permethrin by pyrethroid esterases from resistant and susceptible strains of *Amblyseius fallacis* (Acari: Phytoseiidae). Pestic. Biochem. Physiol. 25: 446–452.

Chiang, C., and A. L. Devonshire. 1982. Changes in membrane phospholipids, identified by Arrhenius plots of acetylcholinesterase and associated with pyrethroid resistance (*kdr*) in house flies (*Musca domestica*). Pestic. Sci. 13: 156–160.

Clark, A. G., and W. C. Dauterman. 1982. The characterization by affinity chromatography of glutathione S-transferases from different strains of house fly. Pestic. Biochem. Physiol. 17: 307–314.

Clark, A. G., and N. A. Shamaan. 1984. Evidence that DDT-dehydrochlorinase from the house fly is a glutathione S-transferase. Pestic. Biochem. Physiol. 22: 249–261.

Clark, A. G., N. A. Shamaan, W. C. Dauterman, and T. Hayaoka. 1984. Characterization of multiple glutathione transferases from the house fly, *Musca domestica* (L). Pestic. Biochem. Physiol. 22: 51–59.

Clark, A. G., N. A. Shamaan, M. D. Sinclair, and W. C. Dauterman. 1986. Insecticide metabolism by multiple glutathione S-transferases in two strains of the house fly, *Musca domestica* (L). Pestic. Biochem. Physiol. 25: 169–175.

Cohen, E., and J. E. Casida. 1986. Effects of insecticides and GABAergic agents on a house fly [^{35}S]t-butylbicyclophosphorothionate binding site. Pestic. Biochem. Physiol. 25: 63–72.

Dauterman, W. C. 1983. Role of hydrolases and glutathione S-transferases in insecticide resistance, pp. 229–247. *In* G. P. Georghiou and T. Saito (eds.), Pest resistance to pesticides. Plenum, New York.

Dauterman, W. C. 1985. Insect metabolism: extramicrosomal, pp. 713–730. *In* G. A. Kerkut and L. I. Gilbert (eds.), Comprehensive insect physiology biochemistry and pharmacology, Vol. 12. Pergamon, Oxford.

de Jersey, J., J. Nolan, P. A. Davey, and P. W. Riddles. 1985. Separation and characterization of the pyrethroid-hydrolyzing esterases of the cattle tick, *Boophilus microplus*. Pestic. Biochem. Physiol. 23: 349–357.

Delorme, R., D. Fournier, J. Chaufaux, A. Cuany, J. M. Bride, D. Auge, and J. B. Berge. 1988.

Esterase metabolism and reduced penetration are causes of resistance to deltamethrin in *Spodoptera exigua* HUB (Noctuidae: Lepidoptera). Pestic. Biochem. Physiol. 32: 240–246.

Devonshire, A. L. 1975. Studies of the carboxylesterases of *Myzus persicae* resistant and susceptible to organophosphorus insecticides. Proc. 8th Brit. Insectic. Fungic. Conf. 1: 67–73.

Devonshire, A. L. 1977. The properties of a carboxylesterase from the peach-potato aphid, *Myzus persicae* (Sulz.), and its role in conferring insecticide resistance. Biochem. J. 167: 675–683.

Devonshire, A. L. 1987. Biochemical studies of organophosphorus and carbamate resistance in house flies and aphids, pp. 239–255. *In* M. G. Ford, D. W. Holloman, B. P. S. Khambay, and R. Sawicki (eds.), Combatting resistance to xenobiotics: biological and chemical approaches. Ellis Horwood, Chichester.

Devonshire, A. L., and G. D. Moores. 1982. A carboxylesterase with broad substrate specificity causes organophosphorus, carbamate, and pyrethroid resistance in peach-potato aphids (*Myzus persicae*). Pestic. Biochem. Physiol. 18: 235–246.

Devonshire, A. L., and G. D. Moores. 1984a. Characterization of insecticide-insensitive acetylcholin-esterase: microcomputer-based analysis of enzyme inhibition in homogenates of individual house fly (*Musca domestica*) heads. Pestic. Biochem. Physiol. 21: 341–348.

Devonshire, A. L., and G. D. Moores. 1984b. Different forms of insensitive acetylcholinesterase in insecticide-resistant house flies (*Musca domestica*). Pestic. Biochem. Physiol. 21: 336–340.

Devonshire, A. L., and R. Sawicki. 1979. Insecticide-resistant *Myzus persicae* as an example of evolution by gene duplication. Nature 280: 140–141.

Devonshire, A. L., G. D. Moores, and C. Chiang. 1983. The biochemistry of insecticide resistance in the peach-potato aphid, *Myzus persicae*, pp. 191–196. *In* J. Miyamoto and P. C. Kearney (eds.), Pesticide chemistry: human welfare and environment, Vol. 3. Pergamon, Oxford.

De Vries, D. H., and G. P. Georghiou. 1981a. Absence of enhanced detoxication of permethrin in pyrethroid-resistant house flies. Pestic. Biochem. Physiol. 15: 242–252.

De Vries, D. H., and G. P. Georghiou. 1981b. Decreased nerve sensitivity and decreased cuticular penetration as mechanisms of resistance to pyrethroids in a (1R)-*trans*-permethrin-selected strain of the house fly. Pestic. Biochem. Physiol. 15: 234–241.

Dowd, P. F., C. C. Gagne, and T. C. Sparks. 1987. Enhanced pyrethroid hydrolysis in pyrethroid-resistant larvae of the tobacco budworm, *Heliothis virescens* (F.). Pestic. Biochem. Physiol. 28: 9–16.

Farnham, A. W. 1977. Genetics of resistance of house flies (*Musca domestica* L.) to pyrethroids. I. Knockdown resistance. Pestic. Sci. 8: 631–636.

Farnham, A. W., A. W. A. Murray, R. M. Sawicki, I. Denholm, and J. C. White. 1987. Characterization of the structure–activity relationship of *kdr* and two variants of *super-kdr* to pyrethroids in the house fly (*Musca domestica* L.). Pestic. Sci. 19: 209–220.

Feyereisen, R., J. F. Koener, D. E. Farnsworth, and D. W. Nebert. 1989. Isolation and sequence of cDNA encoding a cytrochrome P-450 from an insecticide-resistant strain of the house fly, *Musca domestica*. Proc. Natl. Acad. Sci. USA 86: 1465–1469.

Field, L. M., A. L. Devonshire, and B. G. Forde. 1988. Molecular evidence that insecticide resistance in peach-potato aphids (*Myzus persicae* Sulz.) results from amplification of an esterase gene. Biochem. J. 251: 309–312.

Fisher, C. W., and R. T. Mayer. 1984. Partial purification and characterization of phenobarbital-induced house fly cytochrome P-450. Arch. Insect Biochem. Physiol. 1: 127–138.

Forgash, A. J. 1984. History, evolution, and consequences of insecticide resistance. Pestic. Biochem. Physiol. 22: 178–186.

Fournier, D., J.-M. Bride, C. Mouches, M. Raymond, M. Magnin, J.-B. Berge, N. Pasteur, and G. P. Georghiou. 1987. Biochemical characterization of the esterases A1 and B1 associated with organophosphate resistance in the *Culex pipiens* L. complex. Pestic. Biochem. Physiol. 27: 211–217.

Gammon, D. W. 1980. Pyrethroid resistance in a strain of *Spodoptera littoralis* is correlated with decreased sensitivity of the CNS *in vitro*. Pestic. Biochem. Physiol. 13: 53–62.

Gant, D. B., M. E. Eldefrawi, and A. T. Eldefrawi. 1987. Cyclodiene insecticides inhibit $GABA_A$ receptor-regulated chloride transport. Toxicol. Appl. Pharmacol. 88: 313–321.

Georghiou, G. P. 1986. The magnitude of the resistance problem, pp. 14–43. *In* Pesticide resistance: strategies and tactics for management. National Academy Press, Washington, DC.

Georghiou, G. P., and N. Pasteur. 1978. Electrophoretic esterase patterns in insecticide-resistant and susceptible mosquitoes. J. Econ. Entomol. 71: 201–205.

Georghiou, G. P., and N. Pasteur. 1980. Organophosphate resistance and esterase pattern in a natural population of the southern house mosquito from California. J. Econ. Entomol. 73: 489–492.

Georghiou, G. P., N. Pasteur, and M. K. Hawley. 1980. Linkage relationships between organophosphate resistance and a highly active esterase-B in *Culex quinquefasciatus* from California. J. Econ. Entomol. 73: 301–305.

Georghiou, G. P., V. Ariaratnam, M. E. Pasternak, and C. S. Lin. 1985. Organophosphate multiresistance in *Culex pipiens quinquefasciatus* in California. J. Econ. Entomol. 68: 461–467.

Grubs, R. E., P. M. Adams, and D. M. Soderlund. 1988. Binding of [³H]saxitoxin to head membrane preparations from susceptible and knockdown-resistant house flies. Pestic. Biochem. Physiol. 32: 217–223.

Hall, L. M. 1986. Genetic variants of voltage-sensitive sodium channels, pp. 313–324. *In* C. Y. Kao and S. R. Levinson (eds.), Tetrodotoxin, saxitoxin, and the molecular biology of the sodium channel. New York Academy of Sciences, New York.

Hall L. M., and D. P. Kasbekar. 1989. *Drosophila* sodium channel mutations affect pyrethroid sensitivity, pp. 99–114. *In* T. Narahashi and J. Chambers (eds.), Insecticide action: from molecule to organism. Plenum, New York.

Hall, L. M. C., and P. Spierer. 1986. The *Ace* locus of *Drosophila melanogaster*: structural gene for acetylcholinesterase with an unusual 5′ leader. EMBO J. 5: 2949–2954.

Halliday, W. R., and G. P. Georghiou. 1985. Inheritance of resistance to permethrin and DDT in the southern house mosquito (Diptera: Culicidae). J. Econ. Entomol. 78: 762–767.

Hama, H. 1983. Resistance to insecticides due to reduced sensitivity of acetylcholinesterase, pp. 299–331. *In* G. P. Georghiou and T. Saito (eds.), Pest resistance to pesticides. Plenum, New York.

Hammock, B. D., and D. M. Soderlund. 1986. Chemical strategies for resistance management, pp. 111–129. *In* Pesticide resistance: strategies and tactics for management. National Academy Press, Washington, DC.

Hayaoka, T., and W. C. Dauterman. 1982. Induction of glutathione S-transferase by phenobarbital and pesticides in various house fly strains and its effect on toxicity. Pestic. Biochem. Physiol. 17: 113–119.

Hayaoka, T., and W. C. Dauterman. 1983. The effect of phenobarbital induction on glutathione conjugation of diazinon in susceptible and resistant house flies. Pestic. Biochem. Physiol. 19: 344–349.

Hemingway, J. 1982. The biochemical nature of malathion resistance in *Anopheles stephensi* from Pakistan. Pestic. Biochem. Physiol. 17: 149–155.

Hemingway, J. 1983. Biochemical studies on malathion resistance in *Anopheles arabiensis* from Sudan. Trans. R. Soc. Trop. Med. Hyg. 77: 477–480.

Hemingway, J. 1984. The joint action of malathion and IBP against malathion-resistant *Anopheles stephensi*. Bull. World Health Org. 62: 445–449.

Hemingway, J. 1985. Malathion carboxylesterase enzymes in *Anopheles arabiensis* from Sudan. Pestic. Biochem. Physiol. 23: 309–313.

Hemingway, J., C. Smith, K. G. Jayawardena, and P. R. J. Herath. 1986. Field and laboratory detection of the altered acetylcholinesterase resistance genes which confer organophosphate and carbamate resistance in mosquitoes (Diptera: Culicidae). Bull. Entomol. Res. 76: 559–565.

Hodgson, E. 1983. The significance of cytochrome P–450 in insects. Insect Biochem. 13: 237–246.

Hodgson, E. 1985. Microsomal mono-oxygenases, pp. 225–321. *In* G. A. Kerkut and L. I. Gilbert (eds.), Comprehensive insect physiology biochemistry and pharmacology, Vol. 11. Pergamon, Oxford.

Hodgson, E., and A. P. Kulkarni. 1983. Characterization of cytochrome P–450 in studies of insecticide resistance, pp. 207–228. *In* G. P. Georghiou and T. Saito (eds.), Pest resistance to pesticides. Plenum, New York.

Holden, J. S. 1979. Absorption and metabolism of permethrin and cypermethrin in the cockroach and the cotton-leafworm larvae. Pestic. Sci. 10: 295–307.

Jackson, F. R., S. D. Wilson, G. R. Strichartz, and L. M. Hall. 1984. Two types of mutants affecting voltage-sensitive sodium channels in *Drosophila melanogaster*. Nature 308: 189–191.

Kadous, A. A., S. M. Ghiasuddin, F. Matsumura, J. G. Scott, and K. Tanaka. 1983. Difference in the picrotoxinin receptor between the cyclodiene-resistant and susceptible strains of the German cockroach. Pestic. Biochem. Physiol. 19: 157–166.

Kao, L. R., N. Motoyama, and W. C. Dauterman. 1984. Studies on hydrolases in various house fly strains and their role in malathion resistance. Pestic. Biochem. Physiol. 22: 86–92.

Kao, L. R., N. Motoyama, and W. C. Dauterman. 1985. The purification and characterization of esterases from insecticide-resistant and susceptible house flies. Pestic. Biochem. Physiol. 23: 228–239.

Kasbekar, D. P., and L. M. Hall. 1988. A *Drosophila* mutation that reduces sodium channel number confers resistance to pyrethroid insecticides. Pestic. Biochem. Physiol. 32: 135–145.

Lawrence, L. J., and J. E. Casida. 1984. Interactions of lindane, toxaphene, and cyclodienes with brain-specific *t*-butylbicyclophosphorothionate receptor. Life Sci. 35: 171–178.

Lester, H. A. 1988. Heterologous expression of excitability proteins: route to more specific drugs? Science 241: 1057–1063.

Levitan, E. S., P. R. Schofield, D. R. Burt, L. M. Rhee, S. Wisden, M. Koehler, N. Fujita, H. F. Rodriguez, A. Stephenson, M. G. Darlison, E. A. Barnard, and P. H. Seeburg. 1988. Structural and functional basis for GABA_A receptor heterogeneity. Nature 335: 76–79.

Lombet, A., C. Mourre, and M. Lazdunski. 1988. Interaction of insecticides of the pyrethroid family with specific binding sites on the voltage-dependent sodium channel from mammalian brain. Brain Res. 459: 44–53.

Loughney, K., R. Kreber, and B. Ganetzky. 1989. Molecular analysis of the *para* locus, a sodium channel gene in Drosophila. Cell 58: 1143–1154.

Maa, W. C. J., and L. C. Terriere. 1983. Age-dependent variation in enzymatic and electrophoretic properties of house fly (*M. domestica*) carboxylesterases. Comp. Biochem. Physiol. 74C: 461–467.

Malcolm, C. A. 1983a. The genetic basis of pyrethroid and DDT resistance inter-relationships in *Aedes aegypti*. I. Isolation of DDT and pyrethroid resistance factors. Genetica 60: 213–219.

Malcolm, C. A. 1983b. The genetic basis of pyrethroid and DDT resistance inter-relationships in *Aedes aegypti*. II. Allelism of R^{DDT2} and R^{py}. Genetica 60: 221–229.

Malcolm, C. A., and R. J. Wood. 1982a. The establishment of a laboratory strain of *Aedes aegypti* homogeneous for high resistance to permethrin. Pestic. Sci. 13: 104–108.

Malcolm, C. A., and R. J. Wood. 1982b. Location of a gene conferring resistance to knockdown by permethrin and bioresmethrin in adults of the BKPM3 strain of *Aedes aegypti*. Genetica 59: 233–237.

Matsumura, F. 1983. Penetration, binding, and target insensitivity as causes of resistance to chlorinated hydrocarbon insecticides, pp. 367–386. *In* G. P. Georghiou and T. Saito (eds.), Pest resistance to pesticides. Plenum, New York.

Matsumura, F., and S. M. Ghiasuddin. 1983. Evidence for similarities between cyclodiene type insecticides and picrotoxinin in their action mechanisms. J. Environ. Sci. Health B18: 1–14.

Matsumura, F., and G. Voss. 1965. Properties of partially purified malathion carboxylesterase of the two-spotted spider mite. J. Insect Physiol. 11: 147–160.

Milani, R. 1954. Comportamento mendeliano della resistenza alla azione abbattante del DDT: correlazione tran abbattimento e mortalita in *Musca domestica* L. Riv. Parassitol. 15: 513–542.

Miller, T. A., J. M. Kennedy, and C. Collins. 1979. CNS insensitivity to pyrethroids in the resistant *kdr* strain of house flies. Pestic. Biochem. Physiol. 12: 224–230.

Miller, T. A., V. L. Salgado, and S. N. Irving. 1983. The *kdr* factor in pyrethroid resistance, pp. 353–366. *In* G. P. Georghiou and T. Saito (eds.), Pest resistance to pesticides. Plenum, New York.

Miller, L. H., R. K. Sakai, P. Romans, R. W. Gwadz, P. Kantoff, and H. G. Coon. 1987. Stable integration and expression of a bacterial gene in the mosquito *Anopheles gambiae*. Science 237: 779–781.

Moldenke, A. F., D. R. Vincent, D. E. Farnsworth, and L. C. Teeeiere. 1984. Cytochrome P-450 in insects. 4. Reconstitution of cytochrome P-450-dependent monooxygenase activity in the house fly. Pestic. Biochem. Physiol. 21: 358–367.

Motoyama, N., N. Nomura, and W. C. Dauterman. 1980. Multiple factors for organophosphorus resistance in the house fly, *Musca domestica* L. J. Pestic. Sci. 5: 393–402.

Motoyama, N., A. Hayashi, and W. C. Dauterman. 1983. The presence of two forms of glutathione S-transferases with distinct substrate specificity in OP-resistant and -susceptible house fly strains, pp. 197–202. *In* J. Miyamoto and P. C. Kearney (eds.), Pesticide chemistry: human welfare and environment, Vol. 3. Pergamon, Oxford.

Mouches, C., D. Fournier, M. Raymond, M. Magnin, J.-B. Berge, N. Pasteur, and G. P. Georghiou. 1985. Association entre l'amplification de sequences d'ADN, l'augmentation quantititive d'ester-ases et la resistance a des insecticides organophosphores chez des moustiques du complexe *Culex pipiens*, avec une note sur une amplification similaire chez *Musca domestica* L. C. R. Acad. Sci. Ser. III Sci. Vie 301: 695–700.

Mouches, C., M. Magnin, J.-B. Berge, M. de Silvestri, V. Beyssat, N. Pasteur, and G. P. Georghiou. 1987. Overproduction of detoxifying esterases in organophosphate-resistant *Culex* mosquitoes and their presence in other insects. Proc. Natl. Acad. Sci. USA 84: 2113–2116.

Mouches, C., N. Pasteur, J.-B. Berge, O. Hyrien, M. Raymond, B. Robert de St. Vincent, M. de Silvestri, and G. P. Georghiou. 1986. Amplification of an esterase gene is responsible for insecticide resistance in a California *Culex* mosquito. Science 233: 778–780.

Narahashi, T. 1983. Resistance to insecticides due to reduced sensitivity of the nervous system, pp. 333–352. *In* G. P. Georghiou and T. Saito (eds.), Pest resistance to pesticides. Plenum, New York.

Nebert, D. W., and F. J. Gonzales. 1987. P-450 genes: structure, evolution, and regulation. Ann. Rev. Biochem. 56: 945–993.

Nicholson, R. A., and T. A. Miller. 1985. Multifactorial resistance to transpermethrin in field-collected strains of the tobacco budworm *Heliothis virescens* F. Pestic. Sci. 16: 561–570.

Nicholson, R. A., R. J. Hart, and P. O. Osborne. 1980a. Mechanisms involved in the development of resistance to pyrethroids with particular reference to knockdown resistance in house flies, pp. 465–471. *In* Insect neurobiology and pesticide action (Neurotox '79). Society of Chemical Industry, London.

Nicholson, R. A., A. E. Chalmers, R. J. Hart, and R. G. Wilson. 1980b. Pyrethroid action and degradation in the cattle tick (Boophilus microplus), pp. 289–295. *In* Insect neurobiology and pesticide action (Neurotox '79). Society of Chemical Industry, London.

Noda, M., S. Shimizu, T. Tanabe, T. Takai, T. Kayano, T. Ideda, H. Takahashi, H. Nakayama, Y. Kanaoka, N. Minamino, K. Kangawa, H. Matsuo, M. A. Raftery, T. Hirose, S. Inayama, H. Hayashida, T. Miyata and S. Numa. 1984. Primary structure of *Electrophorus electricus* sodium channel deduced from cDNA sequence. Nature 312: 121–127.

Noda, M., T. Ikeda, T. Kayano, H. Suzuki, H. Takeshima, M. Kurasaki, H. Takahashi, and S. Numa. 1986a. Existence of distinct sodium channel messenger RNAs in rat brain. Nature 320: 188–192.

Noda, M., T. Ikeda, H. Suzuki, H. Takeshima, T. Takahashi, M. Kuno, and S. Numa. 1986b. Expression of functional sodium channels from cloned cDNA. Nature 322: 826–828.

Nolan, J., W. J. Roulston, and W. H. Wharton. 1977. Resistance to synthetic pyrethroids in a DDT-resistant strain of *Boophilus microplus*. Pestic. Sci. 8: 484–486.

Omer, S. M., G. P. Georghiou, and S. N. Irving. 1980. DDT/pyrethroid resistance interrelationships in *Anopheles stephensi*. Mosq. News 40: 200–209.

Oppenoorth, F. J. 1982. Two different paraoxon-resistant acetylcholinesterase mutants in the house fly. Pestic. Biochem. Physiol. 18: 26–27.

Oppenoorth, F. J. 1985. Biochemistry and genetics of insecticide resistance, pp. 731–773. *In* G. A. Kerkut and L. I. Gilbert (eds.), Comprehensive insect physiology biochemistry and pharmacology, Vol. 12. Pergamon, Oxford.

Osborne, M. P., and R. J. Hart. 1979. Neurophysiological studies of the effects of permethrin upon pyrethroid resistant (*kdr*) and susceptible strains of dipteran larvae. Pestic. Sci. 10: 407–413.

Osborne, M. P., and A. Smallcombe. 1983. Site of action of pyrethroid insecticides in neuronal membranes as revealed by the *kdr* resistance factor, pp. 103–107. *In* J. Miyamoto and P. C. Kearney (eds.), Pesticide chemistry: human welfare and environment, Vol. 3. Pergamon, Oxford.

Ottea, J. A., and F. W. Plapp, Jr. 1981. Induction of glutathione S-aryl transferase by phenobarbital in the house fly. Pestic. Biochem. Physiol. 15: 10–13.

Ottea, J. A., and F. W. Plapp, Jr. 1984. Glutathione S-transferase in the house fly: biochemical and genetic changes associated with induction and insecticide resistance. Pestic. Biochem. Physiol. 22: 203–208.

Ozoe, Y., M. Eto, K. Mochinda, and T. Nakamura. 1986. Characterization of high affinity binding of [^{3}H]propyl bicyclic phosphate to house fly head extracts. Pestic. Biochem. Physiol. 26: 263–264.

Pasteur, N., and G. Sinegre. 1975. Esterase polymorphism and sensitivity to Dursban organophosphorus insecticide in *Culex pipiens pipiens* populations. Biochem. Genet. 13: 789–803.

Pasteur, N., A. Iseki, and G. P. Georghiou. 1981a. Genetic and biochemical studies of the highly active esterases A′ and B associated with organophosphate resistance in mosquitoes of the *Culex pipiens* complex. Biochem. Genet. 19: 909–919.

Pasteur, N., G. Sinegre, and A. Gabinaud. 1981b. *Est–2* and *Est–3* polymorphisms in *Culex pipiens* L. from southern France in relation to organophosphate resistance. Biochem. Genet. 19: 499–508.

Pauron, D., J. Barhanin, M. Amichot, M. Pralavorio, J.-B. Berge, and M. Lazdunski. 1989. Pyrethroid receptor in the insect Na$^+$ channel: alteration of its properties in pyrethroid-resistant flies. Biochemistry 28: 1673–1677.

Payne, G. T., and D. M. Soderlund. 1989. Allosteric enhancement by DDT of the binding of [^{3}H]batrachotoxinin A-20-α-benzoate to sodium channels. Pestic. Biochem. Physiol. 33: 276–282.

Payne, G. T., R. G. Blenk, and T. M. Brown. 1988. Inheritance of permethrin resistance in the tobacco budworm, *Heliothis virescens* (Lepidoptera: Noctuidae). J. Econ. Entomol. 81: 65–73.

Pedersen, L.-E. K. 1986. The potency of cyclopropane pyrethroid ethers against susceptible and resistant strains of the house fly *Musca domestica*. Experientia 42: 1057–1058.

Picollo de Villar, M. I., L. J. T. van der Pas, H. R. Smissaert, and F. J. Oppenoorth. 1983. An unusual type of malathion-carboxylesterase in a Japanese strain of house fly. Pestic. Biochem. Physiol. 19: 60–65.

Plapp, F. W., Jr. 1986. Genetics and biochemistry of insecticide resistance in arthropods: prospects for the future, pp. 74–86. *In* Pesticide resistance: strategies and tactics for management. National Academy Press, Washington, DC.

Plapp, F. W., Jr., and R. F. Hoyer. 1968. Possible pleiotropism of a gene conferring resistance to DDT, DDT analogs, and pyrethrins in the house fly and *Culex tarsalis*. J. Econ. Entomol. 61: 761–765.

Priester, T. M., and G. P. Georghiou. 1978. Induction of high resistance to permethrin in *Culex pipiens quinquefasciatus*. J. Econ. Entomol. 71: 197–200.

Priester, T. M., and G. P. Georghiou. 1980. Cross-resistance spectrum in pyrethroid-resistant *Culex quinquefasciatus*. Pestic. Sci. 11: 617–624.

Raymond, M., D. Fournier, J. Berge, A. Cuany, J.-M. Bride, and N. Pasteur. 1985a. Single-mosquito test to determine genotypes with an acetylcholinesterase insensitive to inhibition to propoxur insecticide. J. Am. Mosq. Control Assoc. 1: 425–427.

Raymond, M., D. Pasteur, D. Fournier, A. Cuany, J. Berge, and M. Magnin. 1985b. Le gene d'une acetylcholinesterase insensible au propoxur determine la resistance de *Culex pipiens* L. a cet insecticide. C.R. Acad. Sci. Ser. III Sci. Vie 300: 509–512.

Riddles, P. W., P. A. Davey, and J. Nolan. 1983. Carboxylesterases from *Boophilus microplus* hydrolyze *trans*-permethrin. Pestic. Biochem. Physiol. 20: 133–140.

Riskallah, M. R. 1983. Esterases and resistance to synthetic pyrethroids in the Egyptian cotton leafworm. Pestic. Biochem. Physiol. 19: 184–189.

Rossignol, D. P. 1988. Reduction in number of nerve membrane sodium channels in pyrethroid resistant house flies. Pestic. Biochem. Physiol. 32: 146–152.

Saito, T., K. Tabata, and S. Kohno. 1983. Mechanisms of acaricide resistance with emphasis on dicofol, pp. 429–444. *In* G. P. Georghiou and T. Saito (eds.), Pest resistance to pesticides. Plenum, New York.

Salgado, V. L., S. N. Irving, and T. A. Miller. 1983a. Depolarization of motor nerve terminals by pyrethroids in susceptible and *kdr*-resistant house flies. Pestic. Biochem. Physiol. 20: 100–114.

Salgado, V. L., S. N. Irving, and T. A. Miller. 1983b. The importance of nerve terminal depolarization in pyrethroid poisoning in insects. Pestic. Biochem. Physiol. 20: 169.

Salkoff, L., A. Butler, A. Wei, N. Scavarda, K. Giffen, C. Ifune, R. Goodman, and G. Mandel. 1987. Genomic organization and deduced amino acid sequence of a putative sodium channel gene in *Drosophila*. Science 237: 744–749.

Sattelle, D. B., C. A. Leech, S. C. R. Lummis, B. J. Harrison, H. P. C. Robinson, G. D. Moores, and A. L. Devonshire. 1988. Ion channel properties of insects susceptible and resistant to insecticides, pp. 563–582. *In* G. G. Lunt (ed.), Neurotox '88: Molecular basis of drug and pesticide action. Elsevier, Amsterdam.

Sawicki, R. M. 1978. Unusual response of DDT-resistant house flies to carbinol analogues of DDT. Nature 275: 443–444.

Sawicki, R. M. 1985. Resistance to pyrethroid insecticides in arthropods, pp. 143–192. *In* D. H. Hutson and T. R. Roberts (eds.), Progress in pesticide biochemistry and toxicology, Vol. 5, Insecticides. Wiley, New York.

Sawicki, R. M., A. L. Devonshire, A. W. Farnham, K. E. O'Dell, G. D. Moores, and I. Denholm. 1984. Factors affecting resistance to insecticides in house flies, *Musca domestica* L. (Diptera: Muscidae). II. Close linkage on autosome 2 between an esterase and resistance to trichlorphon and pyrethroids. Bull. Entomol. Res. 74: 197–206.

Schnitzerling, H. J., J. Nolan, and S. Hughes. 1983. Toxicology and metabolism of some synthetic pyrethroids in larvae of susceptible and resistant strains of the cattle tick *Boophilus microplus* (Can.). Pestic. Sci. 14:64–72.

Schofield, P. R., M. G. Darlsion, N. Fujita, D. R. Burt, F. A. Stephenson, H. Rodriguez, L. M. Rhee, J. Ramachandran, V. Reale, T. A. Glencorse, P. H. Seeburg, and E. A. Barnard. 1987. Sequence and functional expression of the GABA$_A$ receptor shows a ligand-gated receptor superfamily. Nature 328: 221.

Scott, J. G., and G. P. Georghiou. 1986a. Mechanisms responsible for high levels of permethrin resistance in the house fly. Pestic. Sci. 17: 195–206.

Scott, J. G., and G. P. Georghiou. 1986b. The biochemical genetics of permethrin resistance in the Learn-PyR strain of house fly. Biochem. Genet. 24: 25–37.

Scott, J. G., and G. P. Georghiou. 1986c. Malathion-specific resistance in *Anopheles stephensi* from Pakistan. J. Am. Mosq. Control Assoc. 2: 29–32.

Scott, J. G., and F. Matsumura. 1981. Characteristics of a DDT-induced case of cross-resistance to permethrin in *Blattella germanica*. Pestic. Biochem. Physiol. 16: 21–27.

Scott, J. G., and F. Matsumura. 1983. Evidence for two types of toxic actions of pyrethroids on susceptible and DDT-resistant German cockroaches. Pestic. Biochem. Physiol. 19: 141–150.

Scott, J. G., S. B. Ramaswamy, F. Matsumura, and K. Tanaka. 1986. Effect of method of application on resistance to pyrethroid insecticides in *Blattella germanica* (Orthoptera: Blattellidae). J. Econ. Entomol. 79: 571–575.

Shono, T. 1985. Pyrethroid resistance: importance of the *kdr*-type mechanism. J. Pestic. Sci. 10: 141–146.

Soderlund, D. M., and J. R. Bloomquist. 1989. Neurotoxic actions of pyrethroid insecticides. Ann. Rev. Entomol. 34: 77–96.

Soderlund, D. M., J. R. Sanborn, and P. W. Lee. 1983. Metabolism of pyrethrins and pyrethroids in insects, pp. 401–435. *In* D. H. Hutson and T. R. Roberts (eds.), Progress in pesticide biochemistry and toxicology, Vol. 3. Wiley, New York.

Squires, R. F., J. E. Casida, M. Richardson, and E. Saederup. 1983. [^{35}S]*t*-Butylbicyclophosphorothionate binds with high affinity to brain-specific sites coupled to γ-aminobutyric acid-A and ion recognition sites. Mol. Pharmacol. 23: 326–336.

Sumikawa, K., I. Parker, and R. Miledi. 1984. Partial purification and functional expression of brain mRNAs coding for neurotransmitter receptors and voltage-operated channels. Proc. Natl. Acad. Sci. USA 81: 7994–7998.

Sun, Y.-P., and E. R. Johnson. 1972. Quasi-synergism and penetration of insecticides. J. Econ. Entomol. 65: 349–353.

Talcott, R. E. 1979. Hepatic and extrahepatic malathion carboxylesterases. Assay and localization in the rat. Toxicol. Appl. Pharmacol. 47: 145–150.

Tanaka, K. 1987. Mode of action of insecticidal compounds acting at inhibitory synapse. J. Pestic. Sci. 12: 549–560.

Tanaka, K., J. G. Scott, and F. Matsumura. 1984. Picrotoxinin receptor in the central nervous system of the American cockroach: its role in the action of cyclodiene-type insecticides. Pestic. Biochem. Physiol. 22: 117–127.

Terriere, L. C. 1983. Enzyme induction, gene amplification, and insect resistance to insecticides, pp. 265–297. In G. P. Georghiou and T. Saito (eds.), Pest resistance to pesticides. Plenum, New York.

Tsukamoto, M., T. Narahashi, and T. Yamasaki. 1965. Genetic control of low nerve sensitivity to DDT in insecticide-resistant house flies. Botyu-Kagaku 30: 128–132.

Vincent, D. R., A. F. Moldenke, and L. C. Terriere. 1983. NADPH-cytochrome P–450 reductase from the house fly, *Musca domestica*. Improved methods for purification, and reconstitution of aldrin epoxidase activity. Insect Biochem. 13:559–566.

Vincent, D. R., A. F. Moldenke, D. E. Farnsworth, and L. C. Terriere. 1985. Cytochrome P–450 in insects. 6. Age dependency and phenobarbital induction of cytochrome P–450, P–450 reductase, and monooxygenase activities in susceptible and resistant strains of *Musca domestica*. Pestic. Biochem. Physiol. 23: 171–181.

Wafford, K. A., D. B. Sattelle, I. Abalis, A. T. Eldefrawi, and M. E. Eldefrawi. 1987. γ-Aminobutyric acid-activated ^{36}Cl-influx: a functional *in vitro* assay for CNS γ-aminobutyric acid receptors of insects. J. Neurochem. 48: 177–180.

Waters, L. C., and C. E. Nix. 1988. Regulation of insecticide resistance-related cytochrome P–450 expression in *Drosophila melanogaster*. Pestic. Biochem. Physiol. 30: 214–227.

Waters, L. C., S. I. Simms, and C. E. Nix. 1984. Natural variation in the expression of cytochrome P–450 and dimethylnitrosamine demethylase in *Drosophila*. Biochem. Biophys. Res. Commun. 123: 907–913.

Welling, W., and G. D. Paterson. 1985. Toxicodynamics of insecticides, pp. 603–645. In G. A. Kerkut and L. I. Gilbert (eds.), Comprehensive insect physiology, biochemistry, and pharmacology, Vol. 12. Pergamon, Oxford.

Wheelock, G. D., and J. G. Scott. 1989. Simultaneous purification of a cytochrome P–450 and cytochrome b₅ from the house fly, *Musca domestica* L. Insect Biochem. 19: 481–488.

Wilkinson, C. F. 1983. Role of mixed-function oxidases in insecticide resistance, pp. 175–205. In G. P. Georghiou and T. Saito (eds.), Pest resistance to pesticides. Plenum, New York.

Wu, C.-F., B. Ganetzky, L. Y. Jan, Y.-N. Jan, and S. Benzer. 1978. A *Drosophila* mutant with a temperature-sensitive block in nerve conduction. Proc. Natl. Acad. Sci. USA 75: 4047–4051.

Yasutomi, K. 1983. Role of detoxication esterases in insecticide resistance, pp. 249–263. In G. P. Georghiou and T. Saito (eds.), Pest resistance to pesticides. Plenum, New York.

Yu, S. J., and L. C. Terriere. 1979. Cytochrome P–450 in insects. 1. Differences in the forms present in insecticide resistant and susceptible house flies. Pestic. Biochem. Physiol. 12: 239–248.

Ziegler, R., S. Whyard, A. E. R. Downe, G. R. Wyatt, and V. K. Walker. 1987. General esterase, malathion carboxylesterase, and malathion resistance in *Culex tarsalis*. Pestic. Biochem. Physiol. 28: 279–285.

5

The Role of Population Genetics in Resistance Research and Management

Richard T. Roush and *Joanne C. Daly*

I. Introduction

Pesticide resistance is an evolutionary phenomenon (Dobzhansky 1951), which cannot be fully understood without genetic data. Genetic studies are a major tool in developing improved methods of detecting resistance (Chapter 2), for investigating the mechanisms of resistance (Chapters 3 and 4), and in choosing approaches to manage resistance (Chapters 6, 9, and 10).

The potential long-term significance of the resistance problem was recognized by the early 1950s (Georghiou 1986), but genetic studies of resistance actually began a decade earlier with Dickson's (1941) study of cyanide resistance in the California red scale. Until about the mid-1970s, most studies of pesticide resistance were concerned only with mechanisms and inheritance, but more recent studies have placed considerable emphasis on the ecological genetics of resistance under field conditions (Roush and McKenzie 1987) and the molecular basis of resistance (Chapter 4). Studies of the genetics of resistance in field populations are probably crucial for the success of efforts to manage resistance in the future (National Research Council 1986).

A detailed study of the population genetics of resistance in any specific case involves both a study of the inheritance of resistance, usually made in the laboratory, and of the change in frequency (evolution) of resistance alleles in natural populations. In this chapter, we discuss the genetic basis of resistance, including how the inheritance of resistance can be studied, what dominance in the expression of resistance means theoretically and practically, and whether the currently available data and theory suggest that resistance is inherited primarily at a single locus (monogenic) or more than one (polygenic). We then discuss in more detail the factors that influence the rates of evolution of resistance in the field, particularly in terms of population structure, the initial frequencies of

We thank Bruce Tabashnik and Jay Rosenheim for thoughtful comments on the manuscript and Neil Forrester for the use of Fig. 5.6.

resistant genotypes, and selection in favor of and against resistance. Finally, we describe resistance management tactics from a genetic perspective and suggest new approaches for genetic research that would improve resistance management.

II. Genetic Basis of Resistance

A. Behavioral and Physiological Expression

Resistance may occur through either behavioral or physiological means (Georghiou 1972, Sparks et al. 1989), but very little is known about the potential for the evolution of behavioral resistance in the field or its genetics. Proof of behavioral resistance requires more than simply showing that insects will avoid a pesticide; as in all cases of resistance, there must be proof of genetic differences between strains.

The most persuasive way to provide such evidence is through genetic crosses, but in contrast with the dozens of cases where this has been done for physiological resistance, we are unaware of any case to date of field-evolved behavioral resistance that has been genetically analyzed. Very few studies have even shown that significant levels of behavioral resistance have evolved in the field and that the altered behaviors actually resulted in increased survival of behaviorally resistant individuals. Perhaps the best example successfully demonstrating increased survival through changes in behavior was avoidance of malathion baits by house flies, *Musca domestica* (e.g., Schmidt and LaBrecque 1959). Many other frequently cited cases are ambiguous or doubtful. A commonly cited example is the avoidance of pesticide residues by exophilic (preferring to stay outdoors) behavior in *Anopheles* mosquitoes in Africa, but these observations were made before it became clear that these mosquitoes are composed of a complex of sibling species, some of which are exophilic (Coluzzi et al. 1977). Thus, rather than showing behavioral resistance, this example may have only shown the relative suppression of the nonexophilic species by pesticide use, with an apparent shift to prevalence of the exophilic species.

Because of the paucity of genetic data on field-selected behavioral resistance, this review will concentrate on physiological resistance, where considerably more is known.

B. Issues and Controversies Concerning the Inheritance of Resistance

A complete description of the inheritance of resistance includes an understanding of the number of genes involved, the dominance relationships between genotypes, and the genetic basis of cross-resistance. Studies of the inheritance of resistance have generally used genetic crosses between resistant and susceptible strains in the laboratory. These methods, and their results, have been well documented by many workers, including Tsukamoto (1963, 1983) and Georghiou (1969) and references cited therein. More recently, molecular biology has provided insights about the specific changes in DNA that cause resistance. In general, both molecular and classical studies have documented that high magnitudes of resistance can occur by a genetic change at a single genetic locus (Roush and McKenzie 1987), either by gene amplification or by allelic substitution (Chapter 4).

A more controversial point is whether more than one locus contributes to resistance. Several authors have argued that resistance of sufficient magnitude to cause poor efficacy for any given pesticide in the field usually evolves by only one or two major genes, as summarized by Roush and McKenzie (1987). Because many resistance detection methods (Chapter 2) and more resistance management models (Chapter 6) assume that resistance is monogenic, this may be a particularly important issue. However, it is not the only important one; the dominance of resistance is also very significant to the evolution and management of resistance. Before we can fully describe the meaning and measurement of dominance and weigh the arguments in the controversy over whether the inheritance of resistance is monogenic or polygenic, we must first describe the classical methods by which dominance and the number of genes influencing resistance can be determined. We also discuss the potential applications of genetic techniques that have been rarely applied to the analysis of resistance, including single-pair analysis and quantitative genetics. Along the way, we review the concept of discrimination between genotypes first introduced in Chapter 2 and suggest possible inadequacies of traditional methods for studying the genetics of resistance.

C. Classical Methods for the Analysis of Inheritance

As discussed in Chapter 4, there are at least three kinds of genetic mutations that might cause resistance. One is gene amplification or duplication, as illustrated by *Culex* mosquitoes and the aphid *Myzus persicae*. In other cases, such as for altered acetylcholinesterases, there appear to be changes in particular DNA base pairs that change protein structure. A third possibility is a change in the number of target sites (controversially suggested for the *kdr* gene, Chapter 4) or the amount of enzyme, possibly through changes in genes that regulate the amount of enzyme produced (Oppenoorth 1985). From a population genetic perspective, distinguishing between these possibilities is not always necessary, and has only recently become possible for a few cases at the molecular level. Using one of the best–characterized cases as an example, resistant *Culex* strains that carry the gene amplifications still inherit resistance primarily as a single gene (e.g., Halliday and Georghiou 1985, Chapter 4); strains with different numbers of copies of the amplified gene can be considered as carrying different resistance alleles.

The first step in the genetic analysis of inheritance is to characterize the resistant and susceptible strains to be used. In most cases, the susceptible strain has been protected from pesticide exposure in the laboratory for many years. The resistant strain is usually collected from the field, where it is probably not yet homozygous for the resistance gene(s). To ensure homozygosity for resistance, the strain is ordinarily mass-selected for one to several generations in the laboratory (problems with this approach are discussed in Section II.E.2). Failure to establish homozygosity may lead to results in a later backcross or F_2 that are difficult to interpret. The next step is to cross, typically in groups, the resistant and susceptible strains in both reciprocal directions (resistant females by susceptible males and vice versa), primarily to test the resulting F_1 progeny for evidence of sex linkage or maternal effects, but also to test for homozygosity in the parents, the best indication of which is a homogeneous (linear fit to a probit model) dose-response in the F_1. When sex determination is autosomal, it is necessary to study backcrosses to determine if resistance is linked to a sex–determining allele (e.g., McDonald and Schmidt 1987). If there is no sex linkage, the F_1 progeny are backcrossed to the most phenotypically different parent (e.g., if the F_1s tend to be as susceptible as the susceptible parent, they are backcrossed to the resistant parent). Backcrosses are preferable to F_2s because there are fewer expected phenotypic classes to be distinguished when scoring. The data are then analyzed for the presence of a 1:1 segregation of resistance in the offspring (e.g., Beeman 1983, Halliday and Georghiou 1985, Roush et al. 1986), as illustrated in Fig. 5.1. There are statistical techniques for estimating the number of genes from the genetic variances of the F_2 or backcrosses, but these are dependent on assumptions (see Raymond et al. 1987) that may not be realistic for resistance.

However, there are some common pitfalls that have led to misinterpretations in the literature. One of the most common is the assumption that monogenic inheritance will always show plateaus in the dose–response data (implying segaration for a major gene) or, conversely, that the appearance of plateaus proves monogenic inheritance. As first noted by Tsukamoto (1963), plateaus can only occur when the dose–response lines of the genotypes to be scored in a backcross do not overlap. Plateaus can only occur at those does that kill essentially all individuals of the susceptible genotype(s) but do not affect the resistant genotype(s). Normally, the plateau will occur at 50% mortality, since half of the backcross offspring will be of the susceptible genotype and half will be resistant. On the other hand, there can be no plateau, even when there is only one gene, if the dose responses overlap (Tsukamoto 1963, Georghiou 1969). Although drawing sharply angled inflections has been a common practice, it is also quite incorrect. The proper way to calculate the expected mortality at any given dose in the backcross, without regard to whether the dose–response lines overlap, is to take the average mortality of the F_1 and the backcross parent at that dose (Tsukamoto 1963, Georghiou 1969). To avoid experimenter bias toward simple inheritance, the observed and expected mortalities should be tested for statistically significant deviations against both a hypothesis of monogenic inheritance and alternate hypotheses of more complex or polygenic inheritance (e.g., Roush et al. 1986, Yarbrough et al. 1986). However, there are many possible alternative

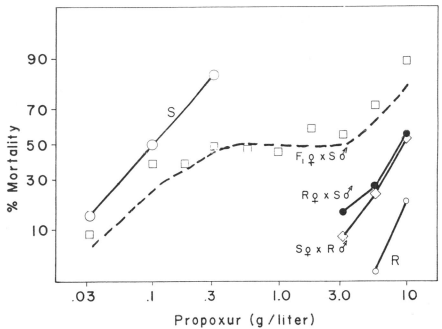

Figure 5.1. Concentration–mortality responses for genetic crosses of carbamate-resistant *Metaseilus occidentalis* when tested with propoxur. The dashed line shows the expected response for a 1:1 segregation of RS and SS females in the blackcross; the expected mortalities at each dose are half of those expected for the RS and half for the SS. (Redrawn from Roush and Plapp 1982a.)

hypotheses and there is no obvious objective way to decide which of them to choose in the absence of prior information on the possible genes available. Multiple tests of alternative hypotheses may also lack statistical power.

Another problem is that the dose responses of resistant and susceptible strains often overlap extensively such that results from a simple backcross are ambiguous with respect to the number of genes involved. In such circumstances, repeated backcrossing may be used to clarify at least whether more than one major gene is involved (Crow 1957, Georghiou 1969), even though it is not very sensitive for detecting minor genes. To consider how repeated backcrossing works, assume for the moment that resistance is controlled by a single gene and that a dose can be chosen that kills essentially all SS (susceptible homozygotes) but causes relatively little mortality of RS (heterozygotes). If the RS survivors of such a dose are backcrossed to SS individuals, there will be about 50% RS and 50% SS in the next generation and in every generation where the process is repeated. That is, the frequency of the resistant phenotype will remain relatively constant (Fig. 5.2). On the other hand, if two genes, R_1 and R_2, are involved, and the parents are homozygous for both, the F_1s will be $R_1S_1R_2S_2$. The first backcross offspring will have four different genotypes in equal frequencies (assuming no linkage):

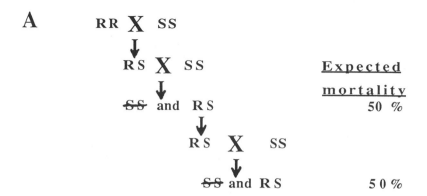

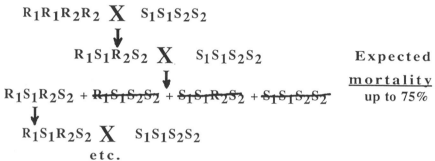

Figure 5.2. Repeated backcrossing where (A) only one major gene is present and (B) two genes are present. Genotypes noted with slashes suffered 100% mortality at the challenge dose. Backcrosses are repeated usually for a total of at least three generations. When resistance is due to a single gene, mortality is a constant 50% after each generation (A). On the other hand, if two genes are involved, the fraction being killed will steadily be less than the 50% expected, as shown in B (adapted from Georghiou 1969). The process would be slightly more complicated for a recessive gene, because each backcrossed generation requires one generation of interbreeding to produce RR homozygotes, at an expected frequency of 25%, that could survive the dose used.

$S_1S_1S_2S_2$, $S_1S_1R_2S_2$, $R_1S_1S_2S_2$, and $R_1S_1R_2S_2$. The dose that kills $S_1S_1S_2S_2$ will also likely kill some $S_1S_1R_2S_2$ and $R_1S_1S_2S_2$, resulting in fewer than 50% survivors each generation over the backcrossing scheme (Fig. 5.2). Repeated backcrossing essentially creates an "isogenic" strain by gradually diluting the fraction of the genome coming from the resistant parent. This method has been frequently applied to the genetic analysis of resistance, perhaps first by Shanahan (1961), who demonstrated that dieldrin resistance in sheep blowfly, *Lucilia cuprina,* was clearly monogenic (Brown 1967). Other examples that provided clear evidence for monogenic inheritance include diazinon resistance in the house fly (Hart

1963) and sheep blowfly (Shanahan 1979) and malathion resistance in *Tribolium castaneum* (Beeman 1983, White and Bell 1988). In a case where there was much poorer discrimination between genotypes, Roush and Wolfenbarger (1985) found by repeated backcrossing that resistance to methomyl in *Heliothis virescens* appeared to be monogenic.

Not only can repeated backcrossing be useful in the genetic analysis of resistance, it can be used to move a major resistance gene into a susceptible genetic background and thereby isolate it from other genes that affect the resistance phenotype. This can be a powerful tool in the separation and identification of resistance mechanisms (Chapter 3), particularly in distinguishing between multiple resistance (the co-occurrence of more than one resistance gene, each conferring resistance to one or more compounds) and cross–resistance (where one resistance gene confers resistance to more than one pesticide; Georghiou 1972). Repeated backcrossing is also essential in the study of fitness disadvantages associated with resistance, as described later in Section III.C.1.

However, the procedure of repeated backcrossing can only help to reveal the number of genes involved if moderate selection pressure is used each generation. The dose used should cause relatively little mortality of the putative heterozygotes. If a dose that kills 75% or more of the backcrosses is used, the most likely genotypes to be killed are the 75% that are most susceptible, leaving $R_1S_1R_2S_2$. These will produce the same frequency of genotypes in the next generation (time and time again), resulting in the lack of apparent dilution of resistance over the backcrossing scheme. The proper dose should cause no more than about 30% mortality of the heterozygotes. It is not a serious problem if this dose also allows some (even 10%) of the susceptibles to survive; later selection would always eliminate them. Better still would be to select only every other generation to moderate selection further and maximize the chances of segregation of genes in the offspring.

Because of the problems outlined above, data showing or failing to show 1:1 segregation in the backcross and even data from repeated backcrossing studies may be viewed as suggestive rather than conclusive. Many cases that have been classified as polygenic (e.g., Thomas 1966) often show somewhat ambiguous inheritance (Georghiou 1969) and vice versa. Far more persuasive is the use of genetic markers to map resistance genes to specific sites on the chromosomes, particularly where specific biochemical mechanisms have been associated with the resistance gene (e.g., citations in Munsterman and Craig 1979, Tsukamoto 1983, Foster et al. 1981). Beeman (1983) confirmed that resistance to malathion in *T. castaneum* was monogenic by crosses to genetic markers. This sort of evidence constitutes the most positive proof, short of cloning the gene, that resistance can occur from a single locus. Genetic markers also provide the strongest method to test whether resistance is monogenic, because they are very sensitive to the detection of minor genes, especially when used with statistical analysis of variance approaches (e.g., McKenzie et al. 1980). The genetics of pesticide resistance has often been successfully investigated through the use of genetic markers even in cases of overlapping dose response curves; the close association of resistance with a specific mutant provides good evidence that there is a tightly linked major gene nearby (examples are cited in Georghiou 1969). The approach is unambiguous even where the homozygous and heterozygous genotypes can only be partially separated (Fig. 5.3).

In many cases, two or more resistant strains may be available and it may be important to determine whether the resistances are caused by different loci. Testing the spectrum of resistance or even linkage to markers is not sufficient for this because there can be two or more resistance alleles at any given locus, each providing different levels of resistance in RS and RR (resistant homozygous) genotypes (e.g., Emeka–Ejiofor et al. 1983, Oppenoorth 1985, Beeman and Nanis 1986). On the other hand, two very closely linked loci can show resistance

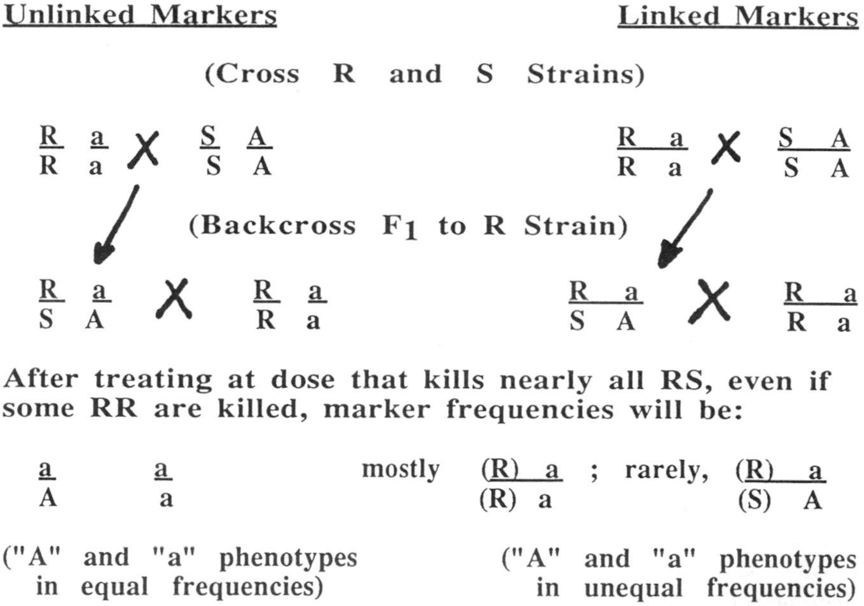

<u>Unlinked Markers</u>   <u>Linked Markers</u>

(Cross   R   and   S   Strains)

$$\frac{R}{R} \frac{a}{a} \quad \times \quad \frac{S}{S} \frac{A}{A} \qquad\qquad \frac{R}{R} \frac{a}{a} \quad \times \quad \frac{S}{S} \frac{A}{A}$$

(Backcross F_1 to R Strain)

$$\frac{R}{S} \frac{a}{A} \quad \times \quad \frac{R}{R} \frac{a}{a} \qquad\qquad \frac{R}{S} \frac{a}{A} \quad \times \quad \frac{R}{R} \frac{a}{a}$$

After treating at dose that kills nearly all RS, even if some RR are killed, marker frequencies will be:

$$\frac{a}{A} \qquad \frac{a}{a} \qquad\quad \text{mostly} \quad \frac{(R)\ a}{(R)\ a} \ ; \ \text{rarely,} \ \frac{(R)\ a}{(S)\ A}$$

("A" and "a" phenotypes   ("A" and "a" phenotypes
  in equal frequencies)   in unequal frequencies)

Figure 5.3. Example of the use of genetic markers to establish the number of genes involved in resistance. Even in cases of overlapping dose response curves, the close association of resistance with a specific mutant provides good evidence that there is a tightly linked major gene nearby (examples are cited in Georghiou 1969). The approach can be effective even where the homozygous and heterozygous genotypes can only be partially separated, as long as both parental strains are homozygous for different alleles, "A" and "a" at a marker locus and linkage is close (e.g., within 20 map units). Because there is little or no recombination in males in many Diptera, such as the house fly, any marker that is on the same chromosome as the resistance gene will be clearly linked to resistance when males are used as the F_1 parent in the backcross. Once linkage is found, the reciprocal backcross can be used to establish a map distance between R and the marker. In this example, it is assumed that there is considerable overlap between the RR and RS genotypes (and even poorer discrimination between SS and RS), necessitating backcrossing of RS to RR.

to similar pesticides. For example, two loci controlling different types of resistance to organophosphorous insecticides in sheep blowfly were too closely linked to be separated in crossing experiments (Raftos and Hughes 1986). One can test whether such genes are inherited at the same locus (often called a test for allelism) by crossing the two resistant strains and backcrossing their offspring to a susceptible strain when resistance is dominant (e.g., Ballantyne and Harrison 1967, Beeman and Nanis 1986, Raftos and Hughes 1986) or, at least in principle, to one of the resistant stains when resistance is recessive. Just as it is possible to have more than one resistance allele, some individuals within the population may carry alleles for greater suspectibility. In just a few generations using family selection (a procedure described in Section III.C.1), McEnroe (1969) was able to select as much as a 10–fold increase in the susceptibility of a strain that was believed to have been completely free of previous selection for resistance.

D. Dominance of Resistance

1. Dominance Is Relative

When resistance is under monogenic control (and there is only one resistant and one susceptible allele), there are just three genotypes, represented here by RR, RS, and SS, as illustrated in Fig. 5.4. Dominance is merely a description of the

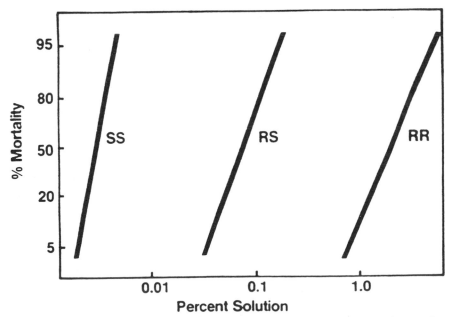

Figure 5.4.   Dose-responses of three genotypes of the seed corn maggot, *Hylemya platura*, when tested with aldrin (Redrawn from McLeod et. al. 1969).

relative phenotypic resemblance between the heterozygotes and the homozygous parents. If the heterozygotes more closely resemble the resistant parent, resistance is described as dominant (although it might also be said that susceptibility is recessive); if the heterozygotes more closely resemble the susceptible homozygotes, resistance is recessive. However, these terms are not absolute. In contrast to many traits discussed in introductory genetics courses, such as eye color in humans (blue eyes recessive to brown), resistance is rarely, if ever, completely recessive or completely dominant. "Incompletely dominant" usually refers to not quite fully dominant, "incompletely recessive" to not fully recessive, "codominant" to nearly exactly intermediate between the parents, and "semidominant" to responses that are between codominant and completely dominant. The resistance shown in Fig. 5.4 might be referred to as codominant or semidominant. Stone (1968) provided a formula to quantify dominance beyond these descriptive terms; Misra (1968) has given methods for calculation of confidence intervals for these estimates. However, the expression and measurement of dominance is more complex than these simple definitions suggest and depends critically on the techniques used. The notion of discrimination between genotypes plays an important role in these concepts.

2. Discrimination Between Genotypes

In Fig. 5.4, the dose–mortality regressions for each genotype are nonoverlapping, so that a concentration of 0.01% would kill virtually all susceptible individuals but very few resistant ones. Similarly, a concentration of a little more than 0.1% can distinguish between heterozygous and homozygous resistant individuals. Such concentrations are called discriminating. The existence of discriminating concentrations or doses is useful for many purposes, including the estimation of gene frequencies in field populations, although this is often difficult, as discussed later and in Chapter 2. Discriminating doses are also important to understanding selection for resistance. Selection proceeds only by discrimination between genotypes. It is obvious that no selection can occur if there is no mortality (or at least fitness reduction) of susceptible homozygotes, but it is perhaps less obvious that no selection can occur if all three genotypes are killed. If all three genotypes are killed, there can be no relative increase of the resistant genotypes in the population.

3. Dominance in Laboratory Assays

Under laboratory conditions, pesticide assays are generally conducted over a range of doses or concentrations (Fig. 5.4). In such "continuous" assays, different resistance mechanisms tend to have characteristic dominances. Nerve insensitivity mechanisms for pyrethroids and DDT, similar to *kdr* (Chapter 4), tend to be recessive or semirecessive in their inheritance (e.g., Halliday and Georghiou

1985, Bull et al. 1988, Farnham et al. 1984, Georghiou 1969, McDonald and Schmidt 1987). Resistance by altered acetylcholinesterases, conferring resistance to organophosphates and carbamates (e.g., Oppenoorth 1985), and changes in the cyclodiene target site (e.g., Emeka–Ejiofor 1983, Georghiou 1969, Oppenoorth 1985, Shanahan 1961, Stone 1962a) tend to be codominant to dominant in their inheritance. Most metabolic mechanisms of resistance tend to be dominant (e.g., Arnold and Whitten 1976, Beeman 1983, Cluck et al. 1985, Ferrari and Georghiou 1981, Georghiou 1969, Halliday and Georghiou 1985, Hughes and Raftos 1985, Raymond et al. 1987, Roush and Plapp 1982a). These patterns can be explained by a reversal of Sewall Wright's physiological theory of dominance (described by Charlesworth 1979). Wright argued as early as the 1920s that most mutations reduce enzymatic activity. If the wild–type allele can produce enough enzyme to compensate for the inactivity of the mutant allele in a heterozygote, the wild–type alle will be dominant. In metabolic resistance to insecticides, the resistance allele produces greater metabolism of the insecticide than the wild–type allele (Chapters 3 and 4) and tends, therefore, to be dominant. Although the exact mechanisms for *kdr* are not yet known (Chapter 4), they presumably require a complete loss of activity to prevent binding by insecticides and, thereby, cause resistance. Thus, just as with the mutants first discussed by Wright, resistance tends to be recessive. The dominance of the effects of resistant genotypes on other characters, such as fitness in the absence of pesticide exposure, does not have to be the same as dominance for toxicological phenotype (Roush and Plapp 1982b). In the case of metabolic resistance, for example, the heterozygotes may benefit from the toxicological fitness advantages of the R allele and the reproductive fitness advantages of the S allele.

1. *Effective Dominance Under Field Exposure*

Dominance of resistance in the field depends on the dose applied. In Fig. 5.4, for example, application of a 0.1% concentration will make the phenotype of resistance recessive by killing the heterozygotes. Similarly, a lower concentration (0.01%) can make resistance effectively dominant (Curtis et al. 1978). The dominance of resistance at any given dose may also depend on life stage, especially whenever some life stages are more susceptible than others.

This prediction that dominance under field conditions is highly dependent on dose has been generally supported by the few studies conducted so far. Effective dominance in the field is often very different from that observed in laboratory assays (see Roush and McKenzie 1987) because of the difference in range of doses that establishes dominance in the laboratory and field. For example, resistance to cyclodienes and lindane is usually codominant (i.e., the responses of the heterozygotes lie roughly halfway between the responses of the resistant and susceptible parental strains) in multiple-dose laboratory experiments (Brown 1967). However, Rawlings et al. (1981) found that resistant heterozygous (RS) anopheline mosquitoes could be nearly completely controlled while some resistant homozygotes (RR) survived in buildings sprayed at moderate concentrations of lindane. Similar results have been obtained in experiments with diazinon resistance in the sheep blowfly. Diazinon resistance is nearly completely dominant in laboratory assays of sheep blowflies (Arnold and Whitten 1976, Shanahan 1979), but under field conditions there was little difference in the viabilities of RS and SS genotypes, even as the diazinon residues decayed to allow considerable survival of the SS homozygotes (McKenzie and Whitten 1982). Similarly, application of high doses of malathion can make resistance effectively recessive in first instar larvae of *Tribolium casteneum,* even though resistance in adults is nearly completely dominant (White and Bell 1988). Thus, these studies show that dominant or codominant resistance can be made effectively recessive by application of a dose of insecticide sufficiently high to kill heterozygotes. On the other hand, even though resistance to pyrethroids in the horn fly, *Haematobia irritans,* appears to be due to a *kdr*-like mechanism and is incompletely recessive (Roush et al. 1986, McDonald and Schmidt 1987), 20–60% of the heterozygotes survive field exposure to pyrethroid–treated ear tags at normal tagging rates (McDonald et al. 1987). Thus, under these field circumstances, pyrethroid resistance is incompletely dominant. These examples demonstrate that, to obtain a complete and realistic appraisal of dominance, resistance must be studied for the appropriate life stages in the field or under conditions that simulate the field.

E. Number of Genes Involved in Resistance: One or Two Genes vs. Many

The results of genetic backcrossing studies, especially those using genetic markers, as summarized in many review articles (e.g., Plapp 1976, Oppenoorth 1985, Chapter 4), suggest that major genes are found in a high percentage of cases of insecticide and acaricide resistance. Major genes have also been found for resistance to some kinds of fungicides and herbicides, particularly those that have very specific modes of action (National Research Council 1986). Even for heavy metal tolerance in plants, previously described as polygenic, new test techniques suggest that at least some cases may actually be mostly due to individual major genes (Macnair 1983). Thus, a single gene is clearly sufficient for resistance to evolve. This does not mean that resistance is always strictly monogenic (Holloway 1986, Via 1986, Tabashnik and Cushing 1989). However, in most cases where the inheritance of resistance in field-collected strains has been well characterized, especially with genetic markers that map resistance to particular loci, resistance of sufficient magnitude to cause control failures in the field appears to be under the control of just one or occasional two loci (Roush and McKenzie 1987).

At least some cases that have been cited as polygenic (e.g., Thomas 1966) are ambiguous or incorrectly analyzed (Georghiou 1969). Nonetheless, there are numerous carefully studied examples where resistance clearly does not conform to a monogenic model (e.g., Liu et al. 1981, cyhexatin resistance in Pree 1987). In addition, there must be a genetic component, perhaps polygenic, to the low levels of resistance often referred to as "vigor tolerance" (Holloway 1986). At least some of the variation observed in pesticide bioassays (Chapter 2) is probably due to genetic variation (Roush and McKenzie 1987, Tabashnik and Cushing 1989). In a population of the rice weevil, *Sitophilus oryzae,* that was believed to be fully susceptible, Holloway (1986) found that 47% of the variation affecting time to knockdown for pirimiphos–methyl was genetic and heritable. In a field-collected strain of diamondback moth, *Plutella xylostella,* that appeared to be at least nearly completely susceptible to fenvalerate (compare WO strain of Tabashnik et al. 1987 to Liu et al. 1981), Tabashnik and Cushing (1989) found a heritability for tolerance of about 20%.

While these examples show that at least some of the variation in pesticide bioassays is genetic, these and other examples also suggest that at least some of the variation is environmental. As noted in Chapter 2, much of the variation in bioassays can be reduced by improved application techniques. For example, changing the formulation increased the slopes of bioassays for cyhexatin resistance in spider mites (Edge and James 1986). In other cases, variation remains intractable. The slope of the concentration–mortality line for carbaryl was very shallow against both resistant and susceptible strains of the predatory mite, *Metasiulus occidentalis* (Roush 1979), suggesting considerable genetic variation within each strain, but vigorous attempts to select for even higher resistance in the resistant strain were unsuccessful. Subsequent studies with the closely related compound propoxur showed that the resistant strain was homozygous for a major gene (Roush and Plapp 1982a) and that propoxur gave a much steeper concentration–response curve than did carbaryl in both the resistant and susceptible strain (Figure 5.5). Thus, although segregation for resistance alleles will increase variation and lower slopes (Roush and Miller 1986), a low slope does not necessarily mean that there is substantial genetic variation.

1. Significant Resistance, Simultaneous or Sequential Selection

At least some of the controversy over the number of genes controlling resistance is due to different interpretations of how the question can be phrased. There is no doubt that at least some cases of resistance are the result of more than one major gene. Polygenic resistance commonly occurs under laboratory selection, and at least some cases of field–evolved resistance clearly show the influences of more than one gene when crossed to fully susceptible strains in the laboratory (Roush and McKenzie 1987). It is probably also true that most cases of monogenic resistance are also influenced by additional minor genes, providing continuous variation (Tabashnik and Cushing 1989). For example, in a particularly careful study, Beeman (1983) and Beeman and Nanis (1986) provided not only clear evidence for a single major gene, but also that the level of resistance was slightly

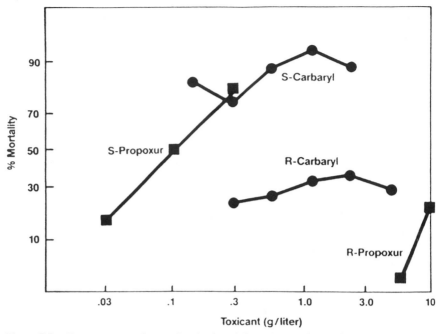

Figure 5.5. Dose–responses for a carbaryl-selected (R) and non–selected (S) strain of *Metaseilus occidentalis*. The R strain was homozygous and cross–resistant to propoxur (Redrawn from Roush 1979 and Roush and Plapp 1982a).

reduced in the genetic background of a susceptible strain. However, the more important question for understanding the evolution of resistance (and determining if monogenic models provide a sufficient description for management) is not whether resistance can be shown to be polygenic in some cases, but whether the evolution of resistance in the field is *significantly* influenced by the nearly *simultaneous substitution* of alleles at more than one locus. The importance of these two distinctions can be illustrated with a few examples, the first three of which relate to the importance of simultaneous substitution.

Pyrethroids were first introduced for control of the diamondback moth on Taiwan in 1976. Populations of diamondback moth on Taiwan already showed widespread resistance to other insecticides, including diazinon (Sun et al. 1978), which conferred cross-resistance to pyrethroids (Liu et al. 1981). Pyrethroids were apparently very useful when first introduced, but their effectiveness had declined significantly within four years (Liu et al. 1981). It seems likely that selection for resistance to pyrethroids was a *sequential* process: (1) selection for a gene or genes providing resistance to diazinon or other previously used insecticides that later weakened the effectiveness of pyrethroids relative to a fully susceptible strain and then (2) selection for an additional gene or genes that further increased pyrethroid resistance. This hypothesis is supported by a genetic analysis of resistance to fenvalerate, one of the pyrethroids. When a field strain that was susceptible to both diazinon (Sun et al. 1978) and pyrethroids was crossed to a strain highly resistant (perhaps 700-fold) to fenvalerate (neither strain was specifically tested for homozygosity), the observed mortalities differed significantly from those expected for monofactorial inheritance (Liu et al. 1981). Synergist studies (Chapter 3) suggested that at least two mechanisms, microsomal monoxygenases and nerve insensitivity (similar to *kdr*), were involved in resistance to pyrethroids (Liu et al. 1981).

A better–documented example is dimethoate resistance in Danish house flies. In a carefully detailed study that also demonstrates the usefulness of genetic markers, Sawicki (1974, 1975) showed that at least six genes controlled or modified resistance to insecticides in a strain that was 120–fold resistant to dimethoate compared to a standard susceptible strain. Two of these genes controlled very little resistance to dimethoate but had been selected by other insecticides or synergists. Of the remaining genes, one on each of chromosomes 3 and 5 contributed only about a 1.5– to 3.8–fold to dimethoate, but had apparently been previously selected by DDT and diazinon. Many populations were highly resistant to DDT and diazinon and were probably nearly fixed (100% allele frequency) for these two genes when dimethoate was introduced (Sawicki 1975). The remaining two genes, both on chromosome 2, were selected to high frequency by dimethoate, to which they contribute a combined 11– to 32–fold resistance (Sawicki 1974, 1975). Thus, when compared to a susceptible strain that had never been exposed to pesticides, the dimethoate–resistant strain showed polygenic inheritance, but if compared to a population collected just before the introduction of dimethoate, loss of the practical efficacy of dimethoate was probably due to substitution at just one or two new loci against a genetic background that already carries other resistance genes.

Even where resistance to a single pesticide becomes polygenic by novel selection of genes by that pesticide alone, it may do so by the sequential addition of genes. In field selection for chlorpyrifos resistance in *Culex pipiens*, for example, resistance was initially monofactorial and associated strictly with a highly active detoxifying esterase, but later evolved to include at least two more factors, an altered acetylcholinesterase and a microsomal monooxygenase (Raymond et al. 1987).

Although it is difficult to study the genetics of resistance in field populations in as much detail as has been done for dimethoate-resistant house flies or chlorpyrifos-resistant *C. pipiens*, even more detailed efforts must be made if the evolution of resistance in field populations is to be fully understood.

The second distinction described above is whether more than one locus significantly influences the evolution of resistance. Secondary genes are often found that further increase the magnitudes of insecticide resistance over those achieved by a major gene. The most common may be genes that intensify resistance by decreasing the rate of penetration of insecticides, such as the *pen* gene of the house fly (Plapp 1976). However, *pen* usually confers little or no resistance in the absence of another resistance mechanism and only two– to six–fold even when another mechanism is present, depending on insecticide, dose, and solvent (Plapp et al. 1979, Oppenoorth 1985). In the case of a dieldrin–resistant strain of house flies, the major gene for dieldrin resistance produces about 700–fold resistance; a minor factor presumed to be *pen* conferred two–fold resistance (Georghiou 1969). Whether *pen* should be considered a resistance gene or a modifier depends on whether there is significant detoxication of the pesticide in susceptible strains; in the absence of such detoxication, *pen* only enhances resistance (Oppenoorth 1985). Although no other low-level resistance gene is as well studied as *pen,* probably because organotin compounds can be used to detect it readily (Farnham et al. 1984), *pen* illustrates that a continuum of effects may exist between major and minor genes.

When a major mechanism can confer 10- to 1000–fold resistance, as is often the case (see citations in Plapp 1976, Oppenoorth 1985, Chapters 3 and 4), genes that contribute much lower effects are probably not important to resistance management. Although the exact magnitudes of resistance that cause control failures vary, they usually appear to be greater than six–fold (Chapter 2). Thus, even though minor genes conferring less than six–fold resistance (including all of those described above) can increase the magnitudes of resistance, they probably cannot in themselves cause a significant control failure. Therefore, it is only the major genes that are of particular importance to resistance management. Even when a gene of only 30–fold is present, minor genes conferring up to three–fold resistance may have virtually no impact on the rate of evolution of resistance (Plapp et al. 1979).

Perhaps the best–studied example illustrating this is the evolution of diazinon resistance in Australian populations of the sheep blowfly. Four populations of sheep blowflies that had already become nearly fixed for a major resistance allele after about 20 years of selection were brought into the laboratory for eight further generations of selection. The level of resistance approximately doubled in each strain and involved genes on at least four chromosomes (McKenzie et al. 1980). Perhaps the most intriguing part of this study is that eight generations of selection were able to produce a polygenic response that more than 20 years of field selection did not. The explanation probably lies in the importance of discriminating between genotypes. As noted earlier in the discussion of effective dominance on treated sheep (Section II.D.4), not even heterozygotes for the major gene, which are about 25–fold resistant to diazinon in the laboratory (Arnold and Whitten 1976), consistently survived diazinon applications on sheep very much better than susceptible homozygotes (McKenzie and Whitten 1982). Under such conditions, it seems very unlikely that alleles of much smaller effect could significantly increase in frequency.

In another example, minor factors for resistance in the horn fly were not measurable until resistance exceeded about 25–fold; with their contributions, resistance eventually reached about 40-to 70–fold (Bull et al. 1988). However, as mentioned in discussing effective dominance, heterozygotes for the major gene, which appears to control nerve insensitivity of a *kdr* type (Roush et al. 1986, Bull et al. 1988), show about 20–60% survival under field exposure (McDonald et al. 1987), even though heterozygotes appear to be only four- to eight–fold resistant (Roush et al. 1986, McDonald and Schmidt 1987). In strains showing 130- to 180–fold resistance, the minor genes appear to confer perhaps two- to four–fold resistance, whereas the major gene controls the remaining 40- to 90–fold (Roush et al. 1986, McDonald and Schmidt 1987).

These two examples may be typical of resistance selection. Recommendations for pesticide application rates often include an "overkill factor" (Farnham et al. 1984) to ensure control even under variable application conditions. Resistance therefore does not always provide complete immunity to the pesticide. Depending on the pesticide and life stage treated, heterozygotes and even resistant homozygotes can be killed by the pesticide application (e.g., Rawlings et al. 1981). Indeed, a useful approach to resistance management is to find such pesticides and life stages (Roush 1989a). Thus, only those genes that provide a high level of resistance confer a generally consistent selective advantage.

In other cases, the modifiers may exist because of prior selection with other compounds, but may still be of only minor importance to control in the field. Even though an esterase gene, selected by earlier use of trichlorphon, produced four- to 10–fold resistance to pyrethroids, neither it nor *pen* appeared to play a detectable role in failures of pyrethroids to control house flies in England, even though both genes were common (Farnham et al. 1984, Sawicki et al. 1984). Control failures were closely associated only with the presence of the nerve insensitivity (*kdr*) gene. Another example is resistance of *Helicoverpa (=Heliothis) armigera* in Australia to pyrethroids. Resistance in the laboratory is influenced by nerve insensitivity (*kdr*–like), reduced penetration, and microsomal monooxygenase (metabolic) genes (Ferris and Gunning 1984). However, in recent years only the microsomal monooxygenase gene has been shown to confer any appreciable resistance to pyrethroids in the field and to respond to selection pressure under field conditions (Forrester 1988b, Section IV.C.1.). The nerve insensitivity gene has declined in frequency since 1986 and appears have a reduced effect on the pyrethroid resistance phenotype since it was first studied in 1983 (R. V. Gunning, pers. comm.). The gene(s) involved in this mechanism may have been selected for by prior use of DDT because *kdr*–like genes often confer much higher resistance to DDT than pyrethroids (Chapter 2).

Holloway (1986) argued that a simple monogenic explanation for resistance was insufficient to account always for the evolution of resistance, but he based his arguments on Fisher's theory of the evolution of dominance, an idea that has been discredited in favor of Wright's physiological theory (Charlesworth 1979), described in Section II.D.3. Contrary to Holloway's assumptions, new resistance alleles are probably rarely completely recessive in toxicological expression; their dominance is largely determined by the mechanism of resistance. Kikkawa (1964), for example, produced a new and nearly completely dominant resistance allele in *Drosophila melanogaster,* at a locus previously shown to control resistance in the field, with the use of X rays. Using ethyl methane sulfonate (EMS), Wilson and Fabian (1986, 1987) have generated mutants in *D. melanogaster* showing as much as 100–fold resistance to methoprene. These mutants appear to be due to changes in the target site, presumably a juvenile hormone receptor, and are incompletely recessive. However, at least 30% of the heterozygotes of the more resistant mutants survive doses that kill 99% of susceptible homozygotes. Six to nine generations of repeated backcrossing can return a resistance gene to an almost completely susceptible genetic background, removing fitness modifiers (McKenzie et al. 1982) and presumably dominance modifiers of the kind invoked by Holloway. Nonetheless, 10 or 11 generations of repeated backcrossing failed to change the nearly complete dominance of resistance to malathion in *T. castaneum* (Beeman and Nanis 1986) or to diazinon in the house fly (Hart 1963), nor did six generations of backcrossing reduce the dominance of dieldrin resistance in the sheep blowfly (Shanahan 1961). Holloway's misconceptions led him to conclude that resistance was likely to be fully recessive (all heterozygotes were killed by pesticide applications) initially and would take hundreds of generations to evolve if monogenic. Since this is much longer than resistance evolution requires in nature, Holloway invoked polygenic inheritance to explain the evolution of resistance. More extensive modelling work shows, however, that even low survival of heterozygotes causes resistance to evolve much more quickly than if resistance is completely recessive and that monogenic inheritance can produce resistance in time periods consistent with field experience (Chapter 6).

2. Conclusions and Research Needs

The inheritance of most traits important to the adaptation of arthropods to their environments, such as developmental time, fecundity, and climatic adaptation, is complex and probably under the control of many genes (Falconer 1981). Pesticide resistance may be unique as an important and widespread adaptation that often shows very simple inheritance (Roush and McKenzie 1987). Another symptom of the unusual nature of resistance is that so many resistance genes seem to be semidominant in expression (Holloway 1986). At least a partial explanation for the unusual genetic characteristics of pesticide resistance is the specificity in mode of action and structure of modern pesticides. Unlike the older inorganic pesticides (such as lead arsenate and hydrogen cyanide), which tend to have multiple sites of action and produced resistance relatively rarely (Georghiou 1983), the more modern synthesis ones usually disrupt a specific target in the nervous system. To accomplish this, the structure of the pesticide must often be precisely correct. Therefore, a subtle change in a single gene coding for a target protein or in the structure or amount of enzyme involved in the metabolism of toxicants can provide high levels of resistance (see Chapter 4, Plapp 1976, Oppenoorth 1985). Although it is probably less likely that an inorganic pesticide can select for monogenic resistance, at least two cases of cyanide resistance appeared to be due to a single gene (Crow 1957, Dickson 1941, Yust et al. 1943). A specific mechanism was implicated in one of these cases (Yust and Shelden 1952).

Polygenic resistance may be significant in some cases, such as for cyhexatin (Pree 1987, Chapter 10), which has an unusual mode of action (Carbonaro et al. 1986), but the available data suggest that the inheritance of resistance is usually fundamentally different from most traits. In those cases that have been amenable to further study due to the availability of chromosomal markers, resistance seems rarely to be measurably influenced by as many as four genes, the only documented example being dimethoate resistance in Danish house flies. In the great majority of examples studied so far, only one or two genes seem to have significant influence on control failures to any given insecticide at any given time or place. The process of selection under field conditions must be further studied if we are to understand how resistance evolves, especially with respect to whether resistance is a monogenic, sequential, or polygenic process. A key element to these studies should be the use of strains of insects collected just prior to the introduction of new compounds. The importance of this for realistic resistance monitoring has already been emphasized (Farnham et al. 1984, Chapter 2).

Unfortunately, field–collected strains will rarely be homozygous for all the genes they carry. Laboratory selection to make the lines homozygous could select genes that might be irrelevant in the field, as demonstrated by laboratory selection for resistance in the sheep blowfly (McKenzie et al. 1980) and for susceptibility in the two–spotted spider mite (McEnroe 1969). Thus, the use of single family lines will be an essential tool for these studies. Establishing the F_1 crosses in single pairs instead of large groups (mass crosses) can be used to determine if the original field–collected parents are homozygous for the genes under study (e.g., Beeman and Nanis 1986). Lack of homozygosity in the parents should be readily detected by segregation in the F_1 where resistance is dominant (e.g., Cluck et al. 1985) but would require single–family backcrosses to the resistant parent when resistance is recessive. Continuing the single pairs into the backcrosses (e.g., Roush and Wolfenbarger 1985) can also help to determine if there are differences in the number of genes carried by each resistant parent.

The use of genetic markers will also be extremely helpful, because they often provide the only convincing way of distinguishing the effects of minor genes. This suggests that good model species for investigating the evolution of resistance will include *Aedes aegypti* (Munsterman and Craig 1979) and other mosquitoes, red flour beetles, *T. castaneum* (Beeman 1983), house flies (Tsukamoto 1983), sheep blowflies (Foster et al. 1981), and *D. melanogaster,* a particularly tractable species (Wilson 1988) that also appears to evolve resistance in the field (Kikkawa

1964; R. T. Roush and R. H. ffrench-Constant, unpubl. data). Electrophoretic markers can be developed for other species, such as *H. virescens*, that do not yet have a sufficient number of visible markers (Heckel et al. 1988), which would make them far more attractive for this work.

F. Quantitative Genetics

Within the last few years, there has been increasing interest in applying quantitative genetics to the study of resistance (Tabashnik and Cushing 1989). Quantitative genetics is a statistical approach developed to analyze cases where phenotypes show polygenic inheritance, cases of such mathematical complexity that simple Mendelian (algebraic) methods could not be applied. However, quantitative genetic approaches can be applied to any trait, and are especially useful for those that show continuous phenotypic variation. The principal model in quantitative genetics is that phenotypic variance is a function of genetic and environmental variances that can partitioned statistically and thereby quantified (Falconer 1981, Via 1986). The techniques of quantitative genetics are potentially very powerful because they make no assumptions about the number of genes controlling trait. Tabashnik and Cushing (1989) and Holloway (1986) have demonstrated how to apply a quantitative approach in describing the inheritance of resistance, as noted earlier in Section II.E. Ferrari et al. (1982) calculated realized heritability estimates from selection studies conducted by Lagunes and Georghiou (Georghiou 1983) on a "synthetic" population of *Culex* mosquitoes produced by incorporating resistance genes from strains resistant to propoxur, temephos, and permethrin at a frequency of 2%. Resistance to both temephos and permethrin was strongly influenced by an independent major gene (Lagunes 1980), Ferrari and Georghiou 1981, Halliday and Georghiou 1985), whereas resistance to propoxur was influenced by at least two genes (Lagunes 1980). The heritabilities were high compared to those generally observed for traits other than resistance (for examples see Falconer 1981) and appeared to be greater for the resistances where single major genes were involved (temephos, 0.404; permethrin, 0.389) than for propoxur (0.254).

Quantitative genetics may be particularly helpful where the inheritance of resistance is clearly under the control of more than one gene or where the inheritance is ambiguous, such as for cyhexatin resistance in mites, as mentioned earlier. Polygenic resistance may become more common if future pesticides are poorly metabolized by arthropods but have multiple modes of action. Although it has been suggested that polygenic resistance may become more common as selection intensity is lowered (Via 1986), this seems likely to occur only if lower rates of pesticides are used in each application, a strategy that does not seem promising if it is the only one followed when major genes are present (Curtis 1985, Roush 1989a; please note that arbitrary increases in doses are not helpful either, as discussed in Section IV.A.1 and Chapter 6).

Quantitative models for resistance management are currently limited by an assumption that selection is weak but could be improved (Via 1986). It would be extremely useful to do so. A critical area for further research is to test whether the assumptions of monogenic versus polygenic inheritance of resistance alter the optimal resistance management strategy for any given situation (Chapter 6). This would help to determine the importance of knowledge of the inheritance of resistance to the practical management of resistance. Because most current models of resistance are monogenic, the limiting factor for this analysis is the development of improved polygenic models.

Quantitative genetic approaches can be applied in any situation (e.g., Ferrari et al. 1982), but Mendelian approaches are often superior, at least for several kinds of investigations, where major genes can be identified, as discussed later in this chapter (Sections III.C and IV.D). Thus, a useful first step in any investigation of resistance is to test, using appropriate F_1 and backcrosses, whether there are any major genes present.

G. Applications of Inheritance Data

Simply knowing the mode of inheritance of resistance does not in itself help to manage resistance. In fact, so many inheritance studies have been conducted over the last 30 years that such data are exciting only for novel chemistries. As noted in Section II.C. and in Chapter 3, one important and perhaps underutilized application of inheritance data is the investigation of resistance mechanisms, particularly in the creation of isogenic lines. Another important use of inheritance data is the development of rapid and efficient resistance monitoring methods based on discriminatory bioassay techniques that can be used both for the management of resistance and for studying the effects of resistance genes on fitness under field conditions (Chapter 2).

III. Influences on the Rates of Evolution of Resistance

Evolution is a change in gene frequencies resulting from the complex interaction of mutation, selection, gene flow, and random genetic drift. A major goal of genetic investigations of resistance is to evaluate the relative significance of these factors and to determine how they might be manipulated to stop or retard the evolution of resistance. Mutation, selection, and random genetic drift are the primary determinants of the frequencies of resistance alleles before the introduction of a novel pesticide. Selection and gene flow are the major influences once pesticide use has become widespread and are the most likely factors that can be manipulated to influence resistance evolution.

A. Initial Allele Frequencies: A Balance Between Mutation and Selection

The frequencies of resistance alleles prior to selection are essentially unknown for any kind of mutation (Roush and McKenzie 1987). McEnroe (1969) estimated that the initial frequency of the oxydemeton resistance factor in a laboratory population of *Tetranychus urticae* was about 4×10^{-4}, but there may be some question as to whether he looked at a single gene. In the absence of other data, initial resistance allele frequencies have been assumed to range from 10^{-2} (Georghiou and Taylor 1977a) to 10^{-13} (Whitten and McKenzie 1982) on the basis of mutation–selection equiplibrium theory. This theory assumes that the frequency, q, of any allele prior to selection in its favor is maintained by an equilibrium between the generation of new alleles by mutation (at a rate represented by u) and selection against the heterozygous genotypes (represented by s), depending on dominance (represented by h). (Resistant homozygotes will be so rare that they can be ignored.) If the fitness disadvantages are incompletely to completely dominant (h ranges from greater than 0 to 1), q is approximately u/hs (Crow and Kimura 1970). Available data on the fitnesses of resistance heterozygotes in the absence of pesticides (summarized later in Section III.C.) suggest that fitness disadvantages, s, range from 0 to 0.5 and that dominance, h, ranges from nearly completely recessive to incompletely dominant (say, 0.05–0.5).

These values for h and s fall into a narrow range. The 10^{-2} to 10^{-13} range in assumptions about initial allele frequencies is due to the absence of data on the mutation rate, u. This has been assumed to range from 10^{-6} to 10^{-17}, depending on the number of nucleotide substitutions that are required (Whitten and McKenzie 1982), but it might be higher (perhaps 10^{-3}, Schimke et al. 1986) for other kinds of mutations, such as gene amplification. In an experiment where *Drosophila melanogaster* were irradiated to enhance mutation, the mutation rate for parathion resistance was about 3×10^{-5} (Kikkawa 1964). Helle showed that the mutation rate for parathion resistance in *T. urticae* was less than 3.3×10^{-7} (Oppenoorth 1985).

Although it has been difficult to measure initial allele frequencies, improved resistance detection techniques may facilitate direct measures of allele frequencies in field populations (Roush and Croft 1986). Accurate estimates of initial allele frequencies may not be terribly important for determining the utility of resistance management tactics (Roush 1989a), but they are still of considerable fundamental interest.

The calculations given here assume populations of infinite size to avoid the loss of resistance alleles by random genetic drift. In finite populations, resistance alleles may be patchily distributed between populations, with some populations losing resistance alleles by the chance effects of sampling and other populations gaining increases in frequencies. Immigration would tend to homogenize populations and make allele frequencies more uniform.

B. Population Structure and Gene Flow

While some populations of major insect pests develop resistance to most insecticides to which they are repeatedly exposed, many other populations of the same or closely related species remain relatively susceptible, even though exposed to similar spray rates and numbers of applications (Roush and McKenzie 1987). Variation in population structure probably contributes to these differences.

Population structure describes the ways in which populations are subdivided into finite (and often small) breeding units as a result of behavior, sex ratios, discrete distribution of resources, population size, the degree to which the population deviates from random mating, and restrictions on dispersal that limit the flow of genes from one location to another. Whether selection occurs before or after the female mates influences the efficacy of selection when it is directed against adults (e.g., Mani 1985). Population size is also important. Although insect populations can be very large, particularly in pest species, pest control regularly reduces the population to small size. Theoretical studies suggest that evolution can occur more rapidly in small, subdivided populations than in large, panmictic ones (Wright 1931 and later, Crow and Kimura 1970); this may be particularly important in cases where resistance is largely recessive (Comins 1977, Curtis et al. 1978, Taylor and Georghiou 1979, Tabashnik and Croft 1982). However, if immigration of susceptible individuals from outside the treated area is great, the frequency of resistance may increase more slowly than predicted by pesticide use (Chapter 6). In fact, there seems to be a consensus that immigration of susceptible individuals into a treated area is one of the more critical influences on resistance development (Chapter 11).

In a number of species, such as house flies (Gibson 1981, Denholm et al. 1985) and greenhouse pests (Dittrich 1975), resistance evolves much more quickly in populations that are isolated by physical barriers or environmental factors from susceptible migrants. There are also striking differences between populations of closely related species (Roush and McKenzie 1987). The Australian noctuid pests *Helicoverpa punctigera* and *H. armigera* are both sprayed with similar insecticides several times each year and have similar generation times. However, only *H. armigera* has developed resistance to a number of insecticides (Wilson 1974, Kay et al. 1983, Gunning et al. 1984). *H. armigera* is thought to be more concentrated in cropping areas in eastern Australia, whereas *H. punctigera* is very common on native hosts, including the semidesert area of central Australia (Zalucki et al. 1986, Gregg et al. 1989). In terms of the management strategies to be discussed in Section IV.A., a higher proportion of the *H. punctigera* than of the *H. armigera* population resides in refugia. Nevertheless, immigration of susceptible individuals appears to be important in diluting resistance in *H. armigera* populations in treated areas (Daly et al. 1988b). Similar phenomena influence the evolution of insecticide resistance in *Helicoverpa* (=*Heliothis*) *zea* and *Heliothis virescens* in North America. *H. virescens* is the species showing the more serious resistance problems (Sparks 1981), apparently because it is concentrated on relatively heavily treated cotton during most of the season, whereas *H. zea* is found on many hosts, such as soybeans and maize, that are only lightly treated. The first field failures associated with pyrethroid resistance in *H. virescens* occurred in fairly isolated areas of western Texas (Plapp and Campanhola 1986, Chapter 9), where all major hosts (including some vegetable crops) had been heavily treated.

On the other hand, dispersal from heavily-treated habitats can worsen resistance in lightly-treated sites, especially where heavily-treated areas have relatively higher pest population densities (perhaps as a result of intense insecticide use) or serve as overwintering sites (e.g., Follett et al. 1985, Denholm et al. 1985, Miller et al. 1985). Under such conditions, it is possible to find relatively high frequencies of resistance even at locations or on hosts that have not been treated (e.g., Scott et al. 1989, Gunning and Easton 1989).

C. Relative Fitness in the Absence of Pesticide Exposure

In contrast to the amount of effort on the mode of inheritance and mechanisms of resistance, studies on the fitness of specific resistance geneotypes in the presence and absence of treatment were rather neglected until the 1970s (Roush and McKenzie 1987). As illustrated by some recent research results, much more work is needed to understand fully the fitness disadvantages associated with resistance.

Resistant strains of arthropods studied in the laboratory often show reproductive disadvantages in developmental time, fecundity, and fertility. In the field, frequencies of resistant individuals usually decline over time in the absence of pesticide use. Unfortunately, however, it is often difficult to associate fitness disadvantages specifically with resistance (Roush and Croft 1986, Roush and McKenzie 1987). Resistant and susceptible strains may differ in fitness attributes independently of resistance (e.g., Heather 1982), and the frequencies of resistant individuals may decline in the field because of dilution by immigration of susceptible individuals (e.g., Gibson 1981). Significant fitness disadvantages seem to be clearly associated only with particular resistance mechanisms, as discussed later in this Section. If so, then at least some cases where resistance is lost in the field may be due to dilution by migration rather than fitness disadvantages.

Relative fitness estimates for the RS heterozygotes compared to SS homozygotes are especially critical because RS individuals will be the most common carriers of resistance during the early stages of a resistance episode (Comins

1977, Roush and Plapp 1982b). There have been only about 20 studies where the fitnesses of RR, RS, and SS genotypes have been compared. Although recent theoretical work suggests that information about fitness disadvantages is less important to the choice of resistance management tactics than was once thought (Roush 1989a), information about fitness disadvantages is important for fundamental questions about the evolution of resistance and may be necessary for the fine-tuning of resistance management programs. Therefore, comparisons of the fitnesses of RR, RS, and SS genotypes should be extended to other species and other resistance mechanisms to ascertain the generality of the conclusions discussed here.

1. Methods

Because of various methodological problems, particularly the confounding effect of immigration, most studies of resistance fitnesses have been conducted in the laboratory rather than in the field. However, recent experiments on the behavioral effects of resistance to cyclodienes (Rowland 1988) and pyrethroids (Chapter 9) suggest that traditional laboratory studies of life history traits (e.g., developmental time, fecundity, and fertility) may overlook important components of fitness. Thus, future studies of fitness should, as much as possible, be conducted in the field or at least under conditions that more closely simulate the field, to provide a more challenging test of the fitness of resistant genotypes.

Two general methods have been used to study fitness disadvantages. Both can be adapted to field use. One approach is to compare fitness components, where such factors as fecundity, developmental time, and mating competitiveness are measured for each genotype. The other method, the "population cage" technique, is to follow changes in genotypic frequencies in replicate heterogeneous populations held for several generations without insecticide treatments. The population cage method is preferable, especially when conducted across a range of environmental conditions (Roush and Croft 1986), but may take several generations before statistically significant differences can be detected.

No matter which method is used to compare fitnesses, it is important to study resistant strains that have evolved in the field. As argued in Section II.E., laboratory strains are not necessarily selected in the same way as in the field and may not be representative of the kind of evolution that is most practically and fundamentally interesting. A comparison of the fitnesses of laboratory–selected and field–selected resistant strains suggests that laboratory–selected strains tend to show more significant fitness disadvantages than field–selected resistant strains (Roush and Croft 1986).

It is also critical to compare all three (RR, RS, and SS) genotypes in a similar genetic background, preferably in a strain that has been repeatedly backcrossed to a strain without previous pesticide exposure, for two reasons. First, as noted in Section II.E., resistant and susceptible strains may differ in many ways unrelated to the resistance genes they carry. Susceptible strains must often be collected from geographic areas different from the resistant strains, but different geographic populations often differ in life history attributes (Whitehead et al. 1985, Hare and Kennedy 1986). Reference susceptible strains have often been maintained in the laboratory for decades. Such strains are often much better adapted to laboratory conditions than field strains, but they may suffer greater inbreeding depression (Roush 1989b). Strains may also differ in their levels of infection by pathogens, which can either increase (Brattsten 1987) or decrease (Hurej et al. 1982) their susceptibility to pesticides. Unless compared in a similar genetic background, there can be no guarantee that the effects measured are truly due to the resistance gene. In genetic terms, the kind of fitness disadvantages that are most useful are those that are due to pleiotropy, which can be defined as multiple phenotypic effects of the same gene (Falconer 1981).

Two approaches can be used to obtain genotypes in a similar genetic background: isolation from a heterogeneous field population by family selection or creation of isogenic lines through backcrossing. Most populations that show resistance are not yet homozygous for resistance. Some families will have higher frequencies of resistance genes than others, but all should have similar complements, on average, of genes (and environmental influences, such as disease) unrelated to resistance. Isolation of resistant homozygotes is usually a simple process of mass selection of the field–collected resistant strain with pesticide exposure for just a few generations (in contrast to selection of an initially susceptible laboratory strain for many generations), especially if selection is applied before mating. However, susceptible strains can be isolated by family selection (e.g., Bogglid and Keiding 1958). To do this, offspring of individual pairs are reared separately and a sample tested for resistance. The more susceptible families are saved and the process is repeated as necessary until a homozygous susceptible strain is produced. This approach is easiest when the phenotype (dose-response regression) of the susceptible genotypes is known from other strains or can be surmised from inspection of data on the field strain for segregation of genotypes. A fairly large sample of families, say at least 10, should be used as a check against sampling effects of the genes that are independent of resistance. A problem with this approach is that some individuals may carry genes for greater susceptibility than is normal (McEnroe 1969), as discussed in Section II.C. The second way to get resistant and susceptible genotypes in a similar genetic background is to move major resistance genes into a susceptible genetic background through repeated backcrossing and mild selection, as also described in Section II.C. and illustrated in Fig. 5.2. This essentially creates an "isogenic" strain by gradually diluting the fraction of the genome coming from the resistant parent.

A second reason for producing isogenic strains is the widely held suspicion that the deleterious effects of resistance genes are minimized by the coadaptation of resistance and more general fitness factors (Roush and McKenzie 1987). This theory has been offered to account for frequent observations that resistance, particularly in laboratory selection programs, would revert toward greater suscep-tibility in the early stages of selection but would gradually become much more stable. Therefore, observations on resistance genotypes in long–established resis-tant strains might underestimate the disadvantages associated with a resistance allele when it is initially rare. Perhaps more important, one theoretical model has suggested that if coadaptation occurs, migration into a treated area may actually increase the rate of resistance evolution (in contrast to most models and observa-tions, as discussed in Section III.B.) by increasing the genetic variation for modifiers (Uyenoyama 1986, Chapter 6). Although coadaptation seems to be a much less common phenomenon in the field than might be supposed (Roush and McKenzie 1987), it warrants further study.

Although either family selection or backcrossing can produce strains with a similar genetic background, only repeated backcrossing is truly suitable for testing whether coadaptation has occurred. Repeated backcrossing dilutes the frequencies of any fitness modifiers coming from the resistant parent. One would predict, therefore, that the apparent fitness of a resistance gene would decline with increasing numbers of backcrossed generations. Five published reports (reviewed in Roush and McKenzie 1987) have adopted this procedure to test for coadapta-tion, but only one case, diazinon resistance in the sheep blowfly (McKenzie et al. 1982), showed this kind of positive evidence for fitness modifiers. However, this is a fairly unusual case, since resistance was maintained at high frequencies for more than 10 years by continuing use of diazinon (Roush and McKenzie 1987). In contrast, repeated backcrossing actually improved fitness of the resistant strain in two other cases (Amin and White 1984, Helle 1965), indicating that the poor fitness of the resistant strain was due to some factor other than resistance. Similar observations have been made during the isolation of other kinds of mutants such as the yellow-eye mutant in *Helicoverpa zea*, which initially had very low viability, but was not different from the laboratory wild type strain after eight generations of outcrossing (Jones et al. 1977).

2. *Applications of Quantitative Genetics*

Techniques from quantitative genetics can be applied to the measurement of fitness disadvantages, but there are some potential pitfalls. Statistical methods are often used to establish genetic correlations between different characters, such as the bristle numbers on the thorax and abdomen of an insect. The principal objective is to predict how change in one character will cause simultaneous changes in other characters, such as how selection for resistance will affect other fitness attributes. There are two causes of such correlations: linkage (in the recombinational sense) and pleiotropy (Falconer 1981). Pleiotropy is the key for understanding resistance and its management. Linkage can be due to historical accidents and may fade as recombination occurs. Strictly speaking, the genes do not even have to be linked on the same chromosome, but need only occur in gametes in greater frequencies than could occur by chance, a phenomenon called linkage disequilibrium (Futuyma 1979). One problem with the use of genetic correlations is that their estimates tend to have large sampling errors and are therefore rarely very precise. A second problem is that genetic correlations are sensitive to gene frequency changes and can change rapidly during the course of selection. These problems limit confidence in applying the theory of correlated responses (Falconer 1981).

A special problem for the use of genetic correlations in studies of resistance is that correlations in pesticide use will tend to cause multiple resistances (linkage disequilibrium for resistance genes) that will confound the identification of true pleiotropy. For example, it is well established that there is no significant cross-resistance between the cyclodienes and pyrethroids used to date (Plapp 1976, Oppenoorth 1985, Chapters 3 and 4). Nonetheless, there are often strong correlations between resistance to pyrethroids and cyclodienes (e.g., Roush et al. 1990), not because of cross-resistance, but because of the fact that pest populations receiving heavy pesticide use are often exposed to many kinds of pesticides. Recent studies of cyclodiene resistance suggest that it may cause significant behavioral disadvantages, as discussed in the next subsection. Thus, one would expect negative correlations between fitness and pyrethroid resistance, not necessarily because of any direct pleiotropy, but because of correlated resistance to pyrethroids. On the basis of a correlations approach alone, one might make recommendations for pyrethroid use that would be wrong as soon as there were enough generations (probably on the order of 5 or 10) of outcrossing to overcome the initial linkage disequilibrium. The principal role of a correlations approach might be to identify traits that appear to be affected by particular kinds of resistances and use these as hypotheses for further work.

In contrast, the isolation of independent resistant genes into a susceptible background by repeated backcrossing and selection with different pesticides, as appropriate to each gene and mechanism being studied, can readily distinguish between linkage and true pleiotropy. The disadvantage of the use of isogenic lines is that it assumes that one can identify and isolate resistance genes, but this does not appear to be a significant disadvantage. Even for cases where resistance is influenced by more than one gene, one can usually isolate the individual genes that cause 90% of the resistance on a log scale. Genes that cause so little resistance that one cannot isolate them probably are not important in the field either. As an example, it has been possible to isolate two of the genes, one controlling microsomal monooxygenases (also known as mixed function oxidases, *mfo*) and the other controlling nerve insensitivity (*kdr*–like), that affect pyrethroid resistance in *H. armigera* with the combined approaches of genetics and toxicology (Daly and Gunning, unpubl. data). Isogenic strains established from field–collected individuals were backcrossed to susceptible individuals to allow segregation of the different resistance genes. Progeny of these crosses were screened electrophysiologically (for *kdr*) and with the metabolic synergist, piperonyl butoxide (for *mfo*, Chapter 3), to locate lines that contained only one or the other resistance gene.

3. Relationships Between Fitness and Resistance Mechanism

Although only a few studies have compared resistant and susceptible genotypes for which the resistance mechanisms were known, some interesting relationships between fitness disadvantages and specific resistance mechanisms seem to have emerged (Roush and McKenzie 1987). (For a review of mechanisms discussed in this Section, see Plapp 1976, Oppenoorth 1985, and Chapters 3 and 4.)

The most serious and consistent disadvantages are usually associated with general esterases. This includes esterases in temephos-resistant *Culex quinquefasciatus* mosquitoes from California (Ferrari and Georghiou 1981, El–Khatib and Georghiou 1985, Hemingway and Georghiou 1984), and organophosphate-carbamate-pyrethroid-resistant green peach aphids, *Myzus persicae* (Beranek 1974, Bauernfiend and Chapman 1985, Eggers-Schumacher 1983, Devonshire and Moores 1982). In some of the strains, the RR genotypes (or most resistant clones in the case of the aphids) appear to be only half as fit as susceptible genotypes. This may be because a relatively large portion of body protein in these strains is devoted to esterases. In highly resistant clones of *M. persicae*, this esterase amounts to 3% of the total body protein and is apparently generated by gene duplication (Devonshire and Moores 1982). Resistant *M. persicae* may be more susceptible to predation than susceptible aphids because of decreased responsiveness to alarm pheromone (Dawson et al. 1983). In the highly resistant strains of *Culex*, the esterases are produced by gene amplification (Mouches et al. 1985, Chapter 4) and may account for 6% of total body protein (Georghiou, pers. com.). Fitness disadvantages may also be associated with high levels of glutathione transferases in house flies (such as the RT strain of Roush and Plapp 1982b), enzymes that may also be produced in increased quantities because of gene duplication (Mouches et al. 1985). Since DDT-dehydrochlorinase is apparently an esterase (Clark and Shamaan 1984, Chapter 4), this may help to explain the strong fitness disadvantages that have been observed in some DDT-resistant strains (examples are cited in Roush and Croft 1986).

Diazinon–resistant sheep blowflies also show significant fitness disadvantages (McKenzie et al. 1982), and the resistance is due to esterases (Hughes and Raftos 1985), but there is no evidence that the esterases are highly amplified. In further contrast, not all *Culex* that are resistant via esterases suffer severe reproductive disadvantages. In chlorpyrifos-resistant *Cx. quinquefasciatus* from Tanzania, for example, generalized esterases are implicated but fitness differences between isogenic RR and SS homozygotes were less than 10% (Amin and White 1984). This may be because the Tanzanian mosquitoes were producing less esterase than the highly resistant California strains; esterase levels (and resistance) varies even among California strains (Mouches et al. 1985).

In contrast, little or no reproductive disadvantages have been associated with the *kdr*–like mechanisms for DDT resistance in house flies (Bogglid and Keiding 1958) and *Boophilus microplus* ticks (Stone 1962a; Schnitzerling et al. 1970), the altered acetylcholinesterase mechanism in *Tetranychus urticae* spider mites (Helle 1965, Smissaert 1964), or malathion–specific carboxylesterases in *Anopheles arabiensis* (Lines et al. 1984, Hemingway et al. 1984) and the flour beetle *Tribolium castaneum* (Beeman 1983, Beeman and Nanis 1986). Similarly, little fitness disadvantage seems to be associated with increased oxidative detoxication in the predatory mite *Metaseiulus occidentalis* (Roush and Hoy 1981, Roush and Plapp 1982a), in house flies (the RD strain in Roush and Plapp 1982b), or in *Helicoverpa armigera* (Section IV.C.1.), consistent with observations that the costs of induced microsomal monooxygenases are small (Neal 1987).

Although now mostly of historical interest, cyclodiene resistance is found in about 62% of all insecticide-resistant arthropods (Georghiou 1986). Greater than 100–fold resistance to dieldrin and endrin in homozygous–resistant strains appears to be inherited as a single gene for an altered–target mechanism throughout all animal taxa studied (Plapp 1976, Kadous et al. 1983, Oppenoorth 1985, Chapter 4), including fish (Yarbrough et al. 1986), and may well be homologous for all insects (Rowland 1988). Although not as severe as the generalized esterase-mediated resistances, cyclodiene resistance seems to be somewhat more deleterious than most mechanisms. For example, on the basis of field data, DDT–resistant phenotypes of *Anopheles culicifacies* were estimated to have relative fitnesses of 62–97% compared to susceptibles. Calculated fitnesses for dieldrin resistance were somewhat lower at 44–79% (Curtis et al. 1978). Similarly, dieldrin resistance declined from a frequency of 50–60% to near 0% in the same *Boophilus* tick strain in which DDT-resistance appeared to be fairly stable (Stone 1962b). These fitness disadvantages are not easily explained by traditional life history studies. Little or no reproductive disadvantages have been observed in laboratory studies of development time or fecundity in field–selected cyclodiene–resistant strains (Emeka–Ejifor et al. 1983, Roush and Croft 1986). On the other hand, the differences may be explained on the basis of behavior. Rowland (1988) has recently provided evidence, using strains of two *Anopheles* species that have been made isogenic by up to 11 generations of repeated backcrossing, that cyclodiene-resistant homozygotes are seriously behaviorally handicapped; homozygous resistant males were significantly less competitive in mating. Although these results differ from those of Gilotra (1965), Gilotra's strains had different geographic origins and were not isogenic. It is important to emphasize the significance of the resistance mechanism, which seems to be a change in a target site (Chapter 4) that may affect nerve function (Rowland 1988), to this fitness disadvantage. Mating competitiveness does not appear to be significantly reduced in predatory mites with a microsomal monooxygenase resistance mechanism (Roush and Hoy 1981, Roush and Plapp 1982a).

Although it may be true that laboratory fitness studies do not stress the resistant insects as much as they might be stressed in nature, the results of the laboratory studies often seem consistent with rates of resistance reversion observed in isolated field populations (i.e., populations where extensive immigration of susceptibles seems unlikely). For example, in the one case where fitness estimates have been obtained in the laboratory and field by the same methods on the same resistance, DDT resistance in *Anopheles stephensi,* the results were quite similar (Curtis et al. 1978, Wood and Bishop 1981).

However, the fact that a major fitness disadvantage for cyclodiene resistance has only recently been identified (Rowland 1988), in spite of examples of such resistances that have been known to be due to a single gene for thirty years (Shanahan 1961), suggests that further efforts to find and accurately measure fitness disadvantages could yet be fruitful. Thus, more rigorous studies on fitness disadvantages should be conducted. Although such studies can make significant progress in the laboratory (e.g., Rowland 1988), field studies have probably been underutilized and may more realistically test the fitnesses of the resistant strains. In particular, very little work has been done on the effects of overwintering stress on resistant genotypes under field conditions. In the few studies that have been made so far, the differences between resistant and susceptible genotypes have been small (Roush and Hoy 1981; Daly et al. 1988b, discussed in Section IV.C.1.), but both studies were concerned only or primarily with microsomal monooxygenase mechanisms. Finally, the inability to show strong fitness disadvantages in the laboratory for most resistance mechanisms suggests that such disadvantages cannot always be assumed.

D. Relative Fitness Under Pesticide Exposure

Selection on resistance genes can occur in a number of ways, some of which can be affected by human activity. In populations exposed to insecticide, the relative fitness of each genotype can vary with pesticide concentration and coverage, age of the individuals treated, and other factors. Lower concentrations and poorer coverage may allow the survival of some susceptibles, thereby raising their fitnesses. The fitnesses of resistant individuals can be reduced by applying higher concentrations, synergists, and mixtures or by spraying a more susceptible life stage.

Part of a generation or some entire generations can remain untreated by insecticide. The same result can be achieved, at least for some resistance loci, by alternating chemical groups being applied or by the use of nonresidual chemicals. A general reduction in selection pressure probably increases the effectiveness of susceptible immigration. Withholding the insecticide also allows for natural selection to operate if there is a fitness deficit associated with the resistant phenotypes in the absence of insecticide.

Although estimates have been made of the relative fitnesses of resistance genes while under selection with pesticides (Curtis et al. 1978), these are probably not very general because of all these influences and will not be discussed here in detail (see Roush and McKenzie 1987). However, these different approaches to pest management form the bases of several resistance management tactics, as discussed later in this chapter and in Chapter 6. Important to the evaluation of tactics that withhold pesticide use is an assessment of how strongly selection acts against resistance genes in the absence of pesticide exposure.

IV. Genetic Perspective on Practical Resistance Management

B. Classification of Tactics

Georghiou (1983) divided resistance management tactics into three categories: management by moderation (lower the mortality caused by any given chemical), by saturation (saturation of the resistance mechanisms of the pest by higher doses or synergists), or by multiple attack (mixing or alternating pesticides). Moderation and saturation had been previously introduced by Sutherst and Comins (1979). Two or more tactics must often be used together in an integrated resistance management (IRM) approach.

In genetic terms, all IRM tactics that show much promise for practical application attempt to alter selection pressures so that increases in the frequencies of resistance genes can be arrested or slowed (Leeper et al. 1986, Chapter 6). Looking at these tactics strictly from the point of view of the how the pesticide is used obscures their underlying genetic similarities. For example, in Georghiou's "management by moderation," there are strategies that reduce selective mortality and strategies that allow for dilution by immigration; alternation (a "multiple attack" tactic) reduces selection pressure, just as does reducing spray frequency ("moderation"). Thus, we will compare resistance management tactics in terms of their effects from a genetic perspective, as summarized in Table 5.1.

Mathematical models are especially useful in comparing these tactics (Chapter 6). However, at least a general idea of how each tactic works and where it can be most effective can be obtained from the simple principle of discrimination between genotypes, introduced earlier with respect to genetic analysis and the effective dominance of resistance. Selection can only occur at concentrations that kill (or reduce the reproduction of) at least some susceptible homozygotes but allow at least some resistant homozygotes to survive. Tactics that reduce this discrimination, particularly the discrimination between susceptible homozygotes and heterozygotes (the most common carriers of resistance genes early in a resistance episode), will generally slow the rate of resistance evolution. This can be illustrated with discussion of the tactics listed in Table 5.1.

Table 5.1. Techniques for Manipulating Selection Pressure in Integrated Resistance Management (IRM) Strategies

A. Reduce fitnesses of resistant individuals when insecticide is applied.
  1. Increase the dose to kill heterozygotes or resistant homozygotes.
  2. Use compounds that confer lower levels of resistance.
  3. Treat the most vulnerable life stage.
  4. Use synergists to supress detoxification mechanisms.
  5. Mix pesticides of differing modes of action and metabolism.

B. Reduce the total amount of selection pressure applied.
  6. Decrease the concentration of insecticide so that some susceptible individuals exposed to the pesticide can survive.
  7. Reduce the number of pesticide applications.
  8. Use pesticides with short residual activity and avoid slow release formulations.
  9. Do not spray all habitats of the pest; use spot treatments.
  10. Rotate pesticides so that not all generations are exposed to the same one, but avoid spatial mosaic treatments.

1. Increase the dose of insecticide until resistance is recessive or neutral

In theory, if the concentration of insecticide is so high that the heterozygotes are killed (i.e., resistance is recessive), if the frequency of resistance is sufficiently low that very few RR individuals can be found in the population, and if relatively high frequencies of susceptible individuals escape exposure altogether and mate with surviving resistant genotypes to assure that they will not produce RR offspring, increases in the frequency of resistance may be very slow (Chapter 6). If the concentration can be increased even higher to kill all resistant homozygotes, the strategy might work even if the gene frequency is high and all the population is treated (greater than 10% in Fig. 5.4). In short, the strategy discriminates poorly between resistant and susceptible genotypes; the R allele is neutral, and selection should not lead to any increases in frequency.

However, there are many practical problems for such a strategy. For those species that tend to mate in the same habitats where they are controlled, pesticide residues may be so persistent that they kill or repel susceptible individuals, thereby inhibiting the mating of susceptible with resistant individuals. In addition, the environmental and economic costs of using a dose high enough to kill the heterozygotes may be great (Tabashnik and Croft 1982, Chapter 6). Consequently, a tactic of increasing the dose applied may be useful only in populations in which resistance is not yet a problem, where doses applied can be tightly controlled and uniform (such as with dipping of cattle or fumigation), and where resistance levels in heterozygotes are likely to be low. Such cases may be rare indeed, since resistance levels of 100- or 1,000–fold have been reported in cattle ticks (Stone and Youlton 1982), one of the few species against which this management tactic might be considered. Nevertheless, this strategy was employed to contain pyrethroid resistance in cattle ticks (Reid 1989).

2. Use compounds that confer lower levels of resistance

Within any group of pesticides that show cross-resistance there are usually some that show lower resistance than others. As discussed in Chapter 2, the ideal compound for resistance monitoring would be one that improves the discrimination between genotypes. In delaying the evolution of resistance, on the other hand, the compound that discriminates most poorly would be favored. Thus, although it may be difficult to increase dose sufficiently to significantly increase mortality of heterozygotes, it may be practical to achieve the same effect by changing the pesticide used (Roush 1989a). Among the cyclodienes and related compounds, for example, dieldrin usually shows high resistance, whereas lindane shows considerably lower resistance (Oppenoorth 1985). In cyclodiene–resistant anopheline mosquitoes, economically feasible concentrations of lindane could make resistance effectively recessive and kill even resistant homozygotes, whereas no reasonable concentration could do this for dieldrin. Under such conditions, resistance can be delayed for considerable periods, as has been predicted for lindane use on these mosquitoes (Rawlings et al. 1981). In at least some aphids, pirimicarb selects less strongly for resistance than does demeton–S–methyl or a deltamethrin–heptenophos mixture (ffrench–Constant et al. 1987). In two species of *Anopheles* mosquitoes, monocrotophos appears to select more strongly for an altered acetylcholinesterase and an oxidase than does pirimiphos-methyl (Hemingway et al. 1988).

3. Treat the most vulnerable lifestage

In a similar fashion, treating the lifestage that shows the lowest levels of resistance can also weaken the discrimination between genotypes. Relative fitnesses vary with the lifestages treated because some resistance mechanisms are more poorly expressed in some lifestages than others (Chapter 2). In other cases, some life stages are so sensitive that they are killed even if resistance is expressed. In the Lepidoptera, genotypically resistant neonate larvae of both the spotted tentiform leafminer (Chapter 10) and *H. armigera* can be killed by fresh deposits of pesticides, even though larger resistant larvae are not. Both pyrethroid-resistant and susceptible *H. armigera* are killed by pyrethroids until the larvae are at least three days of age, suggesting that no selection for resistance occurs in small larvae (Daly et al. 1988a). The current resistance management strategy for *H. armigera* recommends that pyrethroid applications be timed to egg hatch and that an alternative chemical be used if medium–sized larvae are present in appreciable numbers. However, in many lepidopterous pests, the generations are not closely synchronized. Even though the target of insecticide applications may be eggs or small larvae, larger larvae may also be exposed. Strategies aimed at moderation in selection of small larvae may therefore fail if selection occurs against these other life stages. Careful scouting of crops on a regular basis is necessary in these situations.

4. Use synergists to suppress detoxification mechanisms

Synergists, when applied with an insecticide, increase the potency of the insecticide by blocking the action of the resistance mechanism (Chapter 3). This reduces the effect of the mechanism, thereby reducing the discrimination between resistant and susceptible genotypes, much as if the insecticide alone conferred only a low level of resistance. Synergists can only be used effectively if there is a single mechanism for resistance, if resistance does not subsequently develop to the combination of synergist and insecticide (as has happened for house flies, Sawicki 1975), if the synergist is stable under field conditions and has low mammalian toxicity, and if it is cost–effective. Although many synergists are known for different pesticides (Chapter 3), few meet these criteria. Although the synergist piperonyl butoxide is often used in the field, it has usually only given temporary relief from a resistance problem, as in the Colorado potato beetle, *Leptinotarsa decemlineata* (Georghiou 1986), and is not photostable. A few other compounds have fared better even though they may not be synergists in the classical sense of blocking pesticide metabolism. One example is the control of DDT resistance in *H. armigera* in Australia and in *H. virescens* in the United States with toxaphene (camphechlor) (Kay 1977, Goodyer and Greenup 1980, Sparks 1981), where the principal effect may be to enhance residue persistence (Bigley et al. 1981). Another example is the use of chlordimeform for the management of pyrethroid resistance in *H. virescens* in the United States (Chapter 9).

5. Mixtures

Of all the strategies proposed to manage resistance, mixtures may be the most controversial. If resistant alleles at two separate loci exist in the population at very low frequency, it is extremely unlikely that any one individual will carry both alleles. If unexposed susceptible individuals are relatively abundant and mate randomly with the resistant survivors, mixtures must delay resistance (Curtis 1985). Thus, the use of a second insecticide negates much of the advantage of carrying a resistance for the first pesticide, and worsens the discrimination between genotypes. Various laboratory trials have suggested that evolution of resistance could be delayed by mixtures, but such experiments have been misleading in their portrayal of field selection. Field trials usually lack a positive control to show that resistance would have occurred to both compounds in the absence of the mixture (Tabashnik 1989, Chapter 6). Recent models suggest that equal persistence of both pesticides is extremely important to the successful use of mixtures (Roush 1989a). Although a mixture strategy is most effective if implemented while resistances to both pesticides are still rare (Roush 1989a), mixtures are most commonly used as a countermeasure after control failures have already occurred. For example, mixtures of ovicides, such as chlordimeform and methomyl, with larvicides, such as endosulfan, were strongly encouraged in the IRM strategy for *H. armigera* in Australia (Forrester and Cahill 1987) in order to limit further selection for resistance already present to endosulfan and pyrethroids.

6. *Decreasing the dose of application to that which kills most, but not all, of the susceptibles*

Curtis (1985) has shown that this is not a very viable strategy. From a genetic perspective, any dose that allows some susceptibles to survive will always allow even greater survival of the heterozygotes (unless resistance is completely recessive) and an increase in the frequency of resistance. Further, before this strategy can have much of an effort on delaying resistance, a significant fraction of the susceptibles must be allowed to survive, which may not be economically viable in many crops. However, this tactic can be very useful when alternative controls are available. This approach is especially feasible where pesticides are selective (or natural enemies are resistant, Chapter 8) and reduced rates can enhance predation by stabilizing prey densities (Roush 1989a).

7. *Apply insecticides less frequently so that susceptibles have a chance to reproduce and to dilute resistance by interbreeding with resistant individuals*

This can be achieved by using a higher pest population threshold for insecticide application as part of an integrated pest management strategy. Conflicts can arise, however, between the need to control pest numbers in crops and the need to manage insecticide resistance. With *Helicoverpa* spp. in Australia, decisions to spray are determined by the density of eggs laid in the crop. One consequence of raising the threshold at which spraying occurs is that an increased number of older instar larvae may be allowed to develop. If a spray then becomes necessary, there will be stronger discrimination between genotypes than if the larvae were sprayed when younger (as discussed earlier in point 3) and selection may continue to be strong, even though overall selection pressure on neonate larvae is reduced.

8. *Use chemicals with only short residual and avoid slow release formulations so that only the target pest population is killed and subsequent colonizers are not affected*

Persistent pesticides or formulations are functionally equivalent to frequent applications of less persistent materials and may also deter the successful immigration of susceptible individuals. The decay of residues in long residual pesticides can allow survival of resistant individuals that would have been killed by the initial concentration. Thus, persistent chemicals (or formulations) may at some point render an allele that is normally recessive more dominant, thereby increasing the discrimination between genotypes. Resistance appears to occur less frequently when nonpersistent pesticides or formulations are used (Taylor et al. 1983, Chapter 6).

9. Allow refuges for escape of susceptibles

As noted earlier, the recruitment of susceptible genes into a population is important to the delay of resistance; hence, the preservation of refuges for susceptibles to breed is an important approach to the management of resistance. This tactic may often be applied, although inadvertently, in most agricultural crops because 100% kill of susceptibles with registered doses of insecticides is not normally observed (Roush and Croft 1986). However, it is not clear whether these survivors escaped exposure completely or whether they received a dose that was not lethal to them, with important implications for their impact on the development of resistance. If the survivors escaped any exposure, no selection would have occurred among them. On the other hand, if some of them survived exposure to a dose that was lethal to (or reduced reproduction of) their neighbors, some selection might have occurred. The only way to assure that a true refuge has been provided is to leave some habitats completely untreated.

One practical way to achieve this is by treating only those parts of a field that have damaging populations. Spider mites, for example, often build up only in a few areas of a field or orchard; good resistance management would include treating only those areas. Permanent refuges can be created if certain habitats are always left untreated, such as native vegetation. This may be possible for polyphagous pest species, although it must be done in a manner that is compatible with pest management practices that reduce the overall density of the pest. In many insect pests, maintenance of refuges may be difficult, as little is known about the size of the population outside treated areas or the magnitude of migration between areas. Although many insect pests are highly mobile, migration may not be sufficient to retard the spread and increase in resistance.

Another way to provide refuges for susceptible genes is to direct control against adults after they have mated. When resistance is at low frequency and in the absence of significant immigration, selection of unmated females leads to a more rapid rise in the frequency of resistance compared with selection on females that have completed mating. Resistant females that mate before the spray may carry refuges for susceptible genes in the form of sperm, whereas females that mate after application will be mating with resistant males that survived. This approach is most useful for those species that generally only mate once, as is often the case with many Diptera. However, multiple matings are common in insects and in many species, fertilization is usually by the last male to mate (Gwynne 1984).

In some species it may be necessary to preserve refuges for susceptible populations selectively while eradicating reservoirs of resistant populations; burning of trash and cultivation of fields to destroy diapausing pupae may be useful in reducing populations of resistant lepidopterous pests (Fitt and Forrester 1988). In the spider mite, *T. urticae*, Miller et al. (1985) have observed that if the previous season's within–field mite populations had been eradicated, then strawberry plants were colonized by susceptible mites transported into the area on unsprayed nursery plants. In fields where plants are kept for a two–year cycle, vegetation surrounding new plantings can harbor resistant mite populations that reinfest the fields.

10. *Rotate pesticides so that not all generations are exposed to the same one, but avoid spatial mosaic treatments*

Given two or more insecticides with unique modes of action and metabolism, one could alternate their use in either in the short term (rotations) or long term (sequential introduction), or across space (mosaics). All three approaches limit selection pressure on any given locus, but simple genetic simulation models and experiments suggest that rotations limit selection more effectively than mosaics (Curtis 1985, Roush 1989a, Chapter 6).

B. Evaluating the Options

Although these categories may serve to clarify the components of an IRM strategy, they are not easily separable in reality. For example, recommendations to treat only the adult stage (Georghiou 1983) may be based on a combination of considerations, including that it might be the most susceptible life stage (Dittrich et al. 1980), and mating may occur before treatment. Compounds that discriminate strongly between genotypes may also be among the most persistent (e.g., deltamethrin in ffrench–Constant et al. 1987). Further, a well–designed IRM strategy will incorporate a variety of tactics. Thus, IRM strategies devised for a given pest are likely to use a combination of the categories discussed above.

Application of some of these strategies to the real world is likely to incur a number of problems, as has been discussed. Nonetheless, some resistance management programs have already been moderately successful even though they were designed on the basis of a relatively modest amount of information about the expression of resistance and population biology of the pests, as illustrated with the examples given below and in Chapter 11. Although the level of sophistication and success of resistance management programs can be greatly improved by more detailed genetic and ecological information about the resistance episode, it is often possible to choose between various options, at least provisionally, on the basis of relatively little information on pest dispersal and cross–resistance (Roush 1989a).

C. The Genetic Basis of Two Resistance Management Programs.

1. Helicoverpa armigera *in Australia.*

Pyrethroid resistance was first detected in *H. armigera* in Australia in January 1983 after a series of field failures in the Emerald district of central Queensland (Gunning et al. 1984, Daly and Murray 1988). Resistant individuals were subsequently found throughout the species range in eastern Australia, even in areas where synthetic pyrethroids had not been used (Gunning et al. 1984). In response, an IRM strategy was implemented in September 1984 under which the use of pyrethroids was limited to a six–week period in the summer. It applied to all summer field crops in eastern Australia, north of about 30° S, including cotton, sorghum, maize, oilseeds, grain legumes, tomatoes, and tobacco, and for the control of any arthropod pest in these crops (Anonymous 1983, Forrester and Cahill 1987).

The strategy is a conservative one that required little information about the population biology of *H. armigera*. Instead it is based on general notions about which factors influence the evolution of resistance (Daly and McKenzie 1986). It aims to limit selection for resistance to pyrethroids to one of the four to six generations of *H. armigera* per year (pesticide alternation). Growers were also urged to mix pyrethroids with an ovicidal insectide (pesticide mixture). The extent of resistance in untreated populations was unknown when the strategy was first designed. Daly and McKenzie (1986) argued that the strategy was not optimal for minimizing selection for resistance; this could be achieved by restricting pyrethroid usage to time periods or crops where *H. armigera* was a minor pest. However, there were several objectives to the strategy in addition to managing resistance (Daly and McKenzie 1986) that determined the use period in summer.

Intensive programs to monitor the frequencies of resistance phenotypes have been undertaken in northern New South Wales in the major cotton-growing area of the Namoi/Gwydir Valleys (Forrester and Cahill 1987, Forrester 1989), to the east and south of this area in unsprayed maize crops (Gunning and Easton 1989), and in the Emerald Irrigation Area, Queensland (Forrester 1989). These programs measure the frequencies of resistant individuals by a diagnostic dose (Chapter 2), but more recent studies have attempted to measure the frequency of the phenotypes of different resistance mechanisms (Forrester 1988b).

These monitoring programs produced a consistent view of the annual changes that occur in resistance frequencies, illustrated for the Namoi/Gwydir region (Fig. 5.6). Resistance frequencies at the beginning of the cropping season (September-December, called "Stage I") are relatively low; this usually rises in the generation that is treated with pyrethroids in summer (Stage II); in the final generations, which are untreated with pyrethroids (Stage III), the resistance peaks and then declines. The frequency of resistance at the beginning of the next season is

Namoi/Gwydir --- Fenvalerate { % Surviving Discriminating Dose }

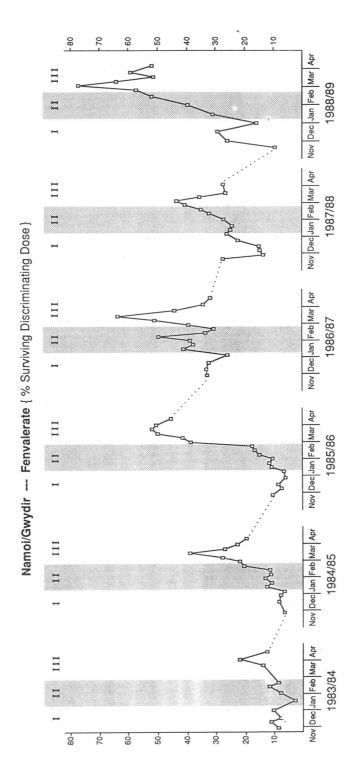

Weekly pyrethroid resistance in *Heliothis armigera* from the Namoi and Gwydir valleys of northern New South Wales for the 6 seasons since the introduction of the Resistance Management Strategy(Stages I, II and III).

generally low (Forrester and Cahill 1987, Forrester 1989), but has been on the increase (Fig. 5.6).

Detailed genetic studies indicate that the factors that affect resistance frequencies and determine their annual periodicity are complex:

1. *More than one gene is involved in resistance.* Three mechanisms are involved in pyrethroid resistance in *H. armigera:* metabolism, nerve insensitivity and decreased penetration (Ferris and Gunning 1984, Gunning pers. comm.) Laboratory studies have since confirmed that the metabolic resistance is associated with mixed function oxidases and appears to be inherited as a single dominant gene (Daly unpubl. data). Nerve insensitivity is recessive and appears to have little effect on the resistance phenotype in the absence of other resistance genes (Daly unpubl. data, Gunning pers. comm.). In the resistant strains derived from the first field failures, the nerve insensitivity mechanism appeared to be the principal mechanism (Ferris and Gunning 1984). Since then, however, this mechanism has declined in frequency (Gunning, pers. comm.) and more recent changes appear to be largely associated with metabolic resistance (Forrester 1988b). Thus, the evolution of resistance seems to be sequential (see Section II.E.1.).

2. *Selection in larvae.* Selective differences in mortality between genotypes occur in the larvae, but vary with larval instar (Daly et al. 1988a). In particular, larvae less than four days old are killed irrespective of genotype. Resistant but not susceptible medium–sized larvae can survive field application rates. Selective mortality may occur in smaller larvae as the insecticide residue is diluted by plant growth or weathering.

3. *Selection in adults.* Selective mortality occurs in adults under laboratory and field conditions (Forrester and Cahill 1987, Daly and Fisk unpubl. data).

4. *Reduced fitness of resistant individuals during diapause.* Daly et al. (1988b) observed a small (ca. 10–15%) but significant decline in the resistance frequency in field collections of diapausing pupae from the beginning of winter to the end of the diapause period that was associated with a significant rise in the mortality of pupae (ca. 10%). The decline in resistance, however, was too small to account for all the drop in resistance from one cropping season to the next.

5. *Immigration of susceptible moths into treated areas and resistant moths into untreated areas.* Resistance frequencies in spring populations of *H. armigera* (Daly et al. 1988b) sampled from weeds and crops other than cotton fall into two groups; some populations have high frequencies (>60%); others were much lower (<35%), similar to the frequencies observed in samples from cotton crops (Forrester 1988a). Relatively high frequencies of resistance, as much as 50%, have been observed in untreated maize crops up to 400 km from the Namoi/Gwydir cotton–producing region (Gunning and Easton 1989). Thus, there appears to be considerable movement of moths between treated and untreated areas that is consistent with light-trap studies (Gregg et al. 1987). The relative importance of immigration of susceptible adults was

not expected as *H. armigera* had been considered to be relatively sedentary and uncommon outside cropping areas. Recent research, prompted by observations of resistance, has shown that *H. armigera* is not as sedentary as was once believed (Daly and Gregg 1985, Farrow and Daly 1986, Gregg et al. 1987).

In summary, as tested in experiments on genetically characterized strains, pyrethroid exposure appears to select for resistance in both adults and larvae. Although the resistance frequency declines during the untreated period, this decline does not seem to be due to fitness disadvantages in resistant individuals, but to the immigration of susceptible moths from untreated and possibly unknown source populations that have lower resistance frequencies. The fact that at least some of the untreated populations show increased frequencies of resistance suggests that these populations may not always suppress resistance and that additional tactics, such as the use of synergists, are needed. Studies on the inheritance of resistance have identified the genes that must be addressed in the resistance management program.

2.   *The cattle tick* Boophilus microplus *in Australia.*

Cattle ticks are important pests of cattle in Australia, Africa, and North and South America. Control of the tick, *Boophilus microplus,* on cattle in Australia illustrates an integrated approach to managing pesticide resistance in an environment where exposure to the pesticide can be controlled. This example also illustrates how toxicological investigations can be combined with genetic principles and information on pest biology to develop a resistance management scheme even where specific details of the genetics of resistance are poorly known.

European cattle, *Bos taurus,* are particularly susceptible to tick infestations. Control can only be maintained with a very extensive cattle dipping program. In Australia, resistance has developed in *B. microplus* to a number of acaricides, including arsenic, DDT, BHC, OPs, carbamates (Wharton 1976, Nolan and Roulston 1979), amidines (Nolan 1981), and the synthetic pyrethroids (Reid 1989). Resistance to many of these acaricides is widespread in Queensland (Roulston et al. 1981, Reid 1989). Strategies to manage acaricide resistance in *Boophilus,* which have been developed over many years, seek to delay the evolution of resistance to new acaricides and to reduce the dependence on acaricides through integrated pest management (Sutherst et al. 1979).

The most important recommendations have been the introduction of tick–resistant cattle, which require dipping for tick control only in autumn, and the regular dipping of cattle at three–week intervals during the control period (Wharton et al. 1969). These recommendations have been supported by extensive studies on the mechanisms of resistance (Nolan et al. 1972, Nolan 1981) and the screening of new groups of acaricides with resistant strains to detect potential cross–resistance. However, studies of the inheritance of resistance (e.g., Stone 1972, Stone 1981) have had only limited impact on IRM; resistance to each new group developed quickly. After resistance became widespread, a new acaricide was introduced (Stone 1981); little knowledge was obtained about how selection operated on individual genes in the field. Nevertheless, inheritance studies have confirmed that resistance to a given chemical group has evolved independently a number of times and can involve different genes in different localities (e.g., Stone et al. 1976).

Various IRM strategies were assessed by Sutherst and Comins (1979), whose models of "management by moderation" and "management by saturation" were the first attempts to categorize and assess a practical IRM strategy in terms of theoretical models of the evolution of resistance (e.g., Comins 1977, Georghiou and Taylor 1977a,b). Sutherst and Comins identified the following approaches on the basis of general genetic considerations:

1. *Reduction in selection pressure by using less acaricide.* During the 1970s, the beef industry in tropical areas of northern Australia rapidly changed over to *Bos indicus* × *B. taurus* hybrid cattle, which are much more resistant to ticks than *B. taurus* (Wharton et al. 1969, Wharton and Norris 1980). This reduced the need to dip in spring and summer. In autumn, when cattle are dipped, tick–susceptible cattle can be culled. Ticks that survive the dipping rarely produce progeny because of poor weather conditions (Sutherst 1983). The size of the tick populations can be reduced further by pasture spelling, which involves withholding cattle from pastures so that ticks are unable to find hosts (the ticks have limited powers of dispersal; Wharton et al. 1969). These strategies produce minimal selection pressure for resistance and prevent build-ups of large pest populations. However, *Bos taurus* continued to be the preferred cattle breed in southern Queensland and extensive and frequent use of acaricides in this area led to the development of high levels of resistance in *B. microplus* (Wharton and Norris 1980). Current research is utilizing molecular biological techniques to develop a commercially available vaccine for cattle that immunizes them against ticks.

2. *Reduction in the relative fitness of resistance individuals.* This proposal has been implemented in a number of ways:
   a. The use of mixtures so that the ticks resistant to one acaricide would be killed by the second acaricide. Chlordimeform and OPs were encouraged for the control of OP-resistant ticks (Roulston et al. 1971).
   b. Eradication of resistant populations. It is assumed that the spread of resistance is primarily by the transportation of cattle. In response to the appearance of a strain highly resistant to OPs (Biarra strain) in a limited area (Brisbane Valley), cattle in the affected area were quarantined and treated at short intervals with increased concentrations of acaricides. Although the initial resistant population was eradicated, resistance had already spread beyond the quarantine area, so further eradication was abandoned (Nolan and Roulston 1979).

    c. Increase the dose of acaricide or use synergists. Laboratory studies with pyrethroids prior to their releases indicated that the recommended dose would select for cross–resistance in DDT–resistant ticks (Nolan and Roulston 1979). Further studies revealed that control could be obtained by increasing the dose of the pyrethroids, or by mixing the pyrethroids with OP compounds that effectively synergized the pyrethroids against DDT–resistant ticks (Nolan and Bird 1977, Nolan 1981). These studies enabled pyrethroids to be introduced for tick control, although DDT resistance was widespread (Roulston et al. 1981). In 1981, deltamethrin and cypermethrin were released pre-mixed with OPs, while cyhalothrin and flumethrin were released, unmixed, but at higher doses (Reid 1989). In recent years, strains with both low–level (1984) and high–level (1986, 1987) resistance to pyrethroids have appeared (Reid 1989). Thus, IRM in *B. microplus* has delayed, rather than prevented, the development of resistance to the pyrethroids. However, without the IRM strategy, it is unlikely that pyrethroids would have been used effectively in field populations of ticks.

    d. Avoidance of selective doses as the acaricide degrades by the use of regular dipping regimes. This was a primary recommendation of Sutherst and Comins (1979) in their "management by saturation". The generation time of these ticks is about three weeks. If repeated dipping is required, Sutherst and Comins recommended that it be undertaken at three week intervals and at high concentrations so that new infestations of ticks would not be exposed to residual (and potentially selective) doses of the acaricide.

In summary, resistance management was based on tactics that would avoid discrimination between resistant and susceptible genotypes independent of any specific information on the expression of resistance. However, it appears that the long–term strategy to manage resistance in ticks will come through better farm management to reduce tick populations without acaricides plus strategic dipping of cattle when tick populations exceed a critical level.

D.   Anticipating the Expression of Resistance

One of the keys to designing or improving a resistance management program for any pesticide is to obtain accurate data on the expression and inheritance of resistance as early as possible when a new pesticide is introduced (or even before the pesticide is introduced), for at least two reasons. First, the efficiency of resistance monitoring programs can be significantly enhanced by the development of more sophisticated test techniques, including biochemical methods, that allow rapid and nearly perfect discrimination between resistant and susceptible individuals (Chapter 2). Second, some knowledge of the genetics (including dominance relationships) and potential mechanisms of resistance (particularly with respect to cross-resistance) can help in the choice of management tactic (Chapter 6).

The importance of anticipating resistance evolution before it occurs has long been recognized, as indicated by the extensive use of artificial selection to investigate the potential for resistance development to new pesticides in laboratory strains of insects and mites. Unfortunately, however, laboratory selection programs have rarely succeeded in providing data that has (or could) significantly improved resistance management. Although it has often been possible to select strains in the laboratory that are resistant, it is usually impossible to get much meaningful information from them. One can conclude that resistance is possible, but we usually strongly suspect that anyway on the basis of the history of resistance evolution. The failure of a laboratory strain to develop resistance is certainly no guarantee that resistance will not develop in the field, as historical examples show (Brown 1967, Nolan and Roulston 1979). When resistance is achieved in the laboratory, its inheritance usually appears to be complex; the mechanisms can usually be described only vaguely as metabolic, occasionally with some low–level insensitivity of the target.

Brown and Payne (1988) have proposed that selection for 40 generations will produce resistance equivalent to that which occurs in nature, but they did not provide any evidence that the genetics and mechanisms of resistance produced in the laboratory were closely related to those that occurred in the field. If resistance was monogenic in these laboratory strains, it should not have taken 40 generations to obtain a significant selection response. During a laboratory selection program for carbaryl resistance in the predatory mite *Metaseulis occidentalis,* for example, a significant selection response had been achieved within 10 generations and near fixation was reached in 18 (Roush and Hoy 1981, Roush and Plapp 1982a), despite poor discrimination between resistant and susceptible genotypes with carbaryl (Fig. 5.5). (This selection program is not a violation of the rule that laboratory selection tends to produce polygenic resistances to truly novel compounds [Roush and McKenzie 1987]; the resistance gene was present in the population at a frequency of about 0.4% [Roush 1979], apparently because of extensive prior selection with carbaryl on the field populations surveyed.) Thus, Brown and Payne (1988) did not prove that laboratory selection was a good model for field selection, only that similar magnitudes of resistance (resistance ratios at the LC_{50}) can occur in the laboratory and field. This supports the proposition that the genetic basis of a trait is not intrinsic to that character, but depends on the initial variation and selection regime applied (Roush and McKenzie 1987).

As noted, laboratory results contrast starkly with those that occur when resistance evolves in the field. Thus, because of our continued dependence on traditional laboratory selection, reliable information on the genetics, expression, and mechanisms of resistance to a truly novel type of pesticide has only been available after resistance had already developed in at least one species because of extensive use of the new compound(s) in the field. (The synthetic pyrethroids might be an exception, but because of their similarities in mode of action with the pyrethrins and DDT, they cannot be considered to be completely novel.) However, given the importance of gathering information on the expression of resistance genes, it seems important to look for a better way.

1. Novel Methods for Anticipating the Expression of Resistance: Quantitative Genetics and Mutagenesis

Quantitative genetic methods may be able to predict short-term changes in resistance (Via 1986), but long-term predictions (more than 10 generations) assume that resistance alleles are common enough in populations prior to control failures that they can be sampled in economically feasible quantitative genetic studies. As discussed in Section III.A., however, major genes for resistance appear to be widespread but rare prior to selection (Roush and McKenzie 1987); normal monitoring often fails to detect resistance until control failures occur (Roush and Miller 1986). Assuming that major genes are initially no more common than 0.001 (which is probably a high estimate) prior to selection, the chance that they would even appear in a sample of as many as 200 individuals used for the quantitative analysis must be very low (Roush and Miller 1986). Even if they had been sampled, their impact on additive variance would probably go undetected, yet the impact of such genes on the final form of resistance would be enormous. In this regard, quantitative genetics suffers the same fundamental problem as classical mass selection in the laboratory.

If the fundamental problem is a poor ability to sample rare alleles, the solution is to develop new methods to sample for such alleles efficiently. One way to do this might be to revise significantly the way selection programs are run in the laboratory. Instead of screening the same strain for 30–40 generations, devote those resources to screening individuals collected directly from the field at a high challenge dose that causes at least 95% mortality. In other words, mimic field selection.

A second possible solution to this problem is suggested by the work of Kikkawa (1964), who used X–rays to induce a parathion resistance allele in *Drosophila melanogaster* that was apparently allelic with a parathion resistance gene found in a resistant strain collected from the field. Kikkawa tested 28,500 flies to isolate the resistant mutant. Since Kikkawa's study, resistance has apparently been induced using the mutagen ethyl methane sulfonate (EMS) in at least three insect species, including *Drosophila* (Pluthero and Threlkeld 1984, Wilson and Fabian 1986), the green peach aphid, *Myzus persicae* (Buchi 1981), and the Colorado potato beetle, *Leptinotarsa decemlineata* (Argentine and Clark 1990). The study by Wilson and Fabian (1986) is particularly interesting because one mutation confers nearly 100–fold resistance to methoprene, apparently by an altered target site. It was isolated from one of 3,490 family lines. Argentine and Clark (1990) used EMS to generate a strain that is about 15–fold resistant to abamectin. Unfortunately, either because of the lack of suitable strains or the inability to conduct genetic crosses (e.g., the parthenogenic green peach aphid), none of these strains except Kikkawa's was compared to field–developed resistant strains.

These studies, particularly those of Kikkawa (1964) and Wilson and Fabian (1986), show that resistance mutagenesis can overcome at least one of the problems of traditional laboratory selection studies: the absence of suitable (major gene) genetic variation. Mutagens have even been identified that increase the frequency of gene amplification, at least in cell lines (Schimke et al. 1986). With further refinements and testing, mutagenesis may provide a more effective way to anticipate resistance development than traditional or quantitative approaches for assessing the potential mechanisms and genetics of resistance.

The use of *Drosophila* as a model organism for resistance mutagenesis and resistance studies is particularly promising; *Drosophila* genes can be used to study altered acetylcholinesterase and reduced neurological sensitivity (Chapter 4).

IV. Summary and Conclusions

Pesticide resistance of practical significance in the field tends to be simply inherited at just one or two loci. Selection acts by discriminating between genotypes. The fitness disadvantages that occur as pleiotropic effects of resistance genes in the absence of pesticide residues do not seem to be universally strong, but vary with resistance mechanism. Thus, the apparent success of pesticide alternation strategies in managing resistance corroborates other evidence suggesting that immigration of susceptible individuals into treated areas is a major influence in delaying resistance.

Although there is an extensive theory of resistance management that shows that there may be advantages to changing application rates or doses, decreasing pesticide persistence, and mixing or alternating pesticides (Chapter 6), few (if any) resistance management programs have been established directly upon considerations of this theoretical framework (Levin et al. 1986). For all practical purposes, all successful current resistance management programs simply seek to reduce the use of a critical insecticide or group of insecticides (Roush 1989a, Chapter 11). Clearly, there is a need for experimental tests of resistance management tactics, especially in the field. The ideal situation for such tests would be in field populations that clearly had resistance potential (so that at least some of the tactics would allow resistance to develop to a statistically significant frequency) and were in similar but suitably isolated habitats.

Genetic investigations can play a key role not only in establishing key pieces of information, such as the effective dominance of resistance and cross-resistance, for the design of such experiments and resistance management programs, but can also help to provide improved resistance detection techniques to monitor the effectiveness of the tactics used. This may be essential to field experiments, where it would be desirable to determine which strategies were failing while the gene frequencies were still so low that economic control failures did not occur. One of the keys to designing or improving a resistance management program for any pesticide is to obtain accurate data on the expression and inheritance of resistance. In some cases, this can be gathered from existing resistant strains, although mutagenesis may prove helpful.

Thus, there are at least four aspects of the genetics of resistance management that deserve considerable further attention: (1) the fitnesses of specific resistance genotypes when treated with pesticides in a manner that simulates field exposure (effective dominance); (2) the fitnesses of specific resistance genotypes in the absence of pesticide exposure (general fitnesses); (3) methods to anticipate accurately the expression of resistance before it evolves in the field (resistance induction); and (4) resolution of technical controversies concerning the theoretical advantages and disadvantages of various resistance management tactics (testing resistance management theory). New approaches and tools for achieving these ends are rapidly becoming available.

References

Anonymous. 1983. Pyrethroid resistance. Aust. Cotton Grower 4(3):4–7.

Amin, A. M., and G. B. White. 1984. Relative fitness of organophosphate-resistant and susceptible strains of *Culex quinquefasciatus* Say (Diptera: Culicidae). Bull. Entomol. Res. 74: 591–598.

Argentine, J. A., and J. M. Clark, 1990. Selection for abamectin resistance in Colorado potato beetle. (Coleoptera: Chrysomelidae). Pest. Sci. 28: 17–24.

Arnold, J. T. A., and M. J. Whitten. 1976. The genetic basis for organophosphorus resistance in the Australian sheep blowfly, *Lucilia cuprina* (Wiedemann) (Diptera, Calliphoridae). Bull. Entomol. Res. 66: 561–568.

Ballantyne, G. H., and R. A. Harrison. 1967. Genetic and biochemical comparisons of organophosphorous resistance between strains of spider mites (*Tetranychus* species: Acari). Entomol. Exp. Appl. 10: 231–239.

Bauernfeind, R. J., and R. K. Chapman. 1985. Non-stable parathion and endosulfan resistance in green peach aphids. (Homoptera: Aphididae). J. Econ. Entomol. 78: 516–522.

Beeman, R. W. 1983. Inheritance and linkage of malathion resistance in the red flour beetle. J. Econ. Entomol. 76: 737–740.

Beeman, R. W., and S. M. Nanis. 1986. Malathion resistance alleles and their fitness in the red flour beetle (Coleoptera: Tenebrionidae) J. Econ. Entomol. 79: 580–587.

Beranek, A. P. 1974. Stable and non-stable resistance to dimethoate in the peach-potao aphid *(Myzus persicae)*. Entomol. Exp. Appl. 17: 381–390.

Bigley, W. S., F. W. Plapp, R. L. Hanna, and J. A. Harding. 1981. Effect of toxaphene, camphene and cedar oil on methyl parathion residues on cotton. Bull. Environ. Contam. Toxicol. 27: 92–94.

Boggild, O., and J. Keiding. 1958. Competition in house fly larvae: experiments involving a DDT-resistant and susceptible strain. Oikos 9: 1–25.

Brattsten, L. B. 1987. Sublethal virus infection destroys cytochrome P-450 in an insect. Experientia 43: 451–454.

Brown, A. W. A. 1967. Genetics of insecticide resistance in insect vectors, pp. 505–552. *In* J. W. Wright and R. Pal (eds.), Genetics of insect vectors of disease. Elsevier, New York.

Brown, T. M., and G. T. Payne. 1988. Experimental selection for insecticide resistance. J. Econ. Entomol. 81: 49–56.

Buchi, R. 1981. Evidence that resistance against pyrethroids in aphids *Myzus persicae* and *Phorodon humuli* is not correlated with high carboxylesterase activity. J. Plant Dis. Protection 88: 631–634.

Bull, D. L., R. L. Harris, and N. W. Pryor. 1988. The contribution of metabolism to pyrethroid and DDT resistance in the horn fly (Diptera: Muscidae). J. Econ. Entomol. 81: 449–458.

Carbonaro, M. A., D. E. Moreland, V. E. Edge, N. Motoyama, G. C. Rock, and W. C. Dauterman. 1986. Studies on the mechanism of cyhexatin resistance in the twospotted spider mite, *Tetranychus urticae* (Acari: Tetranychidae). J. Econ. Entomol. 79: 576–579.

Charlesworth, B. 1979. Evidence against Fisher's theory of dominance. Nature 278: 848–849.

Clark, A. G., and N. A. Shamaan. 1984. Evidence that DDT–dehydrochlorinase from the house fly is a glutathione–S–transferase. Pestic. Biochem. Physiol. 22: 249–261.

Cluck, T., F. W. Platt, Jr., and J. S. Johnston. 1985. Metabolic resistance to insecticides: heterozygosity at the chromosome II locus in house flies, *Musca domestica* (Diptera: Muscidae). J. Econ. Entomol. 78: 1015–1019.

Coluzzi, M., A. Sabatini, V. Petrarca, and M. A. DiDeco. 1977. Behavioral divergence between mosquitoes with different inversion karyotypes in polymorphic populations of the *Anopheles gambiae* complex. Nature 226: 832–833.

Comins, H. N. 1977. The management of pesticide resistance. J. Theor. Biol. 65: 399–420.

Crow, J. F. 1957. Genetics of insect resistance to chemicals. Annu. Rev. Entomol. 2: 227–246.

Crow, J. F., and M. Kimura. 1970. An introduction to population genetics theory. New York: Harper & Row.

Curtis, C. F. 1985. Theoretical models of the use of insecticide mixtures for the management of resistance. Bull. Entomol. Res. 75: 259–265.

Curtis, C. F., L. M. Cook, and R. J. Wood. 1978. Selection for and against insecticide resistance and possible methods of inhibiting the evolution of resistance in mosquitoes. Ecol. Entomol. 3: 273–287.

Daly, J. C., and P. Gregg. 1985. Genetic variation in *Heliothis* in Australia: species identification and gene flow in the two pest species, *H. armigera* (Hübner) and *H. punctigera* Wellengren (Lepidoptera: Noctuidae). Bull. Entomol. Res. 75: 169–184.

Daly, J. C., and J. A. McKenzie. 1986. Resistance management strategies in Australia: the Heliothis and 'Wormkill' programmes, pp. 951–959. *In* Proceedings, British Crop Protection Conference on Pests and Diseases, Brighton, November 1986, The British Crop Protection Council, Surrey.

Daly, J. C., and D. A. H. Murray. 1988. Evolution of resistance to pyrethroids in *Heliothis armigera* (Hübner) (Lepidoptera: Noctuidae) in Australia. J. Econ. Entomol. 81: 984–988.

Daly, J., J. H. Fisk, and N. W. Forrester. 1988a. Selective mortality in field trials between strains of *Heliothis armigera* (Lepidoptera: Noctuidae) resistant and susceptible to pyrethroids: functional dominance of resistance and age class. J. Econ. Entomol. 81:1000–1007.

Daly, J. C., G. P. Fitt, and J. H. Fisk. 1988b. Pyrethroid resistance in pupal and adult *Heliothis armigera*, pp. 73–78. *In* Proceedings Australian Cotton Conference, Surfers Paradise, Queensland.

Dawson, G. W., D. C. Griffiths, J. A. Pickett, and C. M. Woodcock. 1983. Decreased response to alarm pheromone by insecticide-resistant aphids. Naturwissenschaften 70: 254–255.

Denholm, I., R. M. Sawicki, and A. W. Farnham. 1985. Factors affecting resistance to insecticides in house-flies, *Musca domestica* L. (Diptera: Muscidae). IV. The population biology of flies on animal farms in south-eastern England and its implications for the management of resistance. Bull. Entomol. Res. 75: 143–158.

Devonshire, A. L., and G. D. Moores. 1982. A carboxylesterase with broad substrate specificity causes organophosphorus, carbamate and pyrethroid resistance in peach-potato aphids *(Myzus persicae)*. Pest. Biochem. Physiol. 18: 235–246.

Dickson, R. C. 1941. Inheritance of resistance to hydrocyanic acid fumigation in the California red scale. Hilgardia 13: 515–522.

Dittrich, V. 1975. Acaricide resistance in mites. Z. Angew. Entomol. 78: 28–45.

Dittrich, V., N. Luetkemeier, and G. Voss. 1980. OP–resistance in *Spodoptera littoralis:* inheritance, larval and imaginal expression and consequences for control. J. Econ. Entomol. 73: 356–362.

Dobzhansky, T. 1951. Genetics and the origin of species. 3rd ed. Columbia University Press, New York.

Edge, V. E., and D. G. James. 1986. Organo–tin resistance in *Tetranychus urticae* (Acari: Tetranychidae) in Australia. J. Econ. Entomol. 79: 1477–1483.

Eggers-Schumacher, H. A. 1983. A comparison of the reproductive performance of insecticide-resistant and susceptible clones of *Myzus persicae*. Entomol. Exp. Appl. 34: 301–307.

El-Khatib, Z. I., and G. P. Georghiou. 1985. Comparative fitness of temephos-resistant, susceptible, and hybrid phenotypes of the southern house mosquito (Diptera: Culicidae). J. Econ. Entomol. 78: 1023–1029.

Emeka-Ejiofor, S. A. I., C. F. Curtis, and G. Davidson. 1983. Tests for effects of insecticide resistance genes in *Anopheles gambiae* on fitness in the absence of insecticides. Entomol. Exp. Appl. 34: 163–168.

Falconer, D. S. 1981. Introduction to quantitative genetics. 2nd ed. Longman, New York.

Farnham, A. W., K. E. O'Dell, I. Denholm, and R. M. Sawicki. 1984. Factors affecting resistance to insecticides in house-flies, *Musca domestica* L. (Diptera: Muscidae). III. Relationship between the level of resistance to pyrethroids, control failure in the field and the frequency of gene *kdr*. Bull. Entomol. Res. 74: 581–589.

Farrow, R. A. and J. C. Daly. 1987. Long-range movement as an adaptive strategy in the genus *Heliothis* (Lepidoptera: Noctuidae): a review of its occurrence and detection in four pest species. Aust. J. Zool. 35: 1–24.

Ferrari, J. A., and G. P. Georghiou. 1981. Effects of insecticidal selection and treatment on reproductive potential of resistant, susceptible, and heterozygous strains of the southern house mosquito. J. Econ. Entomol. 74: 323–327.

Ferrari, J. A., C. E. Taylor, G. P. Georghiou and A. Lagunes. 1982. Selection with several insecticides in the mosquito *Culex quinquefasciatus:* heritabilities of resistance and genetic correlations. Genetics s100: 23–24.

Ferris, I. G., and R. V. Gunning. 1984. Pyrethroid resistance in *Heliothis armiger,* pp. 137–140. *In* Proceedings 1984 Australian Cotton Growers Research Conference, Toowoomba, Queensland.

ffrench-Constant, R. H., A. L. Devonshire, and S. J. Clark. 1987. Differential rate of selection for resistance by carbamate, organophosphorous and combined pyrethroid and organophosphorous insecticide in *Myzus persicae* (Sulzer) (Hemiptera: Aphididae). Bull. Entomol. Res. 77: 227–238.

Fitt, G. P., and N. W. Forrester. 1988. Overwintering of *Heliothis*–the importance of stubble cultivation. Aust. Cotton Grower 8(4): 7–8.

Follett, P. A., B. A. Croft, and P. H. Westigard. 1985. Regional resistance to insecticides in *Psylla pyricola* from pear orchards in Oregon. Can. Entomol. 117: 565–573.

Forrester, N. W. 1988a. Good news on the resistance front. Aust. Cotton Grower 9(2): 13–14.

Forrester, N. W. 1988b. Field selection for pyrethroid resistance genes. Aust. Cotton Grower 9(3): 48–51.

Forrester, N. W. 1989. Updated insecticide resistance levels. Aust. Cotton Grower 10(1): 32–34.

Forrester, N. W., and M. Cahill. 1987. Management of insecticide resistance in *Heliothis armigera* (Hubner) in Australia, pp. 127–137. *IN* M. G. Ford, D. W. Holloman, B. P. S. Khambay, and R. M. Sawicki (eds.), Combating resistance to xenobiotics: biological and chemical approaches. Ellis Horwood, Chichester, England.

Foster, G. G., M. J. Whitten, C. Konovalov, J. T. A. Arnold, and G. Maffi. 1981. Autosomal genetic maps of the Australian sheep blowfly, *Lucilia cuprina dorsalis* R-D. (Diptera: Calliphoridae) and possible correlations with the linkage maps of *Musca domestica* L. and *Drosophila melanogaster* (Mg.). Genet. Res. 37: 55–69.

Futuyma, D. J. 1979. Evolutionary biology. Sinauer Associates, Sunderland, Mass.

Georghiou, G. P. 1969. Genetics of resistance to insecticides in houseflies and mosquitoes. Exp. Parasitol. 26: 224–255.

Georghiou, G. P. 1972. The evolution of resistance to pesticides. Annu. Rev. Ecol. Syst. 3: 133–168.

Georghiou, G. P. 1983. Management of resistance in arthropods, pp. 769–792. *In* G. P. Georghiou and T. Saito (eds.), Pest resistance to pesticides. Plenum, New York.

Georghiou, G. P. 1986. The magnitude of the resistance problem, pp. 14–43. *In* National Academy of Sciences (ed.), Pesticide resistance: strategies and tactics for management. National Academy Press, Washington, D.C.

Georghiou, G. P., and C. E. Taylor. 1977a. Operational influences in the evolution of insecticide resistance. J. Econ. Entomol. 70: 653–658.

Georghiou, G. P., and C. E. Taylor. 1977b. Genetic and biological influences in the evolution of insecticide resistance. J. Econ. Entomol. 70: 319–323.

Gibson, J. P. 1981. Problems in obtaining a description of the evolution of dimethoate resistance in Danish houseflies *(Musca domestica)*. Pestic. Sci. 12: 565–572.

Gilotra, S. K. 1965. Reproductive potentials of dieldrin-resistant and susceptible populations of *Anopheles albimanus* Wiedemann. Amer. J. Trop. Med. Hyg. 14: 165–169.

Goodyer, G. J., and L. R. Greenup. 1980. A survey of insecticide resistance in the cotton bollworm, *Heliothis armigera* (Hübner) (Lepidoptera: Noctuidae) in New South Wales. Gen. Appl. Entomol. 12: 37–39.

Gregg, P., P. Twine, and G. Fitt. 1987. *Heliothis* in non-cropping areas. Aust. Cotton Grower 8(3): 40–42.

Gregg, P. C., G. McDonald and K. P. Bryceson. 1989. The occurrence of *Heliothis punctigera* Wallengren and *H. armigera* Hübner in inland Australia. J. Aust. Entomol. Soc. 28: 135–141.

Gunning, R. V., and C. S. Easton. 1989. Pyrethroid resistance in *Heliothis armigera* (Hübner) collected from unsprayed maize crops in New South Wales in 1983–1987. J. Aust. Entomol. Soc. 28: 57–61.

Gunning, R. V., C. S. Easton, L. R. Greenup, and V. E. Edge. 1984. Pyrethroid resistance in *Heliothis armigera* (Hübner) (Lepidoptera: Noctuidae) in Australia. J. Econ. Entomol. 77: 1283–1287.

Gwynne, D. T. 1984. Male mating effort, confidence of paternity and insect sperm competition, pp. 117–149. *In* R. L. Smith (ed.), Sperm competition and the evolution of animal mating systems. Academic Press, New York.

Halliday, W. R., and G. P. Georghiou. 1985. Inheritance of resistance to permethrin and DDT in the southern house mosquito (Diptera: Culicidae). J. Econ. Entomol. 78: 762–767.

Hare, J. D., and G. G. Kennedy. 1986. Genetic variation in plant-insect associations: survival of *Leptinotarsa decemlineata* populations on *Solanum carolinense*. Evolution 40: 1031–1043.

Hart, R. J. 1963. The inheritance of diazinon resistance in an Australian strain of *Musca domestica* L. Bull. Entomol. Res. 54: 461–465.

Heather, N. W. 1982. Comparison of population growth rates of malathion resistant and susceptible populations of the rice weevil, *Sitophilus oryzae* (Linnaeus) (Coleoptera: Curculionidae). Queensland J. Agric. Anim. Sci. 39: 61–68.

Heckel, D. G., A. G. Abbott, and T. M. Brown. 1988. Genetic linkage mapping and insecticide resistance in *Heliothis virescens*. Sericologia 28s: 49.

Helle, W. 1965. Resistance in the acarina: mites. Adv. Acarol. 2: 71–93.

Hemingway, J., and G. P. Georghiou. 1984. Differential suppression of organophosphorus resistance in *Culex quinquefasciatus* by the synergists IBP, DEF, and TPP. Pest. Biochem. Physiol. 21: 1–9.

Hemingway, J., M. Rowland, and K. E. Kisson. 1984. Efficacy of pirimiphos methyl as a larvicide or adulticide against insecticide resistant and susceptible mosquitoes (Diptera: Culicidae). J. Econ. Entomol. 77: 868–871.

Hemingway, J., B. C. Bonning, K. G. I. Jayawardena, I. S. Weerasinghe, P. R. J. Herath, and H. Oouchi. 1988. Possible selective advantage of *Anopheles* spp. (Diptera: Culicidae) with the oxidase- and acetylcholineasterase-based insecticide resistance genes after exposure to organophosphates or an insect growth regulator in Sri Lankan rice fields. Bull. Entomol. Res. 78: 471–478.

Holloway, G. J. 1986. A theoretical examination of the classical theory of inheritance of insecticide resistance and the genetics of time to knockdown and dry body weight in *Sitophilus oryzae* (L.) (Coleoptera: Curculionidae). Bull. Entomol. Res. 76: 661–670.

Hughes, P. B., and D. A. Raftos. 1985. Genetics of an esterase associated with resistance to organophosphorus insecticides in the sheep blowfly, *Lucilia cuprina* (Weidemann) (Diptera: Calliphoridae). Bull. Entomol. Res. 75: 535–544.

Hurej, M., P. P. Sikorowski, and H. W. Chambers. 1982. Effects of bacterial contamination on insecticide-treated boll weevils (Coleoptera: Curculionidae). J. Econ. Entomol. 75: 651–654.

Jones, R. L., N. W. Widstrom, and D. Perkins. 1977. Yellow-eye variant of the corn earworm. J. Hered. 68: 264–265.

Kadous, A. A., S. M. Ghiasuddin, F. Matsumura, J. G. Scott, and K. Tanaka. 1983. Difference in the picrotoxinin receptor between the cyclodiene-resistant and susceptible strains of the German cockroach. Pest. Biochem. Physiol. 19: 157–166.

Kay, I. R. 1977. Insecticide resistance in *Heliothis armigera* (Hübner) (Lepidoptera: Noctuidae) in areas of Queensland, Australia. J. Aust. Entomol. Soc. 16: 43–45.

Kay, I. R., L. R. Greenup and C. Easton. 1983. Monitoring *Heliothis armiger* (Hübner) strains from Queensland for insecticide resistance. Queensland J. Agric. Anim. Sci. 40: 23–26.

Kikkawa, H. 1964. Genetical studies on the resistance to parathion in *Drosophila melanogaster*. II. Induction of a resistance gene from its susceptible allele. Botyu-Kagaku 29: 37–42.

Lagunes T., A. 1980. Impact of the use of mixtures and sequences of insecticides in the evolution of resistance in *Culex quinquefasciatus* Say (Diptera: Culicidae). Ph.D. dissertation, Univ. of California, Riverside.

Leeper, J. R., R. T. Roush, and H. T. Reynolds. 1986. Preventing or managing resistance in arthropods, pp. 335–346. *In* National Academy of Science (ed.), Pesticide resistance: strategies and tactics for management. National Academy Press, Washington, D.C.

Levin, B. R., et al. 1986. Population biology of pesticide resistance: bridging the gap between theory and practical applications, pp. 143–156. *In* National Academy of Science (ed.), Pesticide resistance: strategies and tactics for management. National Academy Press, Washington, D.C.

Lines, J. D., M. A. E. Ahmed, and C. F. Curtis. 1984. Genetic studies of malathion resistance in *Anopheles arabiensis* Patton (Diptera: Culicidae). Bull. Entomol. Res. 74: 317–325.

Liu, M-Y., Y-J. Tzeng, and C-N Sun. 1981. Diamondback moth resistance to several synthetic pyrethroids. J. Econ. Entomol. 74: 393–396.

Macnair, M. R. 1983. The genetic control of copper tolerance in the yellow monkey flower, *Mimulus guttatus*. Heredity 50: 283–293.

Mani, G. S. 1985. Evolution of resistance in the presence of two insecticides. Genetics 109: 761–783.

McDonald, P. T., and C. D. Schmidt. 1987. Genetics of permethrin resistance in the horn fly (Diptera: Muscidae). J. Econ. Entomol. 80: 433–437.

McDonald, P. T., C. D. Schmidt, W. F. Fisher, and S. E. Kunz. 1987. Survival of permethrin-susceptible, resistant, and F1 hybrid strains of *Haematobia irritans* (Diptera: Muscidae) on ear-tagged steers. J. Econ. Entomol. 80: 1218–1222.

McEnroe, W. D. 1969. Free genetic variability in the two-spotted spider mite, *Tetranychus urticae* K. (Acarina: Tetranychidae). Massachusetts Exp. Stn. Bull. No. 580. 12 pp.

McKenzie, J. A., and M. J. Whitten. 1982. Selection for insecticide resistance in the Australian sheep blowfly, *Lucilia cuprina*. Experientia 38: 84–85.

McKenzie, J. A., J. M. Dearn, and M. J. Whitten. 1980. Genetic basis of resistance to diazinon in Victorian populations of the Australian sheep blowfly, *Lucilia cuprina*. Aust. J. Biol. Sci. 33: 85–95.

McKenzie, J. A., M. J. Whitten, and M. A. Adena. 1982. The effect of genetic background on the fitness of diazinon resistance genotypes of the Australian sheep blowfly, *Lucilia cuprina*. Heredity 49: 1–9.

McLeod, D., G. R., C. R. Harris, and G. R. Driscoll. 1969. Genetics of cyclodiene-insecticide resistance in the seed-corn maggot. J. Econ. Entomol. 62: 427–432.

Miller, R. W., B. A. Croft, and R. D. Nelson. 1985. Effects of early season immigration on cyhexatin and formetanate resistance of *Tetranychus urticae* (Acari: Tetranychidae) on strawberry in central California. J. Econ. Entomol. 78: 1379–1388.

Misra, R. K. 1968. Statistical tests of hypotheses concerning the degree of dominance in monofactorial inheritance. Biometrics 24: 429–434.

Mouches, C., D. Fournier, M. Raymond, M. Magnin, J.-B. Berge, N. Pasteur, and G. P. Georghiou. 1985. GENETIQUE.-Association entre 1 amplification de sequences d ADN, 1 augmentation quantitative d esterases et la resistance a des insecticides organophosphores chez des moustiques du complexe *Culex pipiens*, avec une note sur une amplification similaire chez *Musca domestica* L. Comptes Rendu Acad. Sci. Paris, Ser. III, 301(16): 695–700.

Munsterman, L. E. and G. E. Craig, Jr. 1979. Genetics of *Aedes aegypti:* updating the linkage map. J. Hered. 70: 291–296.

National Research Council. 1986. Pesticide resistance: strategies and tactics for management. National Academy Press, Washington, D.C.

Neal, J. J. 1987. Metabolic costs of mixed function oxidase induction in *Heliothis zea*. Entomol. Exp. Appl. 43: 175–179.

Nolan, J. 1981. Current developments in resistance to amidine and pyrethroid tickicides in Australia. pp. 109–114. *In* G. B. Whitehead and J. D. Gibson (eds.), Tick Biology and control. Grahamstown, Rhodes Univ., Tick Res. Unit.

Nolan, J., and P. E. Bird. 1977. Co-toxicity of synthetic pyrethroids and organophosphorous compounds against cattle tick *(Boophilus microplus)*. J. Aust. Entomol. Soc. 16: 252.

Nolan, J., and W. J. Roulston. 1979. Acaricide resistance as a factor in the management of acari of medical and veterinary importance, pp. 3–13. *In* J. G. Rodriguez (ed.), Recent advances in acarology, Vol. 2. Academic Press, New York.

Nolan, J., H. J. Schnitzerling, and C. A. Schunter. 1972. Multiple forms of acetylcholinesterase from resistant and susceptible strains of the cattle tick, *Boophilus microplus* (Can.). Pestic. Biochem. Physiol. 2: 85–94.

Oppenoorth, F. J. 1985. Biochemistry and genetics of insecticide resistance, pp. 731–773. *In* G. A. Kerkut and I. I. Gilbert (eds.), Comprehensive insect physiology, biochemistry, and pharmacology, Vol. 12. Pergamon, New York.

Plapp, F. W. 1976. Biochemical genetics of insecticide resistance. Annu. Rev. Entomol. 21: 179–197.

Plapp, F. W., Jr., and C. Campanhola. 1986. Synergism of pyrethroids by chlordimeform against susceptible and resistant *Heliothis*, pp. 167–169. *In* Proceedings, 1986 beltwide cotton production research conference, Dallas, Texas, January 4–8, 1987. National Cotton Council of America, Memphis, Tenn.

Plapp, F. W., Jr., C. R. Browning, and P. J. H. Sharpe. 1979. Analysis of rate of development of insecticide resistance based on simulation of a genetic model. Environ. Entomol. 8: 494–500.

Pluthero, F. G., and S. F. H. Threlkeld. 1984. Mutations in *Drosophila melanogaster* affecting physiological and behavioral response to malathion. Can. Entomol. 116: 411–418.

Pree, D. J. 1987. Inheritance and management of cyhexatin and dicofol resistance in the European red mite (Acari: Tetranychidae). J. Econ. Entomol. 80: 1106–1112.

Raftos, D. A. and P. B. Hughes. 1986. Genetic basis of specific resistance to malathion in the Australian sheep blow fly, *Lucilia cuprina* (Diptera: Calliphoridae) J. Econ. Entomol. 79: 553–557.

Raymond, M., N. Pasteur, and G. P. Georghiou. 1987. Inheritance of chlorpyrifos resistance in *Culex pipiens* L. (Diptera: Culicidae) and estimation of the number of genes involved. Heredity 58: 351–356.

Rawlings, P., G. Davidson, R. K. Sakai, H. R. Rathor, K. M. Aslamkhan, and C. F. Curtis. 1981. Field measurement of the effective dominance of an insecticide resistance in anopheline mosquitos. Bull. WHO 59: 631–640.

Reid, T. J. 1989. Acaricide resistance in Queensland. Information series QI89011, Department of Primary Industries, Queensland Government, Brisbane.

Roulston, W. J., R. H. Wharton, J. Nolan, J. D. Kerr, J. T. Wilson, P. G. Thompson and M. Schotz. 1981. A survey for resistance in cattle ticks to acaricides. Aust. Vet. J. 57: 362–371.

Roulston, W. J., R. H. Wharton, H. J. Schnitzerling, R. W. Sutherst and N. D. Sullivan. 1971. Mixtures of chlorphenamidine with other acaricides for the control of oranophosphorous-resistant strains of cattle tick *Boophilus microplus*. Aust. Vet. J. 47: 521–528.

Roush, R. T. 1979. Selection for insecticide resistance in *Metaseiulus occidentalis* (Nesbitt) (Acarina: Phytoseiidae): Genetic improvement of a spider mite predator. Ph.D. dissertation. Univ. Calif., Berkeley.

Roush, R. T. 1989a. Designing resistance management programs: How can you choose? Pest. Sci. 26: 423–441.

Roush, R. T. 1989b. Genetic considerations in the propagation of entomophagous species, pp. 373–387. *In* R. Baker and P. Dunn (eds.), New Directions in Biological Control, UCLA Symposia on Molecular and Cellular Biology, New Series, Vol. 112. Alan R. Liss, New York.

Roush, R. T., and B. A. Croft. 1986. Experimental population genetics and ecological studies of pesticide resistance in insects and mites, pp. 257–270. *In* National Research Council (ed.), Pesticide resistance: strategies and tactics for management. National Academy Press, Washington, D.C.

Roush, R. T., and M. A. Hoy. 1981. Laboratory, glasshouse, and field studies of artificially selected carbaryl resistance in *Metaseiulus occidentalis*. J. Econ. Entomol. 74: 142–147.

Roush, R. T., and J. A. McKenzie. 1987. Ecological genetics of insecticide and acaricide resistance. Annu. Rev. Entomol. 32: 361–380.

Roush, R. T., and G. L. Miller. 1986. Considerations for design of insecticide resistance monitoring programs. J. Econ. Entomol. 79: 293–298.

Roush, R. T., and F. W. Plapp, Jr. 1982a. Biochemical genetics of resistance to aryl carbamate insecticides in the predaceous mite, *Metaseiulus occidentalis*. J. Econ. Entomol. 75: 304–307.

Roush, R. T., and F. W. Plapp, Jr. 1982b. Effects of insecticide resistance on biotic potential of the house fly (Diptera: Muscidae). J. Econ. Entomol. 75: 708–713.

Roush, R. T., and D. A. Wolfenbarger. 1985. Inheritance of resistance to methomyl in the tobacco budworm (Lepidoptera: Noctuidae). J. Econ. Entomol. 78: 1020–1022.

Roush, R. T., R. L. Combs, T. C. Randolph, J. MacDonald, and J. Hawkins. 1986. Inheritance and effective dominance of pyrethroid resistance in the horn fly. (Diptera: Muscidae). J. Econ. Entomol. 79: 1178–1182.

Roush, R. T., C. W. Hoy, D. N. Ferro, and W. M. Tingey. 1990. Insecticide resistance in Colorado potato beetles (Coleoptera: Chrysomelidae): Influence of crop rotation and insecticide use. J. Econ. Entomol. 83: 315–319.

Rowland, M. 1988. Management of gamma HCH/dieldrin resistance in mosquitoes–a strategy for all insects?, pp. 495–500. *In* 1988 Proceedings, British Crop Protection Conference on Pests and Diseases, Brighton. The British Crop Protection Council, Surrey.

Sawicki, R. M. 1974. Genetics of resistance of a dimethoate-selected strain of houseflies (*Musca domestica* L.) to several insecticides and methyenedioxyphenyl synergists. J. Agr. Food Chem. 22: 344–349.

Sawicki, R. M. 1975. Effects of sequential resistance on pesticide management, pp. 799–808. *In* Proceedings 8th British Insecticide and Fungicide Conference, Nov. 17–20, 1975, Brighton, England, Vol. 1. British Crop Prot. Council, London.

Sawicki, R. M., A. L. Devonshire, A. W. Farnham, K. E. O'Dell, G. D. Moores and I. Denholm. 1984. Factors affecting resistance to insecticides in house–flies, *Musca domestica* L. (Diptera: Muscidae). II. Close linkage on autosome 2 between an esterase and resistance to trichlorphon and pyrethroids. Bull. Entomol. Res. 74: 197–206.

Schimke, R. T., S. W. Sherwood, A. B. Hill, and R. N. Johnston. 1986. Overreplication and recombination of DNA in higher eukaryotes: potential consequences and biological implications. Proc. Natl. Acad. Sci. 83: 2157–2161.

Schmidt, C. H. and G. C. LaBreque. 1959. Acceptability and toxicity of poisoned baits to house flies resistant to organophosphorous insecticides. J. Econ. Entomol. 52: 345–346.

Schnitzerling, H. J., W. J. Roulston, and C. A. Schunter. 1970. The absorption and metabolism of ^{14}C DDT in DDT-resistant and susceptible strains of the cattle tick, *Boophilus microplus*. Aust. J. Biol. Sci. 23: 219–230.

Scott, J. G., R. T. Roush, and D. A. Rutz. 1989. Insecticide resistance of house flies from New York dairies (Diptera: Muscidae). J. Agric. Entomol. 6: 53–64.

Shanahan, G. J. 1961. Genetics of dieldrin resistance in *Lucilia cuprina* (Wied.). Genetica Agaria 14: 307–321.

Shanahan, G. J. 1979. Genetics of diazinon resistance in larvae of *Lucilia cuprina* (Wiedemann) (Diptera: Calliphoridae). Bull. Entomol. Res. 69: 225–228.

Smissaert, H. R. 1964. Cholinesterase inhibition in spider mites susceptible and resistant to organophosphate. Science 143: 129–131.

Sparks, T. C. 1981. Development of insecticide resistance in *Heliothis zea* and *Heliothis virescens* in North America. Bull. Entomol. Soc. Am. 27: 186–192.

Sparks, T. C., J. A. Lockwood, R. L. Byford, J. B. Graves, and B. R. Leonard. 1989. The role of behavior in insecticide resistance. Pest. Sci. 26: 383–399.

Stone, B. F. 1962a. The inheritance of dieldrin-resistance in the cattle tick, *Boophilus microplus*. Aust. J. Agric. Res. 13: 1008–1022.

Stone, B. F. 1962b. The inheritance of DDT-resistance in the cattle tick, *Boophilus microplus*. Aust. J. Agric. Res. 13: 984–1007.

Stone, B. F. 1968. A formula for determining degree of dominance in cases of monofactorial inheritance of resistance to chemicals. Bull. WHO 38: 325–326.

Stone, B. F. 1972. The genetics of resistance by ticks to acaricides. Aust. Vet. J. 48: 345–350.

Stone, B. F. 1981. A review of the genetics of resistance to acaricidal organochlorine and organophosphorous compounds with particular reference to the cattle tick *Boophilus microplus*, pp. 95–102. *In* G. B. Whitehead and J. D. Gibson (eds.), Tick Biology and control. Grahamstown, Rhodes Univ., Tick Res. Unit.

Stone, B. F., J. Nolan and C. A. Schuntner. 1976. Biochemical genetics of resistance to organophosphorous acaricides in three strains of the cattle tick, *Boophilus microplus*. Aust. J. Biol. Sci. 29: 265–279.

Stone, B. F. and N. J. Youlton. 1982. Inheritance of resistance to chlorpyrifos in the Mt Alford strain and to diazinon in the Gracemere strain of the cattle tick *(Boophilus microplus)*. Aust. J. Biol. Sci. 35: 427–440.

Sun, C-N., H. Chi, and H-T. Feng. 1978. Diamondback moth resistance to diazinon and methomyl in Taiwan. J. Econ. Entomol. 71: 551–554.

Sutherst, R. W. 1983. Management of arthropod parasitism in livestock, pp. 41–56. *In* J. D. Dunsmore (ed.), Tropical parasitoses and parasitic zoonoses. 10th Int. Conf., World Association for the Advancement of Veterinary Parasitology, Perth.

Sutherst, R. W., and H. N. Comins. 1979. The management of acaricide resistance in the cattle tick, *Boophilus microplus* (Canestrini) (Acari: Ixodidae), in Australia. Bull. Entomol. Res. 69: 519–537.

Sutherst, R. W., G. A. Norton, N. D. Barlow, G. R. Conway, M. Birley, and H. N. Comins. 1979. An analysis of management strategies for cattle tick *(Boophilus microplus)* control in Australia. J. Appl. Ecol. 16: 359–382.

Tabashnik, B. E. 1989. Managing resistance with multiple pesticide tactics: theory, evidence, and recommendations. J. Econ. Entomol. 82: 1263–1269.

Tabashnik, B. E., and B. A. Croft. 1982. Managing pesticide resistance in crop-arthropod complexes: interactions between biological and operational factors. Environ. Entomol. 11: 1137–1144.

Tabashnik, B. E., and N. L. Cushing. 1989. Quantitative genetic analysis of insecticide resistance: variation in fenvalerate tolerance in a diamondback moth (Lepidoptera: Plutellidae) population. J. Econ. Entomol. 82: 5–10.

Tabashnik, B. E., N. L. Cushing, and M. W. Johnson. 1987. Diamondback moth (Lepidoptera: Plutellidae) resistance to insecticides in Hawaii: intra-island variation and cross-resistance. J. Econ. Entomol. 80: 1091–1099.

Taylor C. E., and G. P. Georghiou. 1979. Suppression of insecticide resistance by alteration of gene dominance and migration. J. Econ. Entomol. 72: 105–109.

Taylor C. E., F. Quaglia, and G. P. Georghiou. 1983. Evolution of resistance to insecticides: a cage study on the influence of migration and insecticide decay rates. J. Econ. Entomol. 76: 704–707.

Thomas, V. 1966. Inheritance of DDT resistance in *Culex pipiens fatigans* Wiedemann. J. Econ. Entomol. 59: 779–786.

Tsukamoto, M. 1963. The log dosage-probit mortality curve in genetic researches of insect resistance to insecticides. Botyu-Kagaku 28: 91–98.

Tsukamoto, M. 1983. Methods of genetic analysis of insecticide resistance, pp. 71–98. *In* G. P. Georghiou and T. Saito (eds.), Pest resistance to pesticides. Plenum Press, New York.

Uyenoyama, M. K. 1986. Pleiotropy and the evolution of genetic systems conferring resistance to pesticides, pp. 207–221. *In* National Academy of Sciences (ed.), Pesticide resistance: strategies and tactics for management. National Academy Press, Washington, D.C.

Via, S. 1986. Quantitative genetic models and the evolution of pesticide resistance, pp. 222–235. *In* National Academy of Sciences (ed.), Pesticide resistance: strategies and tactics for management. National Academy Press, Washington, D.C.

Wharton, R. H. 1976. Tick-borne livestock diseases and their vectors. 5. Acaricide resistance and alternative methods of tick control. World Anim. Rev. 20: 8–15.

Wharton, R. H., K. L. S. Harley, P. R. Wilkinson, K. B. Utech, and B. M. Kelly. 1969. A comparison of cattle tick control by pasture spelling, planned dipping, and tick-resistant cattle. Aust. J. Agric. Res. 20: 783–797.

Wharton, R. H. and K. R. Norris. 1980. Control of parasitic arthropods. Vet. Parasitol. 6: 135–164.

White, N. D. G., and R. J. Bell. 1988. Inheritance of malathion resistance in a strain of *Tribolium castaneum* (Coleoptera: Tenebrionidae) and effects of resistance genotypes on fecundity and larval survival in malathion-treated wheat. J. Econ. Entomol. 81: 381–386.

Whitehead, J. R., R. T. Roush, and B. R. Norment. 1985. Resistance stability and coadaptation in diazinon-resistant house flies (Diptera: Muscidae). J. Econ. Entomol. 78: 25–29.

Whitten, M. J. and J. A. McKenzie. 1982. The genetic basis for pesticide resistance, pp. 1–16. *In* K. E. Lee (ed.) Proc. 3rd. Aust. Conf. Grassland Invertebrate Ecology. South Australia. Govt. Printer, Adelaide.

Wilson, A. G. L. 1974. Resistance of *Heliothis armigera* to insecticides in the Ord Irrigation Area, North Western Australia. J. Econ. Entomol. 67: 256–258.

Wilson, T. G. 1988. *Drosophila melanogaster* (Diptera: Drosophilidae): a model insect for insecticide resistance studies. J. Econ. Entomol. 81: 22–27.

Wilson, T. G., and J. Fabian. 1986. A *Drosophila melanogaster* mutant resistant to a chemical analog of juvenile hormone. Develop. Biol. 118: 190–201.

Wilson, T. G., and J. Fabian. 1987. Selection of methoprene-resistant mutants of *Drosophila melanogaster*, pp. 179–188. *In* J. Law (ed.) Molecular entomology. Alan R. Liss, New York.

Wood, R. J., and J. A. Bishop. 1981. Insecticide resistance: populations and evolution, pp. 97–127. *In* J. A. Bishop and L. M. Cook (eds.), Genetic consequences of man made change. Academic Press, New York.

Wright, S. 1931. Evolution in Mendelian populations. Genetics 16: 97–159.

Yarbrough, J. D., R. T. Roush, J. C. Bonner, and D. A. Wise. 1986. Monogenic inheritance of cyclodiene resistance in mosquito fish, *Gambusia affinis*. Experientia 42: 851–853.

Yust, H. R., and F. F. Shelden. 1952. A study of the physiology of resistance to hydrocyanic acid in the California red scale. Ann. Entomol. Soc. Am. 45: 220–228.

Yust, H. R., H. D. Nelson, and R. L. Busbey. 1943. Comparative susceptibility of two strains of California red scale to HCN, with special reference to the inheritance of resistance. J. Econ. Entomol. 36: 744–749.

Zalucki, M. P., G. Daglish, S. Firempong, and P. Twine. 1986. The biology and ecology of *Heliothis armigera* (Hübner) and *H. punctigera* Wallengren (Lepidoptera: Noctuidae) in Australia: what do we know? Aust. J. Zool. 34: 779–814.

6

Modeling and Evaluation of Resistance Management Tactics

Bruce E. Tabashnik

I. Introduction

Pest resistance to insecticides is a serious worldwide problem. Resistance to one or more pesticides has been documented in more than 440 insect and mite species, with costs of resistance estimated conservatively at $1 billion yearly (Georghiou 1986). Resistance management seeks to slow, prevent, or reverse the evolution of resistance in pests. A secondary goal of resistance management is to promote evolution of resistance in beneficial species, such as natural enemies (see Chapters 8 and 11).

Mathematical models can be useful tools for working toward the goals of resistance management. Models have helped to build a conceptual framework for resistance management and generated hypotheses for experimental tests. Models can provide a relatively fast, safe and inexpensive way to project the consequences of different assumptions about resistance and to weigh the merits of various management options. Models can help to organize data and identify gaps where more research is needed.

There were few modeling studies of insecticide resistance before 1975. Interest in this area has recently exploded, however, and more than 40 research papers on resistance models have been published since 1975 (Table 6.1). Many of the initial modeling studies were discussed in an excellent review by Taylor (1983). More recent reviews include general considerations of modeling and population biology (Pedersen 1984, Levin et al. 1986, Taylor 1986, Riddles and Nolan 1987), population dynamics (May and Dobson 1986), economics (Knight and Norton 1989), pesticide combinations (Curtis 1987, Roush 1989, Tabashnik

I thank M. Caprio, M. Johnson, J. Rosenheim, and R. Roush for their thoughtful suggestions. Special thanks to C. Taylor for providing a preprint of his manuscript. N. Finson and S. Toba provided valuable assistance. This work was supported by USDA grants HAW00947H, USDA/CSRS Special Grants in Tropical and Subtropical Agriculture, USDA Western Regional IPM Grant 88-34103-4059, and a Fujio Matsuda Scholar award through the U.H. Foundation. This is paper no. 3359 of the Hawaii Institute of Tropical Agriculture and Human Resources Journal series.

Table 6.1. Modeling Studies of Insecticide Resistance

Studies	Biological	Operational	Economic
		Factors Emphasized	
Analytical			
MacDonald (1959)	X		
Comins (1977a)	X		
Curtis et al. (1978)	X	X	
Sutherst and Comins (1979)	X	X	X
Taylor and Georghiou (1979)	X		
Cook (1981)	X		
Wood and Mani (1981)	X	X	
Muggleton (1982, 1986)	X		
Mani (1985)	X	X	
Beeman and Nanis (1986)	X		
Comins (1986)	X	X	
Uyenoyama (1986)	X	X	
May and Dobson (1986)	X	X	
Simulation			
Georghiou and Taylor (1977a,b)	X	X	
Greever and Georghiou (1979)	X	X	
Plapp et al., (1979)	X	X	
Maudlin et al. (1981)	X	X	
Curtis (1981)	X	X	
Taylor and Georghiou (1982)	X	X	
Tabashnik and Croft (1982, 1985)	X	X	
Taylor et al. (1983)	X	X	
Knipling and Klassen (1984)		X	
Dowd et al. (1984)	X	X	
Mani and Wood (1984)	X	X	
Sinclair and Alder (1985)	X	X	
Curtis (1985, 1987)	X	X	
Tabashnik (1986b,c)	X	X	
Via (1986)	X	X	
Denholm et al. (1987)	X	X	
Horn and Wadleigh (1988)	X		
Longstaff (1988)	X	X	
Mason et al. (1989)	X	X	
Roush (1989)	X	X	
Optimization			
Hueth and Regev (1974)	X	X	X
Taylor and Headley (1975)		X	X
Gutierrez et al. (1976, 1979)		X	X
Comins (1977b, 1979a,b,c)		X	X
Sarhan et al. (1979)		X	X
Shoemaker (1982)		X	X
Mangel and Plant (1983)		X	X
Lazarus and Dixon (1984)		X	X
Plant et al. (1985)		X	X
Lichtenberg and Zilberman (1986)		X	X
Empirical			
Georghiou (1980, 1986)	X	X	
Tabashnik and Croft (1985)	X		
Rosenheim and Hoy (1986)		X	

1989), and simulation models (Tabashnik 1986a, 1987). This chapter emphasizes recent developments in modeling that have not been thoroughly examined in earlier reviews.

II. Classification of Resistance Models

Modeling studies of insecticide resistance can be classified by four criteria: the basic assumptions, the modeling approach, the factors considered, and the problem addressed. The classification described here is expanded from Tabashnik (1986a).

Assumptions about the mode of inheritance of resistance are a fundamental part of nearly all resistance models. A prevailing view among resistance researchers has been that most economically significant cases of resistance are due to allelic variants at one or two gene loci (Roush and McKenzie 1987, Wood 1981). Because of this view and the tractability of one-locus models, most resistance models have assumed monogenic inheritance, with one allele (R) conferring resistance and another allele (S) conferring susceptibility. Nonetheless, many cases of resistance show patterns that differ significantly from the expectations of monogenic inheritance (Liu et al. 1981, Halliday and Georghiou 1985, Roush et al. 1986, Raymond et al. 1987, Pree 1987, Fournier et al. 1988). Such results as well as theoreticians' interest in polygenic resistance have spurred development of a few nonmonogenic models of resistance.

The second criterion, the modeling approach, is based partly on the level of complexity (Taylor 1983). Four modeling approaches are analytical, simulation, optimization, and empirical. Analytical models aim to analyze general trends by providing a simple mathematical description. Such models do not attempt to capture realistic details; instead they seek to define fundamental principles. Analytical models of the evolution of insecticide resistance usually assume simple population dynamics with discrete generations and no age structure. Population growth is usually determined by some form of the logistic equation.

Compared with analytical models, simulation models are generally more complex and realistic. They often attempt to incorporate specific details to mimic mechanisms that operate in nature. Simulations may include complex population dynamics (age structure, overlapping generations, etc.), temporal and spatial variation in pesticide dose, and other complications. The modeling of complex details typically involves series of coupled equations that cannot readily be solved analytically, so computer programs are used to determined numerical solutions. However, simulations can also be used to project the consequences of simple equations. Simulations are especially valuable for assessing the influence of a large number of factors. Parameter values may be estimated from empirical data and can be systematically varied to assess their influence relative to other factors (Tabashnik 1986a).

Optimization models of insecticide resistance have been used to determine the management strategy that will maximize profit when pest susceptibility to an insecticide is viewed as a nonrenewable natural resource (Hueth and Regev 1974). Dynamic programming techniques have been used to balance the future cost of reduction in pest susceptibility with present losses in crop yield due to the pest (Plant et al. 1985). Biological aspects are usually simplified and viewed as constraints.

Empirical models are based on observed relationships among variables. Unlike the other types of models described earlier, empirical models make no assumptions about causal mechanisms. Such models are derived from data and are likely to have predictive value in the context in which the data were obtained. Although firmly grounded in reality, empirical models may not be useful for extrapolation to conditions that have not been previously studied.

The definition of factors affecting resistance used here differs slightly from earlier descriptions. The genetic, biological, and ecological factors listed by Georghiou and Taylor (1986) are lumped here as biological factors. Operational factors include the insecticide and the way it is applied, as in previous definitions. A new category, economic factors, includes costs associated with pesticide purchase, application, and development, as well as the cost of damage caused by the pest.

Analytical, simulation, and empirical models have been used to examine biological and operational factors, whereas optimization models have emphasized economic and operational factors (Table 6.1). Knight and Norton (1989) review economic analyses in detail. This chapter focuses on biological and operational factors.

The problem addressed by most resistance models is evolution of resistance to one pesticide by one pest species. This first general category includes studies that compare rates of resistance development to different insecticides when each is applied unilaterally (Sinclair and Alder 1985) and those that compare rates of resistance evolution between species when each species is modeled independently (Tabashnik and Croft 1985). A second category includes studies of the effects of multiple insecticide use in mixtures, rotations, or mosaics (Curtis 1985). A third category includes those that incorporate interactions between species, that is, pest–natural enemy interactions (Tabashnik 1986c).

This chapter examines models of resistance to one insecticide, resistance to combinations of two or more insecticides, and resistance in pest–natural enemy systems. Available empirical tests of models are also reviewed. The final segment of the chapter summarizes progress to date and suggests future directions.

III. Resistance to One Insecticide

A. Single-Locus Models

Seminal papers by Comins (1977a), Georghiou and Taylor (1977a,b), and Hueth and Regev (1974) stimulated development of numerous one-locus models employing analytical, simulation, and optimization approaches, respectively (Table 6.1). This section considers predictions and tests of general and specific one-locus models.

1. General Predictions

One-locus models predict that, under most conditions, resistance will evolve faster as

   a. Reproductive potential (particularly generation turnover) increases.

   b. Immigration of susceptible individuals decreases.

   c. Intensity of insecticide use (frequency of application, proportion of population treated, dose, and residue persistence) increases.

a. Reproductive Potential and Generation Time Reproductive potential can be estimated as: (% survival from egg through adult × % of adults that are female × progeny per female)$^{generations/year}$ (Tabashnik and Croft 1985). Most models of resistance development assume that a certain level of selection occurs each generation. Thus, in effect, the generation is the basic time unit and the extent of resistance development depends directly on the number of generations (e.g., May and Dobson 1986). In this scenario the rate of resistance development increases as generation turnover increases.

Comins (1979b) noted, however, that if a pesticide treatment imposes a constant additional daily mortality, then the rate of resistance development is independent of the pest's generation time. He reasoned that pests with a short generation time have a high reproductive rate per unit time and that they require more severe control. Such pests are treated more heavily with insecticides and thus evolve resistance faster. This analysis produces the same general prediction as others (pests with short generation times develop resistance rapidly), but it suggests that selection intensity is crucial, whereas generation turnover and reproductive potential have no direct effect on the rate of resistance development.

Comins' (1979b) reasoning may apply to closed populations, but in the presence of immigration of susceptible individuals, generation turnover and reproductive potential can affect the rate of resistance development directly. Higher generation turnover and reproductive potential increase the population growth rate, which allows the population to increase faster following insecticide treatments. Assuming that the number of immigrants is independent of population size in the treated habitat, higher generation turnover and reproductive potential will decrease the ratio of immigrants to residents in the treated area (i.e., the proportional immigration rate) and thereby cause faster resistance development (Comins 1977a, Taylor and Georghiou 1979, Tabashnik and Croft 1982, May and Dobson 1986). For example, simulations showed that in the presence of immigration by susceptibles, resistance evolved faster as fecundity increased, even though selection intensity was held constant (Tabashnik and Croft 1982, Tabashnik 1986b).

Historical patterns of resistance development in seven species selected by soil applications of aldrin or dieldrin (Georghiou and Taylor 1986) and in 24

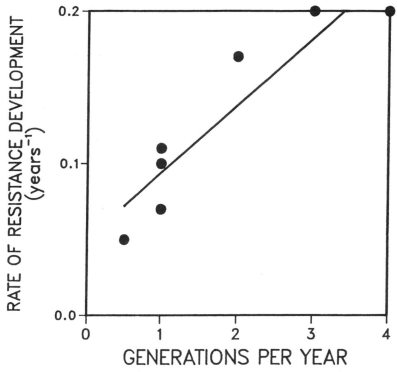

Figure 6.1   Relationship between generations per year and rate of resistance development (rate is the reciprocal of the estimated number of years for resistance development). (**a**) Resistance of soil arthropods to aldrin/dieldrin (after Georghiou 1980). $y = 0.044x + 0.050$, $R^2 = 0.85$, df $= 5$, $P = 0.0031$.

apple insects and mites selected by azinphosmethyl (Tabashnik and Croft 1985) support the idea that generation turnover directly influences resistance development. In both of these cases, the rate of evolution of resistance was directly proportional to the number of generations per year (Fig. 6.1). If one assumes that exposure to insecticides is roughly comparable among species in a particular agroecosystem (or at least not correlated with generations per year) regardless of which pest is targeted, then differences in selection intensity among species are not likely to explain the observed patterns.

May and Dobson (1986) noted that across a wide variety of circumstances and a diversity of organisms, the number of generations required for resistance development was surprisingly narrow (5–100 generations). Most of the pests that are notorious for their ability to develop resistance rapidly have numerous generations annually. An apparent exception to the rule is the Colorado potato beetle, *Leptinotarsa decemlineata,* which has an incredible capacity to evolve resistance, despite having only a few generations yearly (Georghiou 1986). In

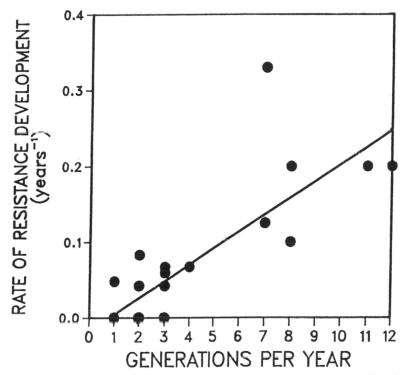

Figure 6.1(b) Resistance of apple arthropods to azinphosmethyl (after Tabashnik and Croft 1985). Excluding species not reported as resistant after 25 years: $y = 0.018\,x + 0.24$, $R^2 = 0.51$, df $= 11$, $P = 0.0059$. Assuming that 11 species not reported as resistant after 25 years (4, 5, and 2 species with 1, 2, and 3 generations per year, respectively) have a zero rate of resistance development: $y = 0.022\,x - 0.019$, $R^2 = 0.67$, df $= 22$, $P < 0.00001$.

field studies of several pest species, however, the Colorado potato beetle showed the greatest tendency to rebound above its equilibrium population size following disturbance (May and Dobson 1986), which indicated its high reproductive potential.

Generation time is normally considered an intrinsic biological factor that is not readily altered by management practices. Longstaff (1988) suggested, however, that cooling of stored grain can slow development of pests, thereby increasing the time per generation and reducing the rate of resistance development. Although Longstaff (1988) notes that the grain storage system is unique in the level of environmental manipulation that can be achieved, the principle of prolonging generation time to manage resistance might be applied to other systems. For example, one could manage resistance in a herbivorous pest by using a crop cultivar that reduces the pest's development rate.

b. Immigration Many single-locus models of resistance show that immigration of susceptible individuals into treated areas can slow resistance development by increasing the frequency of susceptible alleles in a treated population (Comins 1977a, Georghiou and Taylor 1977a, Curtis et al. 1978, Taylor and Georghiou 1979, Tabashnik and Croft 1982). These models also show that resistance evolves more slowly as the ratio of susceptible immigrants to residents in the treated area increases.

Comins (1977a) analyzed a system with a treated area, an untreated area, and movement between the two areas in *both* directions. In this type of system, immigration of susceptible individuals into the untreated area slows resistance development, but emigration of resistant individuals from the treated area speeds resistance development in the untreated area. For example, acaricide resistance in the cattle tick, *Boophilus microplus*, in Australia is thought to spread primarily by emigration (i.e., movement of resistant ticks on transported cattle) (Sutherst and Comins 1979).

If treated and untreated areas have similar population sizes and the proportion of individuals migrating is equal for the two areas, then intermediate levels of migration are optimal for delaying resistance in a two-area system (Comins 1977a). In the special case where emigration from the treated population has neglible impact on the untreated population (i.e., the untreated population is effectively infinite or emigration from the treated population does not occur), the rate of resistance development decreases monotonically as migration increases.

The general prediction that immigration of susceptibles slows resistance development was supported in laboratory experiments with the house fly, *Musca domestica* (Ozburn and Morrison 1963, Taylor et al. 1983); a mosquito, *Anopheles gambiae* (Prasittisuk and Curtis 1982); and the red flour beetle, *Tribolium castaneum* (Wool and Noiman 1983), and it was supported in historical trends seen in apple insects and mites (Tabashnik and Croft 1985). Imai (1987) found that experimental releases of laboratory-reared susceptible house flies into a field population reduced resistance to fenitrothion and diazinon. However, immigration of wild susceptible individuals is difficult to measure in the field, and little direct empirical evidence addresses the importance of this factor in typical field situations (Rawlings and Davidson 1982, Denholm et al. 1985).

Although it is generally believed that immigration can slow resistance development, at least two theoretical studies have found conditions under which the reverse may occur. In a multi-locus analysis of resistance discussed in greater detail later in this chapter, Uyenoyama (1986) reasoned that immigration could speed resistance development by increasing genetic variation and effective population size in a treated area. In addition, on the basis of results from a single-locus resistance model, Maudlin et al. (1981) concluded that the lack of immigration into sprayed areas prevented development of insecticide resistance in the tsetse fly in Nigeria. The tsetse fly, *Glossina* sp., has low reproductive potential and forms small, isolated populations during the dry season in northern Nigeria.

Simulations showed that in the absence of immigration, small tsetse fly populations with a low initial resistance gene frequency were eradicated by intensive insecticide treatment. Immigration into sprayed areas prevented local extinction in the simulations. However, the lack of reports of insecticide resistance in the tsetse fly in the field, even in heavily treated areas where local extinction is unlikely (Maudlin et al. 1981), suggests that other factors restrict resistance development in this pest.

In summary, most one-locus models have emphasized that movement of susceptible individuals from untreated to treated areas (i.e., immigration) can slow resistance development by reducing the frequency of resistance alleles in treated populations. Emigration from treated populations, however, may speed resistance development. Further, migration between small populations can increase effective population size and may thereby promote resistance development in some cases. In addition to direct effects of dispersal on resistance development, interactions between immigration and other factors can be influential (Tabashnik and Croft 1982). Interactions with reproductive potential were discussed in the previous section; interactions with intensity of insecticide use are described later.

  c. Intensity of Insecticide Use Intensity of insecticide use increases with frequency of application, proportion of population treated, dose applied, and residue persistence (i.e., pesticide half-life). There is little doubt that pests that have developed resistance rapidly have been frequently treated with insecticides. For example, bioassays showed that resistance to pyrethroids in the white fly *(Bemisia tabaci)*, pink bollworm *(Pectinophora gossypiella)*, and pear psylla *(Psylla pyricola)* was correlated with the extent of pyrethroid use (Georghiou 1986, T. A. Miller pers. comm., Croft et al. 1989). Resistance in the parasitoid *Aphytis melinus* was correlated with both local (in-grove) and countywide pesticide use history (Rosenheim and Hoy 1986). Resistance researchers also generally agree that, under most conditions, increasing the proportion of the population that is treated will increase the rate of resistance development. In contrast, there is considerable controversy regarding the possibility that resistance can be overwhelmed by a "saturation" (Sutherst and Comins 1979) or "high-dose" (Tabashnik and Croft 1982) strategy.

In principle, one could avoid resistance development by ensuring that all treated insects are killed. As Crow (1952) noted, this is easier said than done. A refinement of this idea is based on results from one-locus models. Several modeling studies have shown that, in theory, the combination of immigration by homozygous susceptible (SS) individuals and use of a dose sufficiently high to kill heterozygous (RS) individuals would suppress resistance development (Comins 1977a, Curtis et al. 1978, Taylor and Georghiou 1979, Tabashnik and Croft 1982). The basic idea is that the few homozygous (RR) resistant survivors in the treated population would mate with the SS immigrants and produce RS offspring that can be killed by insecticide.

Practical considerations and examinations of the assumptions of the models suggest that the potential utility of this high-dose strategy is limited (Tabashnik and Croft 1982, Tabashnik 1986d). Requirements for success of this strategy include immigration by susceptibles, random mating of susceptibles and resistant homozygotes, reproduction by susceptibles in the treated habitat, ability to kill RS heterozygotes with insecticide, low R allele frequency initially, and low reproductive potential. Even when all these conditions are met for a particular target pest species, the high-dose strategy may promote rapid resistance development in nontarget pests, disrupt biological control, increase control costs, and exacerbate the public health hazards and pollution problems caused by pesticides.

Temporal variation in dose, an important component of the intensity of insecticide use, is greatly influenced by pesticide persistence. Experiments with house flies on an English pig farm (Denholm et al. 1983) were consistent with the prediction that nonpersistent insecticides do not rapidly select for resistance (Taylor and Georghiou 1982). Nonpersistent sprays of the pyrethroid bioresmethrin did not cause resistance development, yet the flies developed resistance to the persistent pyrethroid permethrin in the laboratory. The frequency of bioresmethrin sprays was also minimized by treating only when the fly population exceeded a density threshold. The results of this study, however, do not exclude the possibility that the flies had the potential to evolve substantial resistance to permethrin, but not to bioresmethrin. Direct tests of the effect of persistence require comparison between persistent and nonpersistent applications of the same insecticide, rather than comparisons between insecticides.

In a simulation study of grain pests, resistance to dichlorvos was projected to evolve faster than resistance to deltamethrin in some cases, even though the half-life of dichlorvos (two weeks) was assumed to be much less than the half-life of deltamethrin (80 weeks) (Sinclair and Alder 1985). Again, this study did not examine the effect of persistence directly because assumptions about the two insecticides differed in terms of initial dose applied and probit mortality lines as well as persistence. The simulation results were consistent with the rapid development of resistance to dichlorvos observed in some grain pests. Predictions for deltamethrin could not be tested because this compound had not been used in the field.

2. *Empirical Tests of Species-Specific Models*

Some qualitative predictions from general models of the evolution of insecticide resistance have been tested, as described earlier, yet few explicit empirical tests of predictions from species-specific models of resistance have been conducted. There have been two types of validation studies: short-term laboratory experiments (Taylor et al. 1983, Beeman and Nanis 1986) and retrospective comparisons based on historically observed patterns of resistance development in the field (Tabashnik and Croft 1985, Tabashnik 1986b). The following section highlights the approaches and conclusions of validation studies conducted to date.

a. Experimental Tests One of the first explicit empirical tests of a resistance model was a comparison between predicted and observed rates of evolution of resistance to dieldrin in laboratory populations of the house fly (Taylor et al. 1983). Dieldrin resistance in the house fly is monogenic; bioassays can distinguish among three genotypes (Georghiou et al. 1963). Taylor et al. (1983) simulated five different treatment regimes, then compared the predicted resistance gene frequencies and population sizes with those observed in five corresponding experimental cages.

Biological parameters used in the simulations were measured directly from laboratory fly populations. The initial conditions were alike for all cages (90 SS + 10 RS individuals of each sex per cage), but each cage received a different treatment: (A) control—no insecticide and no immigration, (B) slow insecticide decay and no immigration, (C) fast decay and immigration, (D) no decay and no immigration, and (E) no decay and immigration. Immigration was achieved by adding 25 individuals (24 SS + 1 RS) to the appropriate cages three times weekly. Dieldrin was incorporated in the larval medium and acted only on larvae and newly enclosed adults. The initial dieldrin concentration (40 ppm) was the same in treatments B to E, but decay rates corresponding to insecticide half-lives of 1.0 (slow decay) and 0.5 (fast decay) days were mimicked by using decreasing dieldrin concentrations in successive treatments. Each cage was run for 57 days, the equivalent of about four fly generations.

The results showed a strong correlation between predicted and observed values for the final R gene frequency in each treatment (Fig. 6.2). Both the simulations and experiments support earlier predictions that immigration by susceptibles can retard the evolution of resistance, especially when the ratio of immigrants to residents in the treated populations is high (Comins 1977a, Taylor and Georghiou 1979, Tabashnik and Croft 1982). Among the three treatments with immigration, resistance evolved slowest when persistence was lowest (treatment C), but there was little difference between treatments with slow decay (B) and no decay (E). Thus, the rate of resistance development did not decline uniformly as insecticide persistence decreased.

In a similar study, Denholm et al. (1987) found that simulations mirrored results of a selection experiment in which residual treatments of permethrin increased the frequency of the *kdr* allele in a laboratory house fly population. Denholm et al. (1987) emphasized, however, that the relative resistance levels of different *kdr* genotypes depended on the mode of insecticide exposure.

Beeman and Nanis (1986) found that predictions from a simple one-locus model corresponded well with the observed outcome of a single selection with malathion on laboratory colonies of the red flour beetle, *Tribolium castaneum*. Beetles from 29 laboratory colonies established from field collections in 1980 were exposed to filter paper discs impregnated with malathion (Haliscak and Beeman 1983). For each colony, mortality was measured for the parental generation and for the F_1 offspring of survivors treated at the same concentration.

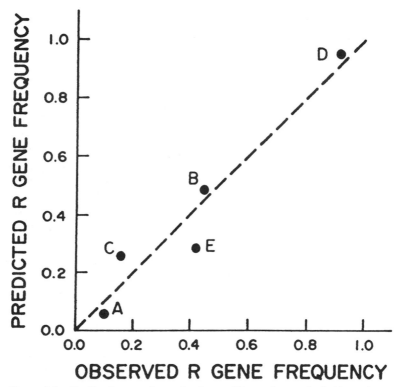

Figure 6.2. Predicted versus observed resistance (*R*) gene frequencies in *Musca domestica* after laboratory selection with dieldrin. Letters indicate treatments (see text). Dashed line shows predicted = observed (after Taylor et al. 1983).

Beeman and Nanis (1986) developed a predictive model based on the following assumptions: (1) Resistance was controlled by two alleles at a single locus, (2) Hardy-Weinberg equilibrium existed before selection, (3) the malathion treatment killed all SS beetles but no RS or RR beetles, and (4) mating after selection was random. Using these assumptions, projected estimates of the frequency of SS beetles (i.e., percent mortality caused by a diagnostic concentration) in the F_1 generation could be calculated from the frequency of SS beetles in the parental generation (Fig. 6.3).

Although the F_1 progeny of one colony (IA-10, the outlier in Fig. 6.3) that were expected to be partially resistant were apparently entirely susceptible (Haliscak and Beeman 1983), the overall correlation between observed and expected SS frequency was significant ($r = 0.63$, $df = 28$, $P = 0.0002$). The very simple prediction that the frequency of SS after selection is correlated with the frequency of SS before selection is, however, even more strongly supported by the data ($r = 0.68$, $P < 0.0001$).

In a similar study, Wool and Noiman (1983) found that observed mortality in

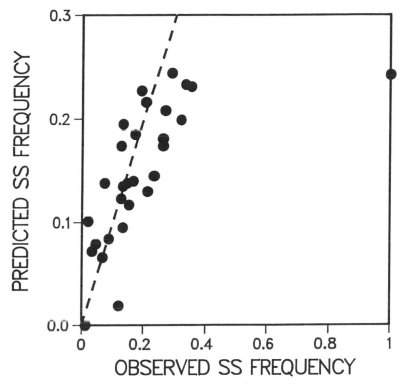

Figure 6.3. Predicted versus observed frequency of susceptible (SS) *Tribolium casta-neum* after one generation of laboratory selection with malathion. Dashed line shows predicted = observed (after Haliscak and Beeman 1983, Beeman and Nanis 1986).

laboratory experiments with *Tribolium castaneum* was consistent with predictions from a one-locus, two-allele model of resistance. They tested the hypothesis that addition of susceptible males to a resistant population will slow or arrest the evolution of resistance to malathion. In initial experiments, the susceptible immigrant males were unintentionally killed by malathion residues before they mated with resistant females. Under these conditions, the rate of resistance development in cages with immigrants was similar to the rate observed in control cages without immigrants. When the protocol was modified to ensure that immigrants had the opportunity to mate before selection with malathion occurred, immigration effectively reduced the frequency of the R allele. Introduction of susceptible males at a ratio of 30 immigrants to 100 residents (in the cage) resulted in a stable level of observed mortality (~25%) that agreed well with the theoretically expected level (~23%).

b. Retrospective Tests In a retrospective study of evolution of resistance to azinphosmethyl, Tabashnik and Croft (1985) tested a model by comparing simulated times versus historically observed times for resistance development in

24 species of apple pests and natural enemies in North America. Some secondary pests and natural enemies have evolved resistance to azinphosmethyl, but it has remained a major pest control tool in apple for more than 30 years because key pests have not evolved resistance to it (Croft 1982).

A survey of fruit entomologists (Croft 1982) was used to estimate population ecology parameters (generations per year, fecundity, immigration, etc.) and the number of years until resistance development for each of 24 apple insect and mite species. The genetic basis of resistance, concentration–mortality lines,and initial R allele frequency (0.001) were assumed to be identical for all 24 species because these parameters were not known for most species. Resistance was controlled by one locus with two alleles, with a concentration–mortality line for each of three genotypes. To simulate historical azinphosmethyl exposure, six sprays at a concentration lethal to 93% of SS, 50% of RS, and 23% of RR were applied yearly. All species were subjected to the same simulated pesticide concentration, spray schedule, and pesticide half-life because all species were present in the same habitat and were exposed to a similar treatment regime in the field. A species was considered resistant in the simulation when the frequency of the R allele exceeded 0.50 and the population size exceeded a threshold. (For pests the threshold was the economic threshold for each species; for all natural enemies, an arbitrary level of 100 individuals per tree was used.)

The results showed a significant rank correlation between predicted and historically observed times to evolve resistance for the 12 pest species ($r_s = 0.65$, $P = 0.022$) and the 12 natural enemy species ($r_s = 0.69$, $P = 0.013$). Thus, ecological differences among species were sufficient to explain a significant portion of observed variation in rates of resistance evolution among apple pests and natural enemies. Simulations of natural enemy resistance development are discussed further in Section V.

The retrospective study of Tabashnik and Croft (1985) suggests that models can predict rates of resistance development in the field. A major limitation of the study, however, is that it relied on estimated values for many important parameters. This problem was addressed, in part, by a sensitivity analysis that showed that many of the model's predictions were minimally affected by substantial variation in some key parameters (immigration, initial population size, and fecundity) that are difficult to estimate yet potentially influential.

In another retrospective study, Tabashnik (1986b) simulated evolution of resistance to the pyrethroid fenvalerate in the diamondback moth, *Plutella xylostella*, a major pest of cabbage and other crucifers. Most of the biological parameters in this one-locus model were based on empirical data from the literature. Predictions from the model agreed reasonably well with observations from the field in Ban-chau, Taiwan, where resistance to fenvalerate was documented less than four years after its introduction (Liu et al. 1981).

Under the assumption of intensive pesticide use (eight sprays per cabbage cycle at a concentration causing 95% mortality of susceptible larvae), sensitivity analyses were done by varying fecundity (four-fold), initial resistance allele frequency (10,000-fold), initial population density (100-fold), and percentage of larvae exposed to insecticide (80–100%). Predicted times for resistance development under these conditions varied from less than one to four years. This range corresponds well with the observed range of times for diamondback moth to develop resistance to fenvalerate in several Asian countries (Thailand, one to two years; Taiwan, three years; Japan, three to four years; Malaysia, four years; [Cheng 1988]). Simulations suggested that spray frequency must be reduced to less than two per cabbage cycle, or spray concentration reduced below LC_{75} (or both), to delay resistance development substantially.

The correspondence between observed and simulated times for resistance development in the diamondback moth study is encouraging, particularly because observed rates of resistance development in the field and many model parameters were based on empirical data. Nonetheless, some important limitations of this work should be recognized. Although the model accurately simulated rapid evolution of resistance under intensive pesticide pressure, its predictions for resistance development under reduced pesticide use were not tested. Experimental verification of these predictions is needed. Convincing demonstration of the utility of this model and other resistance models will require verification of predictions that are truly anticipatory rather than retrospective.

Furthermore, one cannot conclude that the assumptions of a model are correct solely because its predictions are accurate under some circumstances. For example, the diamondback moth model assumed monogenic inheritance of resistance, yet fenvalerate tolerance in this insect may be a polygenic trait (Liu et al. 1981, Tabashnik and Cushing 1989). In addition, the model did not explicitly incorporate cross-resistance, which can be an important factor (Tabashnik et al. 1987). Rigorous testing of resistance models will require comparison of observed versus predicted outcomes across a range of conditions for relevant pesticide use practices and biological parameters. The validity of alternative assumptions can be determined only when those assumptions lead to outcomes that are demonstrably different. Conversely, if alternative models (e.g., monogenic vs. polygenic) produce similar outcomes and recommendations, then it may not be important to distinguish between models for management decisions.

B. Multilocus Models

Few models of the evolution of insecticide resistance have considered resistance due to alleles at more than two loci. In one of the first papers describing this type of model, Plapp et al. (1979) used data from the house fly to develop a simulation model that could incorporate up to six loci for resistance to one insecticide, with each resistance allele conferring 3-, 10-, or 30-fold resistance. They assumed that resistance was multiplicative, so that, for example, 100-fold resistance resulted from the combination of two alleles, conferring 10-fold resistance each. A log-normal dose–morality line was associated with each genotype. In any particular simulation, the slope of the dose–mortality line was identical across genotypes. Effects of a steep slope (2.25, equivalent to relatively low nongenetic variation) were compared with effects of a shallow slope (0.63; relatively high nongenetic variation). Mutation rates were 10^{-4} or 10^{-5}, generations were discrete, and the population was assumed to be extremely large, closed (no immigration), and uniformly exposed to pesticide applications (no refuge).

Plapp et al. (1979) found that resistance developed faster when the dose–mortality lines had a steep slope than when their slope was shallow. They also found that resistance developed quickly when resistance alleles conferred high levels of resistance (10- or 30-fold). Consistent with results from one-locus models, resistance evolved rapidly in nearly all cases when a high concentration ($\geq LC_{90}$ of susceptible individuals) was simulated. When the slope of the dose–mortality line was steep, the rate of resistance development was similar at mutation rates of 10^{-4} and 10^{-5}. With a shallow slope, however, the initial rate of resistance development was substantially slower at the lower mutation rate (10^{-5}).

Simulations contrasting two- and three-locus models at high insecticide concentrations (LC_{90} to LC_{99}) produced some particularly interesting results. If one of the two loci had a semidominant allele conferring 10- or 30-fold resistance, then addition of a third locus with a semidominant allele conferring three-fold resistance had virtually no impact. Conversely, in some cases, addition of this same allele at a third locus substantially increased the rate of resistance development when alleles at the first two loci conferred only three-fold resistance each. Resistance evolved faster when it was due to alleles at two loci, one conferring 10-fold resistance and the other three-fold resistance, than when it was due to alleles at three loci, each conferring three-fold resistance.

Plapp et al. (1979) concluded that the "best insecticides in terms of delaying the onset of resistance are those to which insect populations respond in a heterogeneous manner and further, those to which no single gene confers a high level of resistance." In other words, given similar initial allele frequencies, resistance will evolve more slowly if the selective advantage conferred by resistance alleles is not large (i.e., there is little discrimination among genotypes [see Roush and McKenzie 1987, Chapter 5]). These conclusions are consistent with results from one-locus models. Plapp et al. (1979) implicitly assumed that alleles with major and minor effects were equally rare. Evolution of resistance by selection for several minor genes may occur faster than by selection for a single major gene if the minor genes are initially more common (Lande 1983).

Via (1986) developed polygenic models of insecticide resistance based on the theoretical framework of quantitative genetics (Falconer 1981), in which resistance is viewed as a continuously variable trait influenced by many genes that are not identified individually. Polygenic resistance is apparently less common than monogenic resistance, but this may be due to the intensity of past and current regimes of insecticide application rather than to an inherent bias in genetic potential (Whitten and McKenzie 1982, Via 1986, Tabashnik and Cushing 1989).

Via's (1986) polygenic models assume that variation in tolerance to insecticide within a population is due to a combination of environmental and genetic factors. The observed phenotypic variation is assumed to be continuous and normally distributed. This can be contrasted with single and multilocus models in which there are recognizable discontinuities between the tolerances associated with different genotypes.

The polygenic models of Via (1986) assume that the rate of evolution of resistance to a single insecticide is the product of selection intensity and heritability (the proportion of phenotypic variation that is due to additive genetic variation). Thus, resistance development can be slowed by reducing heritability, selection intensity, or both. In essence, these conclusions parallel those of Plapp et al. (1979) and are consistent with those derived from single-locus models.

In an exceptionally innovative approach, Uyenoyama (1986) developed a theoretical framework based on the idea that evolution of insecticide resistance requires alteration of normal metabolic pathways to detoxify insecticides. A key assumption of her analysis is that in the initial stages of evolution of resistance, increased detoxification capability comes at the expense of normal metabolism. Thus, in the absence of insecticide, resistant individuals will be a disadvantage relative to susceptible individuals.

Uyenoyama (1986) posited that evolution of insecticide resistance entails changes in at least two loci: a regulatory locus controlling the level of synthesis of a key catabolic enzyme and a modifier locus that reduces the pleiotropic cost associated with resistance. This perspective is based partly on biochemical and genetic analyses of microbial resistance and on some studies of insecticide resistance (see Uyenoyama 1986 and references therein).

Recent studies of the evolution of microbial resistance show variation in competitive fitness among mutants of *Escherichia coli* bacteria resistant to virus T4 as well as epistatic modifiers of competitive fitness. Thus, refinement of the resistant phenotype in *E. coli* can occur by selection among resistant genotypes or by selection for epistatic modifiers (Lenski 1988a,b). Direct evidence for the evolution of modifiers of fitness associated with insecticide resistance is scarce (McKenzie et al. 1982, Clarke and McKenzie 1987, Roush and McKenzie 1987, Chapter 5). More empirical work is needed to investigate this possibility.

The most intriguing implication of Uyenoyama's (1986) conceptualization is that, "In contrast to conclusions from single-locus models, migration may have a uniformly detrimental effect as a control strategy opposing the evolution of genetic networks because it promotes the evolution of modifiers of resistance by increasing the effective rate of mutation in the treated area and introducing a preadaptation for resistance into untreated areas." In other words, migration into treated areas may speed resistance development by increasing genetic variation, and migration into untreated areas may promote resistance by spreading resistance alleles. The second prediction is consistent with results from Comins' (1977a) analysis of a two-area system, based on a one-locus model, as previously described. The first prediction, however, is the opposite of expectations from most single-locus models. Direct observations of the effects of migration into treated areas on the rate of resistance evolution and investigations of modifier genes may be needed to determine which of these models is most appropriate.

IV. Resistance To Two Or More Insecticides

The idea of combining two or more pesticides to manage resistance has been reviewed in detail elsewhere (Brown 1977, Georghiou 1983, Ozaki 1983, Leeper et al. 1986, Sawicki and Denholm 1987, Tabashnik 1989, Roush 1989). This section summarizes some of the most recent theoretical and experimental work on insecticide mixtures, rotations, and mosaics.

A. Mixtures

Mixtures of insecticides are applied so that individual are exposed simultaneously to more than one toxicant. Most models of evolutionary response to mixtures of two insecticides assume (1) resistance to each insecticide in the mixture is monogenic, (2) no cross-resistance occurs between insecticides in the mixture, (3) resistant individuals are rare, (4) the insecticides have equal persistence, and (5) some of the population is untreated (Knipling and Klassen 1984, Curtis 1985, Mani 1985, Comins 1986).

If one assumes all the preceding and that resistance for each insecticide is functionally recessive, so that only resistant homozygotes survive exposure (Curtis et al. 1978), then mixtures are predicted to suppress resistance effectively (Mani 1985, Curtis 1985). Under these assumptions, resistance evolves very slowly because doubly resistant homozygotes, the only individuals to survive insecticide exposure, are exceedingly rare. Thus, most survivors are untreated, susceptible individuals. Slight deviations from complete recessiveness, however, would greatly hasten evolution of resistance to mixtures (Mani 1985). Curtis (1985) argued that variation in dosage makes it extremely difficult to achieve absolute recessiveness in practice.

In addition to deviation from complete recessiveness, violations of one or more of the five assumptions listed previously may occur in most field situations. Although many resistances are caused primarily by allelic changes at one locus, nonmonogenic resistance has been reported in many cases (see Via 1986, Tabashnik and Cushing 1989, and references therein). Assuming polygenic inheritance, full-rate mixtures would increase the intensity of selection and the overall rate of resistance development (Crow 1952, Via 1986, Taylor 1989). Cross-resistance is typical among structurally related pesticides and can also occur between compounds from different classes as well (Oppenoorth 1985, Tabashnik et al. 1987, Chapter 4). The importance of equal persistence has also been emphasized (Mac-Donald et al. 1983a, Curtis 1987, Roush 1989); if persistence is unequal, many individuals may be exposed to only one compound in the mixture.

Few experimental tests of mixtures have been performed. Early experimental work (e.g., Burden et al. 1960) suggested that mixtures were not effective for suppressing resistance (see Brown 1977). Review of more recent experimental studies (Table 6.2) shows mixed results. Mixtures of five synthetic insecticides and combinations of fenvalerate with a sticky trap and wasp parasite did not effectively slow resistance development in laboratory studies with the house fly (Pimentel and Burgess 1985). In another laboratory study with the house fly, a mixture of dichlorvos and permethrin slowed development of resistance to permethrin, but not to dichlorvos (MacDonald et al. 1983a).

B. Rotations

In a rotation, two or more insecticides are alternated in time so that each individual is exposed to only one compound, but the population experiences more than one insecticide over time. The use of rotations is based on the assumption that the frequency of individuals resistant to one insecticide will decline during application of an alternate insecticide. Such a decline can occur if there is negative cross-

Table 6.2. Experimental Tests of Insecticide Mixtures, Rotations, and Mosaics (1976–1985)

Reference	Lab (L) or Field (F)	Tactics: Insecticides
	Musca domestica	
Pimentel and Bellotti (1976)	L	mosaic: citric acid, copper sulfate, magnesium nitrate, potassium hydroxide, ammonium phosphate, sodium chloride
Pimentel and Burgess (1985)	L	mixture: arsenite, dieldrin, malathion rotation: methomyl, fenvalerate, sticky trap, parasite
MacDonald et al. (1983a)	L	mixture, rotation: dichlorvos, permethrin
MacDonald et al. (1983b)	F	rotation: dichlorvos, permethrin
	Culex quinquefasciatus	
Georghiou et al. (1983)	L	rotation: diflubenzuron, permethrin, propoxur, temephos

resistance, a substantial fitness cost associated with resistance, or immigration of susceptible individuals. Georghiou (1986) stated that, "Despite the search for pairs of compounds with negatively correlated resistance, none has been discovered that would have the potential for field application." Fitness costs associated with resistance (i.e., the disadvantage of resistant individuals compared with susceptible individuals in the absence of pesticide) are generally moderate or nil (Roush and McKenzie 1987, Chapter 5).

Results from some models assuming no substantial fitness disadvantage and no cross-resistance suggest that rotations would not effectively suppress resistance (Knipling and Klassen 1984, Comins 1986, Via 1986, Curtis 1987). Recent modeling work by Roush (1989), however, suggests that rotations may be more effective than other multiple pesticide use tactics. The impact of immigration on rotations has not been fully explored.

Rotations of various insecticides have shown some success in slowing resistance development in laboratory experiments with the house fly and the southern house mosquito, *Culex quinquefasciatus* (Table 6.2). One of the few (perhaps only) field tests suggested that rotation of dichlorvos and permethrin did little to slow resistance development in the house fly on Canadian farms (MacDonald et al. 1983b). Regular sanitation as a substitute for insecticide use appeared to be the most effective way to manage resistance (MacDonald et al. 1983b).

C. Mosaics

Mosaics use a spatial patchwork of insecticide applications so that adjacent areas are treated simultaneously with different insecticides. Verbal arguments favoring mosaics have been made (Byford et al. 1987), but models have suggested that mosaics would generally not slow evolution of resistance. Curtis (1985) used a two-locus model (assuming one locus for each of two insecticides in a mosaic) to show that mosaics would not be effective unless susceptible individuals were released into the treated area. Based on an analytical model, Comins (1986) concluded that mosaics would actually increase the rate of resistance development. Using polygenic models, Via (1986) showed that a spatial mosaic could be effective if there was negative cross-resistance, or strong positive cross-resistance and an intermediate optimum for insecticide tolerance. Because negative cross-resistance is rare and fitness in the presence of insecticide usually increases monotonically with tolerance (i.e., no intermediate optimum exists), the conditions for success of a mosaic described by Via (1986) are probably not met in most field situations.

The only experimental test of a mosaic, a laboratory study with the house fly, provided some support for the use of mosaics. Pimentel and Bellotti (1976) found that resistance to a mosaic of six toxic organic salts did not evolve in 32 generations, whereas substantial resistance to each salt used singly evolved in 7–10 generations. A rigorous test would require 60 generations of mosaic selection versus 10 generations of single compound selection, but this comparison was not possible (see Tabashnik 1989). A more important consideration is that mosaics requiring six different types of pesticide are probably not feasible.

In summary, theoretical models and available data suggest that the effectiveness of mixtures, rotations, and mosaics requires special conditions that are not generally met in the field. Experimental work on pesticide combinations has been very limited—consisting primarily of laboratory experiments with the house fly. Much more experimental work is needed to assess rigorously the potential of multiple pesticide use tactics (Mani 1985, Taylor 1989). Consideration of factors that are usually not included in models—such as negative effects on beneficials and other nontarget organisms, increased cost, and safety hazards—suggests that reducing pesticide use through integrated pest management may be more productive than attempts to optimize pesticide combinations (Tabashnik 1989).

V. Resistance in Natural Enemy–Pest Systems

Documented cases of insecticide resistance are rare in arthropod natural enemies compared to those in pests (Georghiou 1986, Chapter 8). Two of the hypotheses proposed to explain this disparity, the preadaptation and food limitation hypotheses (Georghiou 1972, Croft and Strickler 1983, Tabashnik and Johnson 1990), have been partially evaluated using models. This section reviews the hypotheses and the implications of results from models of resistance development in natural enemies.

The basic idea of the preadaptation hypothesis is that herbivorous pests detoxify plant allelochemicals and are thus better preadapted to detoxify pesticides than are entomophagous natural enemies (Croft and Morse 1979). In simulations of resistance development in 12 natural enemies of apple pests, Tabashnik and Croft (1985) incorporated the preadaptation hypothesis six ways, none of which substantially improved the overall correspondence between simulated and observed times for resistance development in natural enemies. Lowering the initial R allele frequency or the LC_{50} for homozygous susceptible (SS) natural enemies by 10- or 100-fold had little impact on the rate of evolution. Conversely, reducing the LC_{50} of all three presumed genotypes (SS, RS, RR) by 10- or 100-fold delayed resistance for more than 25 years in all natural enemies, even those reported to be resistant in the field after 8–15 years.

The essence of the food limitation hypothesis is that natural enemies do not evolve resistance readily because pesticides greatly reduce their food supply by killing susceptible prey or hosts (Huffaker 1971, Georghiou 1972). Tabashnik and Croft (1985) incorporated this hypothesis in simulation projections by assuming that a natural enemy began to evolve resistance only after its host or prey became resistant. Thus, the observed time to develop resistance for the host or prey was added to the originally predicted time for the respective natural enemy. This modification of the predictions improved their correspondence with observed rates of resistance development, but it oversimplified the dynamic interactions between pests and natural enemies.

Other models have been used to project the potential impact of food limitation on the population dynamics and resistance evolution in natural enemies (Tabashnik and Johnson 1990). The Lotka–Volterra equations of predator–prey population growth show that equivalent mortality will suppress a predator population more than its prey because the predator's birth rate and the prey's death rate are proportional to the product of the population sizes of both species (Wilson and Bossert 1971). In contrast, the predator's death rate and the prey's birth rate are directly proportional to their respective population sizes but are not affected by the population size of the other species. Thus, a pesticide treatment that kills 90% of predator and prey populations reduces the predator's birth rate and the prey's death rate by a factor of 100 but reduces the predator's death rate and the prey's birth rate only by a factor of 10. More refined models also show that natural enemy populations are more severely suppressed by pesticides than are pest populations, even if the immediate mortality is similar for both populations (e.g., Waage et al. 1985).

May and Dobson (1986) noted that after pesticide applications, natural enemy populations recover slowly because food is scarce, but pests rebound readily due to lack of predators. They reasoned that, in effect, natural enemies show undercompensating density-dependence, whereas pests show overcompensating density-dependence. Undercompensating density-dependence reduces the average size of treated natural enemy populations, thereby increasing the impact of immigration of susceptible individuals. Thus, in the presence of immigration, pests develop resistance faster than natural enemies.

Recent simulation studies included evolutionary potential for resistance in both predator and prey, as well as coupled predator–prey population dynamics (Tabashnik 1986c). The key assumption of these simulations was that low prey density reduced the predator's rates of consumption, survival, and fecundity. Predator functional response and the effects of food shortage on predator survival and fecundity were based, in part, on experimental data from mites (Dover et al. 1979). Even though the predator and prey were assumed to have equal intrinsic tolerance and equal genetic potential for evolving resistance, intensive pesticide use caused rapid resistance development in the pest (prey), but either suppressed resistance development or caused local extinction of the natural enemy (predator). These theoretical results imply that food limitation is sufficient to account for pests' ability to evolve pesticide resistance more readily than natural enemies.

The successful selection of pesticide-resistant natural enemies in the laboratory, where food is not limited, provides indirect support for the idea that food limitation retards evolution of resistance in natural enemies (Croft and Strickler 1983, Hoy 1985, Chapter 8). The general trend that natural enemies evolve resistance only after their host or prey develops resistance (Georghiou 1972, Croft and Brown 1975, Tabashnik and Croft 1985) is also consistent with the food limitation hypothesis. However, patterns of variation in resistance to five insecticides in field populations of *Aphytis melinus,* a parasite of California red scale *(Aonidiella aurantii),* suggested that food limitation was not a key factor in this system (Rosenheim and Hoy 1986). Direct empirical tests of predictions from models of resistance development in natural enemies have not yet been performed.

VI. Conclusions and Future Directions

Modeling studies of insecticide resistance have been useful for evaluating the relative importance of different factors, identifying qualitative trends, and suggesting potential resistance management tactics. Models have shown that resistance evolves faster as reproductive potential, particularly generation turnover, increases. Models have identified dispersal as a key factor that may have major direct effects on resistance development, as well as significant interactions with other factors. To some extent, models have provided quantitative verification of common sense about pesticide resistance; under most conditions, the best way to slow resistance development is to reduce pesticide use, particularly the number of treatments and the proportion of the population that is treated. This approach has been termed management by "moderation" or the "low pesticide use" strategy (Sutherst and Comins 1979, Tabashnik and Croft 1982).

Models have also fostered consideration of management tactics that are based on the idea of overwhelming pest evolutionary responses by "saturation" (high dose) and "multiple attack" (combinations) (Comins 1977b, Sutherst and Comins 1979, Taylor and Georghiou 1979, Wood and Mani 1981, Georghiou 1983, Knipling and Klassen 1984, Mani 1985). Careful examination of such strategies has shown that they require special assumptions that are violated in most field situations (Sutherst and Comins 1979, Tabashnik and Croft 1982, Curtis 1985, Tabashnik 1989, Roush 1989). Strategies that rely on intensive insecticide use are often "high risk" because their success requires special conditions, and their failure may have disastrous effects, including extremely rapid resistance development in the target pest (Sutherst and Comins 1979, Rawlings et al. 1981, Tabashnik and Croft 1982). Even when such tactics suppress resistance in a particular pest, they may still be undesirable because of their negative side effects, including disruption of biological control, increased environmental hazards, high cost, and promotion of resistance in secondary pests (Tabashnik and Croft 1982, Tabashnik 1986, Taylor 1989, Johnson and Tabashnik 1990).

I urge modelers and other resistance management workers to view tactics in the broad context of pest management, human safety, and environmental quality. In this context, it is clear that resistance management is a subcomponent of integrated pest management (Chapter 11), which seeks to minimize pesticide use by fully exploiting nonchemical control strategies. We need to identify "soft-fail" tactics that have the potential to be especially effective under favorable assumptions yet are no worse than alternative tactics under unfavorable assumptions (Comins 1986).

Nearly all the resistance models developed thus far assume that inheritance of resistance is monogenic. Although this assumption may be appropriate for most cases, nonmonogenic resistance is often reported (see Roush and McKenzie 1987, Tabashnik and Cushing 1989, and references therein). Limited studies conducted to date show that some of the general trends predicted by monogenic and polygenic models are similar (Plapp et al. 1979, Via 1986, Tabashnik 1989). Uyenoyama's (1986) polygenic model of pleiotropic costs suggests that migration generally favors evolution of resistance, whereas in monogenic models, immigration of susceptibles has been considered a major factor opposing resistance development. Models should be used to compare systematically the implications of monogenic versus polygenic inheritance across a range of assumptions about biological and operational conditions. This type of theoretical work is needed to determine if the choice of an optimal management strategy depends on the mode of inheritance of resistance. Field experiments are also needed to measure dispersal and its impact on resistance development.

Most models of the evolution of resistance are deterministic. They typically assume that the initial frequency of a resistance allele is sufficiently high ($>10^{-5}$) that resistance is present in local populations. In the earliest stages of resistance development, however, the presence or absence of resistance in finite, local populations may be affected by stochastic events (Whitten and McKenzie 1982). More detailed theoretical analysis of this "stochastic phase" of resistance development would be useful.

Recent efforts to test models have slightly narrowed the gap between theoretical and empirical work in resistance management, but previous pleas for more experimental work (Taylor 1983, Tabashnik 1986a) are still largely unanswered. Empirical comparison of the effectiveness of different management tactics is almost entirely lacking. Long-term experiments under field or fieldlike conditions are the most rigorous approach to this issue. Retrospective studies that combine information on pesticide use with bioassay data from numerous field populations can also be informative (e.g., Rosenheim and Hoy 1986).

In conclusion, the most important challenges for modeling and evaluation of resistance management are to test tactics empirically, to consider the potential impact of nonmonogenic resistance and stochastic events, and to view resistance management in the broad context of integrated pest management. With the achievement of these goals, resistance management can help to promote safer, saner pest management.

References

Beeman, R. W., and S. M. Nanis. 1986. Malathion resistance alleles and their fitness in the red flour beetle (Coleoptera: Tenebrionidae). J. Econ. Entomol. 79: 580–587.

Brown, A. W. A. 1977. Epilogue: resistance as a factor in pesticide management, pp. 816–824. *In*

Proceedings, XV International Congress of Entomology. Entomological Society of America. College Park, Md.

Burden, G. S. , C. S. Lofgren, and C. N. Smith. 1960. Development of chlordane and malathion resistance in the German cockroach. J. Econ. Entomol. 53: 1138–1139.

Byford, R. L., J. A. Lockwood, and T. C. Sparks. 1987. A novel resistance management strategy for horn flies (Diptera: Muscidae). J. Econ. Entomol. 80: 291–296.

Cheng, E. Y. 1988. Problems of control of insecticide-resistant *Plutella xylostella*. Pestic. Sci. 23: 177–188.

Clark, G. M., and J. A. McKenzie. 1987. Developmental stability of insecticide resistant phenotypes in blowfly; a result of canalizing natural selection. Nature 325: 345–346.

Comins, H. N. 1977a. The development of insecticide resistance in the presence of immigration. J. Theor. Biol. 64: 177–197.

Comins, H. N. 1977b. The management of pesticide resistance. J. Theor. Biol. 65: 399–420.

Comins, H. N. 1979a. Analytic methods for the management of pesticide resistance. J. Theor. Biol. 77: 171–188.

Comins, H. N. 1979b. The management of pesticide resistance: models, pp. 55–69. *In* M. A. Hoy and J. J. McKelvey, Jr. (eds.), Genetics in relation to insect management. Rockefeller Foundation, New York.

Comins, H. N. 1979c. The control of adaptable pests, pp. 217–226. *In* G. A. Norton and C. S. Holling (eds.), Pest management: proceedings of an international conference. Oxford. Pergamon, Oxford.

Comins, H. N. 1986. Tactics for resistance management using multiple pesticides. Agric. Ecosystems Environ. 16: 129–148.

Cook, L. M. 1981. The ecological factor in assessment of resistance in pest populations. Pestic. Sci. 12: 582–586.

Croft, B. A. 1982. Arthropod resistance to insecticides: a key to pest control failures and successes in North American apple orchards. Entomol. Exp. Appl. 3: 88–110.

Croft, B. A., and A. W. A. Brown. 1985. Responses of arthropod natural enemies to insecticides. Annu. Rev. Entomol. 20: 285–335.

Croft, B. A., and J. G. Morse. 1979. Recent advances in natural-enemy pesticide research. Entomophaga 24: 3–11.

Croft, B. A., and K. Strickler. 1983. Natural enemy resistance to pesticides: documentation, characterization, theory and application, pp. 669–702. *In* G. P. Georghiou and T. Saito (eds.), Pest resistance to pesticides. Plenum, New York.

Croft, B. A., E. C. Burts, H. E. van de Baan, P. H. Westigard, and H. W. Riedl. 1989. Local and regional resistance to fenvalerate in *Psylla pyricola* Foerster (Homoptera: Psyllidae) in western North America. Can. Entomol. 121: 121–129.

Crow, J. F. 1952. Some genetic aspects of selection for resistance. National Res. Council Publ. 219: 72–75.

Curtis, C. F. 1981. Possible methods of inhibiting or reversing the evolution of insecticide resistance in mosquitoes. Pestic. Sci. 12: 557–564.

Curtis, C. F. 1985. Theoretical models of the use of insecticide mixtures for the management of resistance. Bull. Entomol. Res. 75: 259–265.

Curtis, C. F. 1987. Genetic aspects for selection for resistance, pp. 150–161. *In* M. G. Ford, D. W. Hollman, B. P. S. Khambay, and R. M. Sawicki (eds.), Combating resistance to xenobiotics: biological and chemical approaches. Horwood, Chichester, England.

Curtis, C. F., L. M. Cook and R. J. Wood. 1978. Selection for and against insecticide resistance and possible methods of inhibiting the evolution of resistance in mosquitoes. Ecol. Entomol. 3: 273–287.

Denholm, I., A. W. Farnham, K. O'Dell, and R. M. Sawicki. 1983. Factors affecting resistance to insecticides in house-flies, *Musca domestica* L. (Diptera: Muscidae). I. Long-term control with bioresmethrin of flies with strong pyrethroid-resistance potential. Bull. Entomol. Res. 73: 481–489.

Denholm, I., R. M. Sawicki, and A. W. Farnham. 1985. Factors affecting resistance to insecticides in house-flies, *Musca domestica* L. (Diptera: Muscidae). IV. The population biology of flies on animal farms in south-eastern England and its implications for the management of resistance. Bull. Entomol. Res. 75: 143–158.

Denholm, I., R. M. Sawicki, and A. W. Farnham. 1987. Laboratory simulation of selection for resistance, pp. 138–149. *In* M. G. Ford, D. W. Holloman, B. P. S. Khambay and R. M. Sawicki (eds.), Combating resistance to xenobiotics: biological and chemical approaches. Horwood, Chichester, England.

Dover, M. J., B. A. Croft, S. M. Welch, and R. L. Tummala. 1979. Biological control of *Panonychus ulmi* (Acarina: Teranychidae) by *Amblyseius fallacis* (Acarina: Phytoseiidae) on apple: a prey–predator model. Environ. Entomol. 8: 282–292.

Dowd, P. F., T. C. Sparks, and F. L. Mitchell. 1984. A microcomputer simulation program for demonstrating the development of insecticide resistance. Bull. Entomol. Soc. Am. 30: 37–41.

Falconer, D. S. 1981. Introduction to quantitative genetics, 2nd ed. Longman, London.

Fournier, D., M. Pralavario, A. Cuany and J. Berge. 1988. Genetic analysis of methidathion resistance in *Phytoseiulus persimilis* (Acari: Phytosiidae). J. Econ. Entomol. 81: 1008–1013.

Georghiou, G. P. 1972. The evolution of resistance to pesticides. Annu. Rev. Ecol. Syst. 3: 133–168.

Georghiou, G. P. 1980. Insecticide resistance and prospects for its management. Residue Rev. 76: 131–145.

Georghiou, G. P. 1983. Management of resistance in arthropods, pp. 769–792. *In* G. P. Georghiou and T. Saito (eds), Pest resistance to pesticides. Plenum, New York.

Georghiou, G. P. 1986. The magnitude of the resistance problem, pp. 14–43. *In* Pesticide resistance: strategies and tactics for management. National Academy of Sciences, Washington, D.C.

Georghiou, G. P., and C. E. Taylor. 1977a. Genetic and biological influences in the evolution of insecticide resistance. J. Econ. Entomol. 70: 319–323.

Georghiou, G. P., and C. E. Taylor. 1977b. Operational influences in the evolution of insecticide resistance. J. Econ. Entomol. 70: 653–658.

Georghiou, G. P., and C. E. Taylor. 1986. Factors influencing the evolution of resistance, pp. 157–169. *In* Pesticide resistance: strategies and tactics for management. National Academy of Sciences, Washington, D.C.

Georghiou, G. P., R. B. March, and G. E. Printy. 1963. A study on the genetics of dieldrin-resistance in the housefly (*Musca domestica* L.). Bull. WHO 29: 155–165.

Georghiou, G. P., A. Lagunes, and J. D. Baker. 1983. Effect of insecticide rotations on evolution of resistance, pp. 183–189. *In* J. Miyamoto (ed.), IUPAC pesticide chemistry, human welfare and the environment, Pergamon, Oxford.

Greever, J., and G. P. Georghiou. 1979. Computer simulations of control strategies for *Culex tarsalis* (Diptera: Culicidae). J. Med. Entomol. 16: 180–188.

Guttierez, A. P., U. Regev, and H. Shalet. 1979. An economic optimization model of pesticide resistance: alfalfa and Egyptian alfalfa weevil—an example. Environ. Entomol. 8: 101–107.

Guttierez, A. P., U. Regev, and C. G. Summers. 1976. Computer model aids in weevil control. Calif. Agric. April: 8–9.

Haliscak, J. P., and R. W. Beeman. 1983. Status of malathion resistance in five genera of beetles infesting farm-stored corn, wheat, and oats in the United States. J. Econ. Entomol. 76: 717–722.

Halliday, W. R., and G. P. Georghiou. 1985. Inheritance of resistance to permethrin and DDT in the southern house mosquito (Diptera: Culicidae). J. Econ. Entomol. 78: 762–767.

Horn, D. J., and R. W. Wadleigh. 1988. Resistance of aphid natural enemies to insecticides, pp. 337–347. *In* A. K. Minks and P. Harrewijn (eds.), Aphids, their biology, natural enemies, and control, Vol. B. Elsevier, Amsterdam.

Hoy, M. A. 1985. Recent advances in genetics and genetic improvement of the Phytoseiidae. Annu. Rev. Entomol. 30: 345–370.

Hueth, D., and U. Regev. 1974. Optimal agricultural pest management with increasing pest resistance. Am. J. Agric. Econ. 56: 543–552.

Huffaker, C. B. 1971. The ecology of pesticide interference with insect populations, pp. 92–107. *In* J. E. Swift (ed.), Agricultural chemicals—harmony or discord for food, people, and the environment. Univ. Calif. Div. Agric. Sci. Public., Berkeley.

Imai, C. 1987. Control of insecticide resistance in a field population of houseflies, *Musca domestica*, by releasing susceptible flies. Res. Popul. Ecol. 29: 129–146.

Johnson, M. W., and B. E. Tabashnik. 1990. Enhanced biological control through pesticide selectivity. *In* T. W. Fisher et al. (eds.), Principles and application of biological control. University of California Press, Berkeley (in press).

Knight, A. L., and G. W. Norton. 1989. Economics of agricultural pesticide resistance in arthropods. Annu. Rev. Entomol. 34: 293–313.

Knipling, E. F., and W. Klassen. 1984. Influence of insecticide use patterns on the development of resistance to insecticides—a theoretical study. Southwest. Entomol. 9: 351–368.

Lande, R. 1983. The response to selection on major and minor mutations affecting a metrical trait. Heredity 50: 47–65.

Lazarus, W. F., and B. F. Dixon. 1984. Agricultural pests as common property: control of the corn rootworm. Am. J. Agric. Econ. 66: 456–465.

Leeper, J. R., R. T. Roush, and H. T. Reynolds. 1986. Preventing or managing resistance in arthropods, pp. 335–346. *In* Pesticide resistance: strategies and tactics for management. National Academy of Sciences, Washington, D.C.

Lenski, R. E. 1988a. Experimental studies of pleiotropy and epistasis in *Escherichia coli*. I. Variation in competitive fitness among mutants resistant to virus T4. Evolution 42: 425–432.

Lenski, R. E. 1988b. Experimental studies of pleiotropy and epistasis in *Escherichia coli*. II. Compensation for maladaptive effects associated with resistance to virus T4. Evolution 42: 433–440.

Levin, B. R., J. A. Barrett, E. C. Cruze, A. P. Dobson, F. Gould, J. H. Greaves, D. Heckel, R. M. May, H. T. Reynolds, R. T. Roush, B. E. Tabashnik, M. Uyenoyama, S. Via, M. J. Whitten, and M. S. Wolfe. 1986. Population biology of pesticide resistance: bridging the gap between theory and practical applications, pp. 143–156. *In* Pesticide resistance: strategies and tactics for management. National Academy of Sciences, Washington, D.C.

Lichtenburg, E., and D. Zilberman. 1986. The econometrics of damage control: why specification matters. Am. J. Agric. Econ. 68: 261–273.

Liu, M. Y., Y. J. Tseng, and C. N. Sun. 1981. Diamondback moth resistance to several synthetic pyrethroids. J. Econ. Entomol. 74: 393–396.

Longstaff, B. C. 1988. Temperature manipulation and the management of insecticide resistance in stored grain pests: a simulation study for the rice weevil, *Sitophilus oryzae*. Ecol. Modelling 43: 303–313.

MacDonald, G. 1959. The dynamics of resistance to insecticides by anophelines. Riv. Parassitol. 20: 305–315.

MacDonald, R. S., G. A. Surgeoner, K. R. Solomon, and C. R. Harris. 1983a. Effect of four spray regimes on the development of permethrin and dichlorvos resistance in the laboratory by the house fly (Diptera: Muscidae). J. Econ. Entomol. 76: 417–422.

MacDonald, R. S., G. A. Surgeoner, K. R. Solomon, and C. R. Harris. 1983b. Development of resistance to permethrin and dichlorvos by the house fly (Diptera: Muscidae) following continuous and alternating insecticide use on four farms. Can. Entomol. 115: 1555–1561.

Mangel, M., and R. E. Plant. 1983. Multiseasonal management of an agricultural pest. I. Development of the theory. Ecol. Modelling 20: 1–19.

Mani, G. S. 1985. Evolution of resistance in the presence of two insecticides. Genetics 109: 761–783.

Mani, G. S., and R. J. Wood. 1984. Persistence and frequency of application of an insecticide in relation to the rate of evolution of resistance. Pestic. Sci. 15. 325–336.

Mason, G. A., B. E. Tabashnik, and M. W. Johnson. 1989. Effects of biological and operational factors on evolution of insecticide resistance in *Liriomyza* (Diptera: Agromyzidae). J. Econ. Entomol. 82: 369–373.

Maudlin, I., C. H. Green, and F. Barlow. 1981. The potential for insecticide resistance in *Glossina* (Diptera: Glossinidae)—an investigation by computer simulation and chemical analysis. Bull. Entomol. Res. 71: 691–702.

May, R. M., and A. P. Dobson. 1986. Population dynamics and the rate of evolution of pesticide resistance, pp. 170–193. *In* Pesticide resistance: strategies and tactics for management. National Academy of Sciences, Washington, D.C.

McKenzie, J. A., M. J. Whitten, and M. A. Adena. 1982. The effect of genetic background on the fitness of diazinon resistance genotypes of the Australian sheep blowfly, *Lucilia cuprina*. Heredity 49: 1–9.

Muggleton, J. 1982. A model for the elimination of insecticide resistance using heterozygous disadvantage. Heredity 49: 247–251.

Muggleton, J. 1986. Selection for malathion resistance in *Oryzaephilus surinamensis* (L) (Coleoptera: Silvanidae): fitness values of resistant and susceptible phenotypes and their inclusion in a general model describing the spread of resistance. Bull. Entomol. Res. 76: 469–480.

Oppenoorth, F. J. 1985. Biochemistry and genetics of insecticide resistance, pp. 731–773. *In* G. A. Kerkut and L. I. Gilbert (eds.), Comprehensive insect physiology, biochemistry, and pharmacology, Vol. 12. Pergamon, Oxford.

Ozaki, K. 1983. Suppression of resistance through synergistic combinations with emphasis on planthoppers and leafhoppers infesting rice in Japan, pp. 595–613. *In* G. P. Georghiou and T. Saito (eds.), Pest resistance to pesticides. Plenum, New York.

Ozburn, G. W., and F. O. Morrison. 1963. The effect of diluting a colony of DDT resistant houseflies with non-resistant houseflies. Phytoprotection 44: 32–36.

Pedersen, O. C. 1984. Models of pesticide resistance dynamics. Acta Agric. Scand. 34: 145–152.

Pimentel, D., and A. C. Bellotti. 1976. Parasite–host population systems and genetic stability. Am. Nat. 95: 65–79.

Pimentel, D., and M. Burgess. 1985. Effects of single versus combinations of insecticides on the development of resistance. Environ. Entomol. 14: 582–589.

Plant, R. E., M. Mangel, and L. E. Flynn. 1985. Multiseasonal management of an agricultural pest II: The economic optimization problem. J. Environ. Econ. Man. 12: 45–61.

Plapp, F. W., Jr., C. R. Browning, and P. J. H. Sharpe. 1979. Analysis of rate of development of insecticide resistance based on simulation of a genetic model. Environ. Entomol. 8: 494–500.

Prasittisuk, C., and C. F. Curtis. 1982. Further study of DDT resistance in *Anopheles gambiae* Giles (Diptera: Culicidae) and a cage test of elimination of resistance from a population by male release. Bull. Entomol. Res. 72: 335–344.

Pree, D. J. 1987. Inheritance and management of cyhexatin and difocol resistance in the European red mite (Acari: Tetranychidae). J. Econ. Entomol. 80: 1106–1112.

Rawlings, P., and G. Davidson. 1982. The dispersal and survival of *Anopheles culicifacies* Giles (Diptera: Culicidae) in a Sri Lankan village under malathion spraying. Bull. Entomol. Res. 72: 139–144.

Rawlings, P., G. Davidson, R. K. Sakai, H. R. Rathor, M. Aslamkhan, and C. F. Curtis. 1981. Field measurement of the effective dominance of an insecticide resistance in anopheline mosquitos. Bull. WHO 59: 631–640.

Raymond, M., N. Pasteur, and G. P. Georghiou. 1987. Inheritance of chlorpyrifos resistance in *Culex pipiens* L. (Diptera: Culicidae) and estimation of the number of genes involved. Heredity 58: 351–356.

Riddles, P. W., and J. Nolan. 1987. Prospects for the management of arthropod resistance to pesticides. Int. J. Parasitol. 17: 679–688.

Rosenheim, J. A., and M. A. Hoy. 1986. Intraspecific variation in levels of pesticide resistance in field populations of a parasitoid, *Aphytis melinus* (Hymenoptera: Aphelinidae): the role of past selection pressures. J. Econ. Entomol. 79: 1161–1173.

Roush, R. T. 1989. Designing resistance management programs: how can you choose? Pestic. Sci. 26: 423–441.

Roush, R. T., and J. A. McKenzie. 1987. Ecological genetics of insecticide and acaricide resistance. Annu. Rev. Entomol. 32: 361–380.

Roush, R. T., R. L. Combs, T. C. Randolph, and J. A. Hawkins. 1986. Inheritance and effective dominance of pyrethroid resistance in the horn fly (Diptera: Muscidae). J. Econ. Entomol. 32: 361–380.

Sarhan, M. E., R. E. Howitt, and C. V. Moore. 1979. Pesticide resistance externalities and optimal mosquito management. J. Environ. Econ. Man. 6: 69–84.

Sawicki, R. M., and I. Denholm. 1987. Management of resistance to pesticides in cotton pests. Trop. Pest Manag. 33: 262–272.

Shoemaker, C. A. 1982. Optimal integrated control of univoltine pest populations with age structure. Oper. Res. 30: 40–61.

Sinclair, E. R., and J. Alder. 1985. Development of a computer simulation model of stored product insect populations on grain farms. Agric. Syst. 18: 95–113.

Sutherst, R. W., and H. N. Comins. 1979. The management of acaricide resistance in the cattle tick *Boophilus microplus* (Canestrini) (Acari: Ixodidae), in Australia. Bull. Entomol. Res. 69: 519–537.

Tabashnik, B. E. 1986a. Computer simulation as a tool for pesticide resistance management, pp. 194–206. *In* Pesticide resistance: strategies and tactics for management. National Academy of Sciences, Washington, D.C.

Tabashnik, B. E. 1986b. Model for managing resistance to fenvalerate in the diamondback moth (Lepidoptera: Plutellidae). J. Econ. Entomol. 79: 1147–1451.

Tabashnik, B. E. 1986c. Evolution of pesticide resistance in predator–prey systems. Bull. Entomol. Soc. Am. 32: 156–161.

Tabashnik, B. E. 1986d. Insect resistance. Science 234: 802.

Tabashnik, B. E. 1987. Computer-aided management of insecticide resistance, pp. 215–218. *In* Proc. 1987 Beltwide Cotton Production Research Conferences, National Cotton Council of America, Memphis.

Tabashnik, B. E. 1989. Managing resistance with multiple pesticide tactics: theory, evidence, and recommendations. J. Econ. Entomol. 82: 1263–1269.

Tabashnik, B. E., and B. A. Croft. 1982. Managing pesticide resistance in crop–arthropod complexes: interactions between biological and operational factors. Environ. Entomol. 11: 1137–1144.

Tabashnik, B. E., and B. A. Croft. 1985. Evolution of pesticide resistance in apple pests and their natural enemies. Entomophaga 30: 37–49.

Tabashnik, B. E., and N. L. Cushing. 1989. Quantitative genetic analysis of insecticide resistance: variation in fenvalerate tolerance in a diamondback moth (Lepidoptera: Plutellidae) population. J. Econ. Entomol. 79: 189–191.

Tabashnik, B. E., and M. W. Johnson. 1990. Evolution of pesticide resistance in natural enemies. *In* T. Fisher et al. (eds.), Principles and application of biological control, University of California Press, Berkeley (in press).

Tabashnik, B. E., N. L. Cushing, and M. W. Johnson. 1987. Diamondback moth (Lepidoptera: Plutellidae) resistance to insecticides in Hawaii: intra-island variation and cross-resistance. J. Econ. Entomol. 80: 1091–1099.

Taylor, C. E. 1983. Evolution of resistance to insecticides: the role of mathematical models and computer simulations, pp. 163–173. *In* G. P. Georghiou and T. Saito (eds.), Pest resistance to pesticides. Plenum, New York.

Taylor, C. E. 1986. Genetics and evolution of resistance to insecticides. Biol. J. Linn. Soc. 27: 103–112.

Taylor, C. E. 1989. On the use of more than one insecticide to control resistance: theory and computer simulation (in manuscript).

Taylor, C. E., and J. C. Headley. 1975. Insecticide resistance and the evolution of control strategies for an insect population. Can. Entomol. 107: 237–242.

Taylor, C. E., and G. P. Georghiou. 1979. Suppression of insecticide resistance by alteration of gene dominance and migration. J. Econ. Entomol. 72: 105–109.

Taylor, C. E., and G. P. Georghiou. 1982. Influence of pesticide persistence in evolution of resistance. Environ. Entomol. 11: 746–750.

Taylor, C. E., F. Quaglia, and G. P. Georghiou. 1983. Evolution of resistance to insecticides: a cage study on the influence of immigration and insecticide decay rates. J. Econ. Entomol. 76: 704–707.

Uyenoyama, M. K. 1986. Pleiotropy and the evolution of genetic systems conferring resistance to pesticides, pp. 207–221. *In* Pesticide resistance: strategies and tactics for management. National Academy of Sciences, Washington, D.C.

Via, S. 1986. Quantitative genetic models and the evolution of pesticide resistance, pp. 222–235. *In* Pesticide resistance: strategies and tactics for management. National Academy of Sciences, Washington, D.C.

Waage, J. K., M. P. Hassel, and H. C. J. Godfray. 1985. The dynamics of pest–parasitoid–insecticide interactions. J. Appl. Ecol. 22: 825–838.

Whitten, M. J., and J. A. McKenzie. 1982. The genetic basis for pesticide resistance, pp. 1–16. *In* K. E. Lee (ed.),. Proc. 3rd Australasian Conf. Grassland Invert. Ecol., South Aust. Gov. Print., Adelaide.

Wilson, E. O., and W. H. Bossert. 1971. A primer of population biology. Sinauer Associates, Sunderland, Mass.

Wood, R. J. 1981. Insecticide resistance: genes and mechanisms, pp. 53–96. *In* J. A. Bishop and L. M. Cook (eds.), Genetic consequences of man made change. Academic, New York.

Wood, R. J., and G. S. Mani. 1981. The effective dominance of resistance genes in relation to the evolution of resistance. Pestic. Sci. 12: 573–581.

Wool, D., and S. Noiman. 1983. Integrated control of insecticide resistance by combined genetic and chemical treatments: a warehouse model with flour beetles (Tribolium; Tenebrionidae, Coleoptera). Z. Agnew. Entomol. 95: 22–30.

7

The Effect of Agrochemicals on Vector Populations

George P. Georghiou

I. Introduction

Crop losses due to the action of herbivorous arthropods, parasitic fungi, nematodes, molluscs and noxious weeks have been estimated to represent at least one third of production (Cramer 1967). Losses from insects alone were stated to be from as low as 12% of potential production (Anonymous 1974) to several times that much (Pimentel et al. 1978). Since modern plant protection chemicals offer the most practical means of reducing crop losses, they are being used extensively throughout the world, and it is expected that demand for these will continue to rise as developing countries strive to increase their agricultural production and to improve their economic standards. The world market for pesticides in 1985 was estimated at $13,778 million and was expected to increase to $15,759 million by 1990 (Table 7.1) (Anonymous 1985).

Strange as it may seem, intensive efforts to control pests on crops by insecticides have in a number of cases diminished man's ability to control adequately insect vectors of human disease in the same environment. The ready availability of pesticides, often accompanied by inadequate controls, has led to excesses in their use, especially on nonfood crops. These abuses have not only complicated

Table 7.1. World Market of Pesticides (in millions of U.S. 1984 dollars)*

	1980	1985†	1990‡
Herbicides	4,891	6,331	7,183
Insecticides	3,916	4,268	4,815
Fungicides	2,199	2,537	2,947
Others	559	642	804
Total	11,565	13,778	15,759

*Data from Anonymous (1985).

†Estimated.

‡Projected.

agricultural pest control by the selection of resistant strains of pests and the suppression of beneficial insects, but also altered, in some cases profoundly, the susceptibility levels of vectors of human diseases.

Insects of medical importance, especially mosquitoes, are often found breeding in agricultural habitats and are hence exposed to the insecticides employed in agriculture. It is estimated that 90% of all insecticides produced by industry are used for agricultural purposes, with the treatment of cotton and rice receiving the greatest share of these chemicals in some countries (WHO 1986). Aerial application of insecticides, especially by ultralow volume, is known to result in some drift into surrounding areas, even under optimal meteorological conditions (Yates et al. 1978). These treatments may be expected to have a suppressive effect on mosquitoes in that environment. Such suppression, when occurring repeatedly, could lead to development of resistance to insecticides.

During the past several years, reports from various parts of the world have indicated that mosquito resistance has been more severe in areas where crops are treated frequently with insecticides (Busvine and Pal 1969, WHO 1976, 1980, 1986). Although some of the evidence is circumstantial, an increasing body of information points to a direct cause–effect relationship between the use of insecticides in agriculture and serious problems in mosquito control. The writer established a close correlation between the type and quantities of insecticides applied in cotton-growing areas of El Salvador and Nicaragua and resistance to insecticides in *Anopheles albimanus* (Georghiou 1972). The dilemma arising from this problem has been discussed in historical perspective by Garcia-Martin and Najera-Morrondo (1972); in economic and sociological terms, by Chapin and Wasserstrom (1981); and with reference to malaria resurgence, by Sharma and Mehrotra (1986). The subject has been reviewed by the writer on several occasions (Georghiou 1975, 1982) and by the WHO Expert Committee on Vector Biology and Control (WHO 1976, 1980, 1986). The present paper represents an extension and updating of the material of those earlier reviews.

II.   Toxicity of Agrochemicals to Vectors

Most agricultural insecticides are nonspecific; they are toxic to agricultural pests as well as to vectors and their predators. The data in Table 7.2 indicate that some widely used agricultural insecticides, such as parathion, methyl parathion, DDT, and dieldrin, are highly toxic to mosquito larvae. In fact, the toxicity of parathion and dieldrin exceeds that of chlorpyrifos, the most commonly used mosquito larvicide. Numerous data in the published literature confirm the considerable potential of agricultural insecticides to suppress mosquitoes, larvae as well as adults, and their predators, through direct toxicity (reviews by Mulla et al. 1979, Mulla and Mian 1981).

Other pesticides (e.g., herbicides, fungicides, and molluscicides) appear to possess limited toxicity to insects at the concentrations in which they are employed in the field (Georghiou et al. 1974b). However, certain fungicides and herbicides

Table 7.2. Relative Toxicity of Various Insecticides to Larvae of *Anopheles albimanus**

	LC_{50} (mg/ml)
Propoxur	0.39
Carbaryl	0.89
Malathion	0.085
Parathion	0.0031
Methyl parathion	0.0065
Fenthion	0.023
Fenitrothion	0.025
Dichlorvos	0.11
DDT	0.01
Dieldrin	0.003
Chlorpyrifos	0.006
Temephos	0.011

*Data from Georghiou (1970), Ariaratnam and Georghiou (1971).

are synergistic of insecticides against mosquito larvae. For example, the fungicide kitazin, which is used against rice blast in the Far East, synergizes organophosphates, making them more effective against resistant strains (Hemingway and Georghiou 1984). The herbicide diquat and related one-electron transfer agents were found to produce significant synergism of propoxur and fenthion to *A. albimanus* and *Culex quinquefasciatus,* possibly by inhibition of microsomal detoxication systems. However, the concentrations of herbicide needed for synergism far exceed those applied for weed control (Georghiou et al. 1974b), and herbicides and insecticides are seldom applied together.

Entomopathogenic fungi are capable of limiting the population levels of a number of pest species. Fungicides affect only minimally the action of entomopathogens. Livingston et al. (1978) showed that benomyl and other fungicides at recommended dosages did not suppress significantly an epizootic of *Entomophthora gammae* on several lepidopterous pests of soybeans. The incidence of *Nomuraea rileyi* was suppressed in a low-density population of *Anticarsia gemmatalis* (Lepidoptera), but when larval density increased, no difference was detected in the incidence of *N. rileyi* infection in treated and untreated plots (Livingston et al. 1978).

The impact of agricultural fungicides on fungi that are parasitic on vectors does not appear to have been adequately investigated. Thus, the following discussion applies only to agricultural insecticides and their role in the selection of insecticide-resistant mosquito populations.

III. Cases of Vector Resistance Due to Agrochemical Applications

A partial list of cases of insecticide resistance in mosquitoes alleged to have been caused or aggravated by agricultural insecticides is given in Table 7.3. The large number of *Anopheles* species may be due to the fact that the susceptibility of anophelines has been followed more closely in connection with malaria control programs. It is suspected, however, that many other cases also exist, especially in *Culex* and *Aedes,* that remain unexamined. It is of interest to note that the majority of cases on record involve dieldrin or DDT, undoubtedly because resistance to dieldrin is known to develop readily and is easily detected in heterozygotes, and because DDT has enjoyed wider and longer use than other classes of insecticides. However, selection by organophosphates (OP), carbamates, or both has also occurred (e.g., *A. albimanus, A. sinensis, A. sacharovi, A. stephensi, C. quinquefasciatus, C. pipiens, Ae. nigromaculis,* etc.). Cases of resistance in pests of man other than mosquitoes, attributable to agricultural application of insecticides are rare. Body lice in Burundi were reported to have developed resistance to malathion as a result of contamination of workers treating coffee trees and of the storage of insecticides in houses, though it was observed that in some cases agricultural insecticides had been used for lice control (WHO 1986).

IV. Evidence for Implication of Agrochemicals in Vector Resistance

The evidence for implication of agricultural insecticides in vector resistance is in most cases indirect and circumstantial. Many of the available reports point out either that the problem of vector resistance exists in agricultural areas that are heavily treated with insecticides or that mosquito populations are more resistant in agricultural than in nonagricultural areas even when both areas have received an equal number of treatments by public health authorities. Although some of this information may be convincing enough, definitive evidence comes only from a small number of documented cases, especially those involving *A. albimanus* in Central America (Georghiou et al. 1971, 1973, 1974a; Georghiou 1972; Hobbs 1973; Ariaratnam and Georghiou 1974, 1975; Ayad and Georghiou 1975, 1979; Bailey et al. 1981). In this case, most of the relevant evidence was obtained during the early stages of emergence of OP and carbamate resistance (1970–1973), that is, during the "genesis" of resistance and before its spread beyond "focal" areas. The geographical distribution of *A. albimanus* and the occurrence of resistance to DDT, organophosphates, and carbamates in 1985 are shown in Figure 7.1.

The evidence of the implication of agricultural insecticides is discussed under the following categories:

A. Appearance of vector resistance prior to the application of chemicals against the vector.

B. Higher resistance of vectors in agricultural areas than in nonagricultural.

C. Correlation between intensity of insecticide use on crops and degree of resistance in vectors.

Table 7.3. Cases in Resistance to Insecticides in Mosquitoes Presumed to Have Been Caused or Aggravated by Indirect Selection Pressure by Agricultural Insecticides

Species	Country/area	Crop	Insecticide Resistance	Reference
Anopheles aconitus	Indonesia (Java)	various, rice	dieldrin, DDT	12 43
A. albimanus	El Salvador, Nicaragua	cotton, rice	parathion, methyl parathion, malathion, fenitrothion, propoxur, carbaryl	20, 24, 25
	El Salvador	cotton	DDT	46
	Mexico, Guatemala, El Salvador, Honduras, Nicaragua	cotton	DDT, dieldrin	various (in 12)
A. culicifacies	India (Andhra Pradesh, Madhya Pradesh)	various	malathion	54, 55
A. gambiae	Ivory Coast	coffee, cocoa	dieldrin	29
	Nigeria	ground nuts	dieldrin	16
	Ghana	cocoa	dieldrin	Coker 1956 (in 29)
	Mali	cotton	dieldrin	12
	Burkina Faso	cotton	DDT	Hamon et al. 1968 (in 12)
	Sudan, Ethiopia, Togo, Senegal	various	DDT	55
A. maculipennis	Romania, Turkey	various	dieldrin	Duport, 1965* (in 12, 47, 48)
A. melanoon subalpinus	Turkey	various	dieldrin	47
A. melas	Zaire	bananas	DDT	30
A. pharoensis	Egypt	cotton	dieldrin	Zahar and Thymakis, 1962* (in 12)
			DDT	Zahar, 1965* (in 12)
	Sudan	various	dieldrin, DDT	32
A. quadrimaculatus	USA	cotton	dieldrin	39
	Mexico	cotton	DDT, dieldrin	38
A. rufipes	Mali	cotton	dieldrin	Hamon, 1968* (in 12)
A. sacharovi	Greece, Turkey	cotton, rice	DDT, dieldrin	11, 47, 48, 61
A. sinensis	China	rice	DDT, malathion	2
A. stephensi	Pakistan	cotton	malathion	23
Aedes aegypti	Tahiti	coconut	dieldrin	40
Ae. nigromaculis	USA	various	DDT, dieldrin, OP	various (in 12)
Culex quinquefasciatus	USA	various	OP	28
Culex tritaeniorhynchus	Rep. of Korea, Japan	rice	several	58

*Communication to WHO.

187

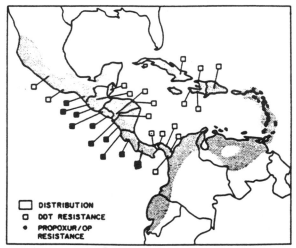

Figure 7.1. Distribution of *Anopheles albimanus* and resistance to DDT and propoxur/organophosphates (WHO 1980).

D. Fluctuations in resistance of vectors in parallel with periods of agricultural spraying.

E. Correspondence between the spectrum of resistance of vectors and types of insecticides applied to crops.

F. Temporary suppression of mosquito population densities following application of agricultural sprays.

A. Appearance of Vector Resistance Prior to the Application of Chemicals Against Vectors

Dieldrin resistance was present in *A. sacharovi* and *A. maculipennis* in Turkey in many agricultural areas in which neither BHC nor dieldrin had been used for vector control (Ramsdale 1973). In the delta area of Egypt, where cotton was treated extensively with toxaphene and DDT, marked resistance to dieldrin and incipient resistance to DDT were noted in *Anopheles pharoensis* in 1959 prior to the commencement of residual house spraying (Zahar and Thymakis, unpublished report to WHO, 1962). Similar occurrences of resistance were reported for dieldrin in *Anopheles maculipennis* in Romania (M. Duport, unpublished report to WHO, 1965), *Anopheles aconitus* in the Malang district of East Java (Brown and Pal 1971), in *Anopheles gambiae* in Mali (Hamon et al. 1961), and in the lower Volta region of Ghana (W. Z. Coker, cited in Hamon and Garrett-Jones, 1963).

B. Higher Vector Resistance in Agricultural than in Nonagricultural Areas

Frequently cited are reports of the presence of mosquito resistance in areas with industrialized agriculture and its absence in areas with subsistence agriculture, although both had received residual house spraying. Thus, in Greece, Belios (1961) found higher dieldrin and DDT resistance in *A. sacharovi* in the cotton and rice growing area of Laconia than in Etolia and Euboea, a fact he attributed to strong selection pressure on the larvae by agricultural insecticides.

In Turkey, Ramsdale (1973, 1975) pointed out that, although DDT had been widely used for more than 20 years in public health and agriculture, the incidence of DDT resistance was not related to the duration of DDT house spraying operations: More than 10 years of regular house treatment had not affected the susceptibility of *A. sacharovi* or *A. maculipennis* in the southeastern part of the country. However, remarkable DDT resistance had developed in the cotton-growing district of Manan, Adana (M. H. Holstein, cited by Brown and Pal 1971). Similarly, in Pakistan it was noted that the incidence of malathion resistance in *A. stephensi* coincides to a large degree with the geographic distribution of cotton cultivation (Fig. 7.2, Georghiou 1986).

Considerable differences in the susceptibility to OP and carbamate insecticides in field strains of *A. albimanus* from different areas were reported by Georghiou

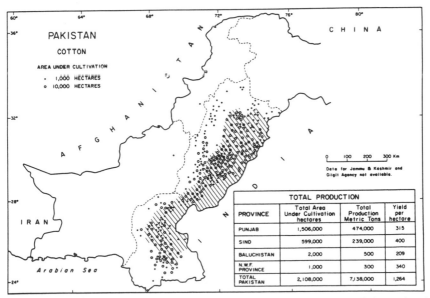

Figure 7.2. Distribution of cotton cultivation and malathion resistance in *Anopheles stephensi* in Pakistan (Georghiou 1986). (Shaded area: malathion resistance; circles and dots: area under cotton cultivation.)

et al. (1971): Strains from Haiti, where agricultural use of these insecticides was minimal, were of normal susceptibility, whereas strains from the cotton- and rice-growing areas to El Salvador showed remarkable levels of OP and carbamate resistance. Such differences were apparently not due to the extreme geographical separation of these populations, since a strain from the isolated area of Texisti-peque, Santa Ana, El Salvador, which had not experienced commercial use of OP or carbamate insecticides, was equally susceptible to propoxur and DDT, as was the strain from Haiti (Georghiou 1972).

C. Correlation Between Intensity of Insecticide Use on Crops and Degree of Resistance in Vectors

The implication of the involvement of agricultural insecticides in the development of resistance by mosquitoes was enhanced by the demonstration in 1971 of correlations between the intensity of pest control operations on cotton and rice in areas of El Salvador and Nicaragua and the degree of OP and carbamate resistance (Georghiou 1972). In these areas cotton crops are treated with insecticides at frequent intervals, with as many as 30 applications being made during the six-month growing season. As a result, resistance of cotton pests to chemicals has been an extremely serious problem. According to Smith (1968), problems of resistance to insecticides in cotton insects in Central America "have usually started first in El Salvador."

In the course of an extensive field survey of susceptibility of *A. albimanus* in Nicaragua, Honduras, El Salvador, and Guatemala in 1971, the writer (Georghiou 1972) obtained evidence of significant resistance to carbamates (propoxur) and to OP's (malathion and parathion) in the Department of La Paz, El Salvador, and in the Sebaco Valley, Nicaragua. Lower resistance levels, or absence of resistance, were obtained elsewhere in these countries (Georghiou 1972). Examination of the available agronomic and pest control information in El Salvador indicated that the higher resistance in La Paz was in agreement with the more intensive chemical pest control that was practiced on cotton and rice in that department as compared to other departments of the country. No information could be obtained on the quantities of insecticides used per acre in each department. However, calculations from data provided by the Ministry of Agriculture indicated that 26% of the country's cotton acreage was found in the Department of La Paz. Here the average holding per cotton grower was 81 hectares as compared to 23 hectares per grower in the remainder of the country. There were indications that the larger the holding, the greater the tendency to apply insecticide treatments on a fixed schedule rather than discriminately when and where needed. Eighty-seven percent of the cotton acreage of El Salvador was treated by aircraft, a practice that may result in frequent contamination of mosquito breeding habitats. In the Department of La Paz, 95% of the acreage was treated by air, and approximately one fifth of this was treated by ultra-low-volume (ULV) sprays.

As with cotton, rice cultivation in El Salvador was also more intensive in La Paz. During 1969–1970, 53.6% of the rice acreage of the country was in the central part of the coastal plain, including La Paz, as compared to 25.5% in the western and 20.9% in the eastern parts.

Intensive use of insecticides, especially on rice and cotton, was also practiced in Nicaragua. Complete reliance on chemical control of rice pests led to a substantial increase in the number of insecticide applications made and in the variety of compounds used, as the rice pests were being made resistant to a wider and wider range of compounds. On one large farm at La Concepcion, Nicaragua, on which rice growing began in 1963, the following treatments had been applied up to 1972 (G. P. Georghiou, Report to Pan American Health Organization, Washington, DC):

Year	Insecticide Applications	Total
1963	none	0
1964	carbaryl (4)	4
1965	carbaryl (6)	6
1966	carbaryl (3); carbaryl + methyl parathion (4)	7
1967	monocrotophos (2); carbaryl + methyl parathion (7)	9
1968	monocrotophos (3); endrin (2); disulfoton ethyl/methyl parathion (2)	8
1969	perthane (1); monocrotophos (2); endrin (2); naled (3)	8
1970	perthane (1); naled (2); "benfucarb" (3); methamidophos (2)	8
1971	naled (3); "benfucarb" (3); methamidophos (3); diazinon (1)	10
1972	changed to sorghum	

As in the case of OP and carbamate resistance, DDT resistance has occurred at higher levels in the cotton-growing areas of Nicaragua, as indicated in Figure 7.3.

Additional evidence of the impact of agricultural insecticides is revealed by a study of OP multiresistance in populations of *C. quinquefasciatus* in California (Georghiou et al. 1975). Strains collected from dairy waste drains of two farms located six miles apart in the intensely agricultural Central Valley revealed significant differences in the levels of OP resistance. Both breeding sites had experienced similar larvicidal treatments that, in 1973 and 1974, consisted exclusively of chlorpyrifos. However, examination of the official records of agricultural insecticide applications within a three-mile radius of each farm indicated that during 1971–1974 approximately twice as large a quantity of OP and carbamate insecticides had been applied in the area of the more resistant population.

D. Fluctuations of Vector Resistance in Parallel with Periods of Agricultural Spraying

A study conducted over a two-year period (1970–1972) in the cotton-growing area of El Salvador indicated that the susceptibility levels of *A. albimanus* showed seasonal fluctuations in parallel with the period of spray applications to cotton.

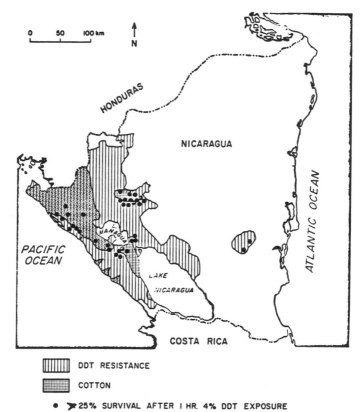

Figure 7.3. Geographical distribution of DDT resistance in *Anopheles albimanus* and area under cotton cultivation, Nicaragua, 1970 (Georghiou 1982).

Sampling was done in June and February of each year, that is, at the beginning and end of the cotton-growing season. Resistance to parathion, methyl parathion, malathion, fenitrothion, carbaryl, and propoxur was found to rise during the spray period and to decline somewhat during the nonspray period (Fig. 7.4). With reference to a susceptible strain, the resistance levels observed (at the LC_{50}) were malathion 117×, parathion 158×, methyl parathion 144×, fenitrothion 45×, propoxur >1,000×, and carbaryl 443× (Georghiou et al. 1973).

E. Correspondence Between the Spectrum of Resistance in Vectors and Types of Insecticides Applied to Crops

The El Salvador studies of 1970–1973 have also indicated that the spectrum of multiresistance in *A. albimanus* can be traced to the types of insecticides applied to cotton. Parathion and methyl parathion were the principal insecticides used on this crop for over a decade. Since the geographical distribution of *A. albimanus*

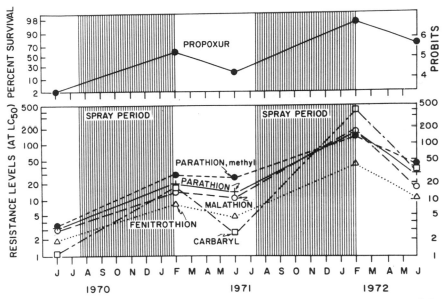

Figure 7.4. Fluctuations in resistance levels toward organophosphates and carbamates in *Anopheles albimanus* with reference to alternating agricultural spray and nonspray periods, El Salvador, 1970–1971 (Georghiou et al. 1973).

in this country coincides to a large extent with the agricultural area, data on insecticide imports into the country were used as an approximate indicator of the degree of selection pressure contributed by each chemical. Calculations from official records revealed that during the 10-year period 1961–1970, 51.22% of the insecticides imported were OP's; 46.37%, organochlorines, and 2.41%, carbamates. Methyl parathion and parathion constituted the bulk of the OP's (93.3%), whereas carbaryl was the most common carbamate (88.1%) (Georghiou 1972, Georghiou et al. 1973).

It is obvious that methyl parathion and parathion have had the greatest impact on *A. albimanus,* as reflected by the high resistance levels to these compounds. The elevated resistance to malathion may have resulted from the relatively limited malathion treatments, with additional selection of malathion-resistant genotypes by other OP's. Carbamate resistance may be the consequence of carbaryl applications and to a lesser extent of propoxur, such resistance being supported and enhanced further by OP selection pressure. This suggestion has been validated by biochemical tests, which revealed that OP (parathion) and carbamate (propoxur) resistance in this mosquito is due to the selection of a variant acetylcholinesterase that is far less sensitive to inhibition by these compounds (Ayad and Georghiou 1975, Hemingway and Georghiou 1983). This property is encoded by a single gene that may also be responsible for the high resistance toward the other OP's and carbamates (Georghiou et al. 1975). Results obtained on *A. albimanus* (Ayad and Georghiou 1975) and *C. pipiens* (Raymond et al. 1986) indicate that the mechanism of reduced sensitivity of acetylcholinesterase confers higher resistance toward carbamates than organophosphates.

F. Temporary Suppression of Vector Population Densities in Sprayed Areas

That agricultural applications do indeed suppress the density of mosquito populations was clearly demonstrated by Hobbs (1973) in El Salvador. Adult *A. albimanus* density fluctuations were measured weekly from February through December, 1972, within a 100-square-kilometer cotton-growing area, and in a comparable area well removed from cotton fields. Adult densities began to build up in May in both areas following the onset of the rainy season. However, whereas in the noncotton area the density remained relatively high throughout the rainy season, in the cotton area it declined abruptly in mid-August and remained low until December. This decline coincided with the increased application of insecticides to cotton as revealed by the number of flights of pest control aircraft per week (Fig. 7.5). In the same study, 16 larval breeding sites, representing the principal larval habitats within the cotton area, were negative at the end of August and remained so until December. The single exception that remained continuously positive was a cattle watering pond, situated the farthest from cotton fields (Hobbs 1973).

As was to be expected, continued severe selection pressure eventually resulted in the evolution of such high resistance that the mosquito population could no longer be suppressed effectively by agricultural sprays. This was demonstrated by a study conducted five years later in the same area of El Salvador (Bailey et al. 1981). Standardized collections were made at weekly intervals from June 18, 1977, to March 25, 1979, at several sites within the cotton area as well as at sites located at least 1 kilometer away from cotton fields (Figs. 7.6, 7.7). The data show that the adult mosquito population within the cotton area was suppressed only mildly by the first treatments and that its density recovered steadily, so that by the end of the spraying season it had reached a level similar to that observed in the non–cotton area. Roughly similar results were obtained by larval counts. It was thus obvious that through continued yearly selection and intermingling of the population of the sprayed and nonsprayed areas a uniformly resistant population had evolved whose density could no longer be drastically suppressed by the agricultural pest control treatments. A similar situation was reported for *C. tritaeniorhynchus* in the Republic of Korea and in Japan. Here widespread use of insecticides on rice was introduced in 1966. This led to good control of the species up to the end of the 1970s, when it developed resistance to all insecticides used for agricultural purposes. As a result, populations of this species could no

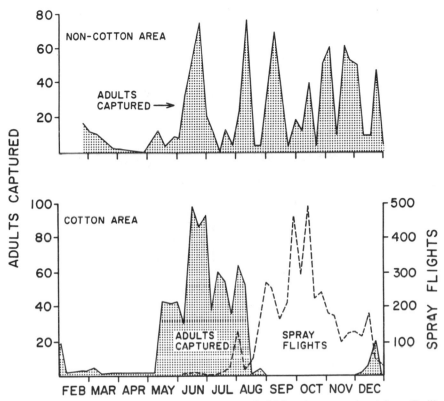

Figure 7.5. Adult densities of *Anopheles albimanus* in Melara (cotton area) and Santa Emilia (noncotton area), El Salvador, with records of cotton spray flights by week, 1972. (Data from Hobbs 1973.)

longer be controlled by agricultural pesticide applications (WHO 1986).

V. Discussion

The available evidence leaves little doubt that agricultural insecticides, when applied over a wide area, are capable of exerting strong selection pressure against mosquito populations. Such selection could be the result of decimation of the adult population, suppression of larvae by contamination of breeding habitats, or both. The consequences of such selection have been shown to be commensurate with the extent and frequency with which such exposure occurs in a given area.

In the most serious case studied, El Salvador (as well as throughout the Pacific coastal zone of Nicaragua, Honduras, Guatemala, and southwestern Mexico), selection was shown to occur with regularity from August through December, as

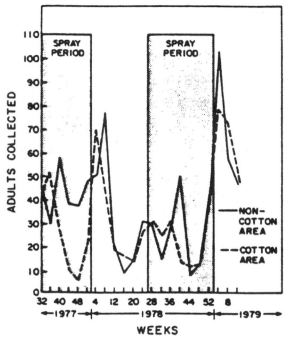

Figure 7.6. Adult densities of *Anopheles albimanus* in cotton-
and non–cotton-growing areas in El Salvador, 1977–1979 (data
from Bailey et al. 1981).

evident in the large number of spray flights carried out and the nearly complete
suppression of mosquito populations during that period. Such selection resulted
in the development of resistance, which was found to rise annually to higher
levels in concert with the seasonal spray–nonspray periods. Resistance was
quantitatively congruent with the intensity of agricultural operations in each area
and qualitatively indicative of the types of compounds that were employed in the
largest quantities in agriculture. The resultant multiresistance has considerably
reduced the efficacy of residual applications of propoxur, malathion, and fenitrothion with a concomitant resurgence of malaria transmission.

Continued selection of a vector population by agricultural sprays, over several
years, may eventually result in the population being fairly homogeneously resistant over a sufficiently large area so that population immigration no longer causes
significant fluctuations in resistance. At this stage, the advantage of agricultural
sprays (i.e., the temporary suppression of the vector population, and hence
reduction in the incidence of transmission), is reduced.

The possibility that in certain cases the development of resistance in mosquitoes
may be caused entirely or in part by the application of insecticides in public health
programs cannot be excluded. For example, it has been suggested that in the

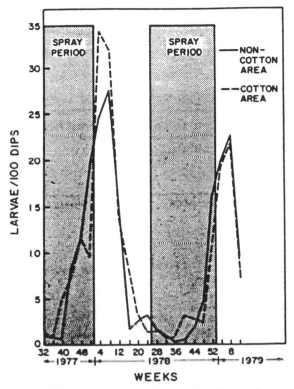

Figure 7.7. Larval densities of *Anopheles albimanus* in cotton- and non–cotton-growing areas in El Salvador, 1977–1979 (data from Bailey et al. 1981).

Sudan, *Anopheles arabiensis* became resistant in the adult stage specifically to malathion, which is used in house spraying, and not to the OP's used on cotton crops (WHO 1986). In most cases, however, residual house spraying alone, as applied for malaria control, exerts selection pressure on no more than 50% of the population, even under a rigorous program, since mostly female mosquitoes enter houses. This proportion would be even smaller in rural, sparsely populated areas, and where the vector is partly exophagic, as with *A. albimanus*. Under these conditions, the possibility of resistance due to these treatments alone is reduced. In India, *Anopheles minimus,* a highly endophilic species that is not directly exposed to the impact of agricultural insecticides has not developed resistance after 20 years of house spraying (WHO 1986). Where larvicides are used, especially in combination with adulticides, a higher degree of selection pressure is exerted, particularly if the same or related chemicals are used. Exposure to both agricultural and public health treatments enhances further the prospects for resistance development. In the case of *C. quinquefasciatus* referred to earlier, it was concluded that the large variety of insecticides applied to crops had predisposed the population to respond readily to the specific selection pressure applied by chlorpyrifos at the breeding sites.

In searching for alternatives to alleviate this problem, two points of concern must be borne in mind. First, in many developing countries, the introduction and use of insecticides in agriculture remains unrestricted and unregulated. New compounds are often introduced before they have been full tested and licensed in developed countries. In contrast, their use in official public health projects, especially for residual house spraying, is preceded by exhaustive testing, requiring a number of years for completion (e.g., WHO Insecticide Testing Program, Wright 1971). Since the same compounds are usually candidates for both agriculture and public health, their earlier availability and use in agriculture jeopardizes their subsequent effectiveness against mosquitoes, because of the resulting resistance. Addressing itself to the broader issue of pesticide marketing in developing countries, FAO prepared an International Code of Conduct on the Distribution and Use of Pesticides that was adopted by a FAO international conference in 1985 (FAO 1986). This voluntary code does not resolve the prioritization question, but serves as a point of reference until such time as developing countries have established adequate national regulatory infrastructures to cope with the problem.

A second point of concern is the exposure of mosquitoes to the multitude of insecticides applied in agriculture. Exposure to compounds of varied chemical structure appears to select in favor of several pathways of detoxication (or "site insensitivity"), thus preparing the mosquito population to respond more readily to the specific selection that is subsequently applied against it.

In view of the dilemma imposed by the indirect exposure and selection of vectors by agrochemicals, a number of alternatives have been considered on various occasions. These include:

1. Collaboration between agricultural and public health agencies and industry, for the purpose of identifying chemicals with unique modes of action that offer the greatest promise of prolonged effectiveness against vector mosquitoes, and reserving these primarily for public health use.

To be effective, any agreement of this nature must be established at the highest level of authority, preferably among FAO, WHO, and industry, and must be implemented by governments.

The discovery of *Bacillus thuringiensis israelensis* (BTI) offers an opportunity for testing the feasibility of such measures. In this connection it should be pointed out that current biotechnology research aimed at broadening the spectrum of toxicity of *B. thuringiensis* so as to include both agricultural pests and vectors would appear to be undesirable from the resistance point of view. Also undesirable would be the introduction of a single toxin-producing gene of BTI into algae that are found in mosquito breeding habitats as a means of suppressing mosquito production. There is a growing concern that unless the entire complement of BTI toxin genes (now known to be at least four, encoding the 135, 125, 65 and 28 KD proteins) is utilized, the risk for rapid development of resistance would be high.

2. Reduction of indirect selection against mosquitoes by agricultural insecticides by the introduction of comprehensive pest management practices.

It is realized that agricultural exports provide a considerable share of the foreign exchange earnings of developing countries. Thus, a reduction in the extent of insecticide usage can be expected only when alternative effective pest control measures have been convincingly demonstrated to the growers. For a discussion of progress and of the practical considerations in this area, the reader is referred to the proceedings of a recent WHO/FAO/UNEP conference (Anonymous 1987). Since specific research is necessary to provide comprehensive pest management programs for each crop, progress in this area may be slow. Nevertheless, it is believed that the present excesses in insecticide usage can be reduced through closer collaboration between the public health and agricultural services and by an intensive program of research in agricultural pest management.

3. Greater emphasis on the use of comprehensive mosquito control measures.

The need for supplementing insecticide applications with other measures designed to reduce mosquito populations is widely acknowledged (Garcia-Martin and Najera-Morrondo 1972, Mulhern 1972, Schliessman 1974). Such measures could contribute to reduction of mosquito population densities below threshold levels, at least during part of the season, thereby reducing the number of insecticide applications needed. Since much mosquito breeding occurs in irrigation and runoff water, the cooperation of agriculture in effecting the necessary engineering improvements (Junkert and Townzen 1973) (e.g., drainage) and/or agronomic modifications (e.g., intermittent vs. continuous irrigation of rice fields) would be required.

VI. Summary

Reports implicating agricultural insecticides in the development or intensification of resistance in mosquitoes are available from different areas of the world, especially where large acreage is devoted to the cultivation of cotton or rice. The evidence is particularly convincing in the Pacific coast zone of Central America where multiple resistance in the vector Anopheles albimanus toward organochlorine, organophosphorus and carbamate insecticides has impeded malaria eradication efforts.

Evidence of the implication of agricultural insecticides consists of (1) appearance of mosquito resistance prior to application of chemicals against vectors,

(2) higher mosquito resistance in agricultural than non-agricultural areas, (3) correlation between intensity of insecticide use on crops and degree of resistance in mosquitoes, (4) fluctuations of mosquito resistance in parallel with periods of agricultural spraying, (5) correspondence between spectrum of mosquito resistance and types of insecticides applied to crops, and (6) temporary suppression of mosquito population densities in sprayed areas.

Measures considered for coping with the problem include (1) collaboration between appropriate international agencies and governments for the purpose of identifying chemicals with unique modes of action which offer the greatest promise of prolonged effectiveness against vector mosquitoes, and reserving these primarily for public health use, (2) introduction of comprehensive pest management practices in agriculture, and (3) greater emphasis on supplemental mosquito control measures.

References

1. Anonymous. 1974. A Hungry World: The Challenge to Agriculture. Summary Report by University of California Food Task Force. Division of Agric. Sciences, Univ. of Calif., Berkeley 68 pp.
2. Anonymous. 1980. Report on the WHO Technical Visit on Vector Biology and Control to the People's Republic of China. WHO mimeo. *CHN/VBC/001*, 47 pp.
3. Anonymous. 1985. A look at world pesticide markets. *Farm Chemicals* 148: 26–34.
4. Anonymous. 1987. Report of Seventh Meeting, WHO/FAO/UNEP Panel of Experts on Environmental Management for Vector Control (PEEM), World Health Organization, VBC/87.2, 72 pp.
5. Ariaratnam, V. and G. P. Georghiou. 1971. Selection for resistance to carbamate and organophosphorus insecticides in *Anopheles albimanus*. *Nature* 232: 642–644.
6. Ariaratnam, V. and G. P. Georghiou. 1974. Carbamate resistance in *Anopheles albimanus:* cross resistance spectrum and stability of resistance. *Bull. WHO* 51: 655–659.
7. Ariaratnam, V. and G. P. Georghiou. 1975. Carbamate resistance in *Anopheles albimanus:* penetration and metabolism of carbaryl in propoxur-selected larvae. *Bull. WHO* 52: 91–96.
8. Ayad, H. and G. P. Georghiou. 1975. Resistance to organophosphates and carbamates in *Anopheles albimanus* based on reduced sensitivity of acetylcholinesterase. *J. Econ. Entomol.* 69: 295–297.
9. Ayad, H. and G. P. Georghiou. 1979. Resistance pattern of *Anopheles albimanus* Wied. following selection by parathion. *Mosquito News* 39:121–125.
10. Bailey, D. L., P. E. Kaiser, and R. E. Low. 1981. Population densities of *Anopheles albimanus* adults and larvae inside and outside cotton-growing areas in El Salvador. *Mosquito News* 41:151–154.
11. Belios, G. D. 1961. WHO Unpublished Working Paper *WHO/Mal.* 307.
12. Brown, A. W. A. and R. Pal. 1971. Insecticide Resistance in Arthropods. *WHO Monograph* Ser. 38, 491 pp. Geneva, Switzerland.
13. Busvine, J. R. and R. Pal. 1969. The impact of insecticide resistance on control of vectors and vector-borne diseases. *Bull. WHO* 40:371–444.
14. Chapin, G. and R. Wasserstrom. 1981. Agricultural production and malaria resurgence in Central America and India. *Nature* (London) 293: 181–185.
15. Cramer, H. H. 1967. Plant protection and world crop production. *Pfl. Nachr. Bayer.* 20: 1–524.
16. Elliott, R. 1959. Insecticide resistance in populations of *Anopheles gambiae* in West Africa. *Bull. WHO* 20: 777–796.
17. FAO. 1986. International code of conduct on the distribution and use of pesticides. Food and Agriculture Organization of the United Nations, Rome, 28 pp.

18. Garcia-Martin, G. and J. A. Najera-Morrondo. 1972. The interrelationships of malaria, agriculture and the use of pesticides in malaria control. *Bol. Ofic. Sanit. PanAmer.* 6: 15–23.
19. Georghiou, G. P. 1970. Considerations on the relationship of larval and adult tolerance to insecticides in mosquitoes. *Proc. & Papers, Calif. Mosquito Control Assoc.* 38: 55–59.
20. Georghiou, G. P. 1972. Studies on resistance to carbamate and organophosphorus insecticides in *Anopheles albimanus*. *Am. J. Trop. Med. Hyg.* 21: 797–806.
21. Georghiou, G. P. 1975. Implications of agricultural insecticides in the development of resistance by mosquitoes. *World Health Organization VBC/EC/75.3*, 13 p. (Also in Proc. UC/AID Conference *The Agromedical Approach to Pesticide Management"* Guatemala City, Jan. 1976. Univ. of Calif., Berkeley, mimeo, pp. 24–41).
22. Georghiou, G. P. 1982. The implication of agricultural insecticides in the development of resistance by mosquitoes with emphasis on Central America. pp. 95–121. *In* Resistance to Insecticides Used in Public Health and Agriculture. *Proc. Int. Workshop, Colombo, Sri Lanka. Natl. Science Council, Sri Lanka.*
23. Georghiou, G. P. 1986. A review of insecticide resistance in malaria vectors in Pakistan and recommendations for future action. Unpubl. Report to U.S. Agency for International Development, Washington, D.C., 44 pp.
24. Georghiou, G. P., V. Ariaratnam, and S. G. Breeland. 1971. *Anopheles albimanus:* Development of carbamate and organophosphorus resistance in nature. *Bull. WHO.* 46:551–554.
25. Georghiou, G. P., S. G. Breeland, and V. Ariaratnam. 1973. Seasonal escalation of organophosphorus and carbamate resistance in *Anopheles albimanus* by agricultural sprays. *Environ. Entomol.* 2: 369–374.
26. Georghiou, G. P., V. Ariaratnam, H. Ayad, and B. Betzios. 1974a. Present status of research on resistance to carbamate and organophosphorus insecticides in *Anopheles albimanus*. *WHO/VBC/74.508*, 9 pp.
27. Georghiou. G. P., A. L. Black, R. I. Krieger, and T. R. Fukuto. 1974b. Joint action of diquat and related one-electron transfer agents with propoxur and fenthion against mosquito larvae. *J. Econ. Entomol.* 67: 184–186.
28. Georghiou, G. P., V. Ariaratnam, M. E. Pasternak, and Chi Lin. 1975. Organophosphorus multiresistance in *Culex pipiens fatigans* Wied. in California. *J. Econ. Entomol.* 68: 461–467.
29. Hamon, J. and C. Garrett-Jones. 1963. La résistance aux insecticides chez des vecteurs majeurs du paludisme et son importance operationnelle. *Bull. WHO.* 281–324.
30. Hamon, J., and J. Mouchet. 1961. La résistance aux insecticides chez les insectes d'importance medicale. *Med Trop.* 21: 565–596.
31. Hamon, J., M. Eyraud, B. Diallo, A. Dyemkouma, H. Bailly-Choumara, and S. Ouanou. 1961. Les moustiques de la République du Mali. *Ann. Soc. Ent. France.* 130: 95–129.
32. Haridi, A. M. 1966. Report in *WHO Inf. Circ. Insect Resist.* No. 58–59, p. 10.
33. Hemingway, J., and G. P. Georghiou. 1983. Studies on the acetylcholinesterase of *Anopheles albimanus* resistant and susceptible to organophosphate and carbamate insecticides. *Pestic. Biochem. Physiol.* 19: 167–171.
34. Hemingway, J., and G. P. Georghiou. 1984. Differential suppression of organophosphorus resistance in *Culex quinquefasciatus* by the synergists IBP, DEF and TPP. Pestic. Biochem. Physiol. 21: 1–9.
35. Hobbs, J. H. 1973. Effects of agricultural spraying on *Anopheles albimanus* densities in a coastal area of El Salvador. *Mosquito News* 33: 420–423.
36. Junkert, R., and K. R. Townzen. 1973. Biological and engineering evaluation of an irrigated pasture mosquito problem in Stanislaus County, California, and recommendations for its alleviation. *Calif. Vector News.* 20: 1–9.
37. Livingston, J. M., W. C. Yearian, and S. Y. Young. 1978. Effect of insecticides, fungicides, and insecticide-fungicide combinations on development of lepidopterous larval populations in soybeans. *Environ. Entomol.* 7: 823–828.
38. Martinez Palacios, A. 1959. Resistencia fisiologica a dieldrin y DDT de *Anopheles albimanus* en Mexico. *Bol. Comm. Nac. Errad. Palud.*, Mexico, 3: 31–32.

39. Mathis, W., H. F. Schoof, K. D. Quarterman, and R. W. Fay. 1956. *Public Health Rpts.* 71: 876–878.

40. Mouchet, J. and J. Laigret. 1967. La résistance aux insecticides chez *Aedes aegypti* à Tahiti. *Med. Trop.* 27: 685–692.

41. Mulhern, T. D. 1972. An approach to comprehensive mosquito control. *Calif. Vector Views* 19: 61–64.

42. Mulla, M. S. and L. S. Mian. 1981. Biological and environmental impacts of the insecticides malathion and parathion on non-target biota in aquatic ecosystems. *Residue Reviews* 78: 101–135.

43. Mulla, M. S., G. Majori, and A. A. Arata. 1979. Impact of biological and chemical mosquito control agents on non-target biota in aquatic ecosystems. *Residue Reviews* 7: 121–173.

44. O'Connor, C. T., and Arwati. 1974. Insecticide Resistance in Indonesia. Unpubl. Document, *WHO/VBC/*74.505, 8 pp.

45. Pimentel, D., J. Krummel, D. Gallahan, J. Hough, A. Merrill, I. Schreiner, P. Vittum, F. Koziol, E. Back, D. Yen, and S. Fiance. 1978. Benefits and costs of pesticide use. *BioScience* 28: 772, 778–784.

46. Rachou, R. G., G. Lyons, M. Moura-Lima, and J. A. Kerr. 1965. Synoptic epidemiological studies of malaria in El Salvador. *Am. J. Trop. Med. Hyg.* 14:1–62.

47. Ramsdale, C. D. 1973. Insecticide resistance in the anophelines of Turkey. Abstract, *9th Intern. Congr. Trop. Med. Malar.* 1: 260–261.

48. Ramsdale, C. D. 1975. Insecticide resistance in the *Anopheles* of Turkey. *Trans. Roy Soc. Trop. Med. Hyg.* 69: 226–235.

49. Raymond, M., D. Fournier, J.-M. Bride, A. Cuany, J. Bergé, M. Margnin and N. Pasteur. 1986. Identification of resistance mechanisms in *Culex pipiens* (Diptera: Culicidae) from southern France: Insensitive acetylcholinesterase and detoxifying oxidases. *J. Econ. Entomol.* 79: 1452–1458.

50. Sharma, Y. P., and Mehrotra, K. N. 1986. Malaria resurgence in India: A critical study. Soc. Sci. Med. 22:835–845.

51. Schliessmann, D. J. 1974. Technical and economic justification for the use of comprehensive measures in malaria control and eradication. Unpublished document, *WHO/MAL/*74.835, 7 pp.

52. Smith, R. F. 1968. Second Session of *FAO Panel of Experts on Integrated Pest Control*, Rome, 19–24 Sept. 1968. (mimeo.) pp. 39–42.

53. Subha Rao, Y. 1979. Susceptibility status of *Anopheles culicifacies* to DDT, dieldrin and malathion in village Mangapeta, District Warangal, Andhra Pradesh. *J. Com. Dis.* 2: 41–43.

54. Watal, B. L., G. C. Joshi, and M. Das. 1981. Role of agricultural insecticides in precipitating vector resistance. *J. Com. Dis.* 13: 71–74.

55. WHO. 1973. Review of susceptibility tests of malaria vectors to insecticides from 1 July 1970 to 31 December 1971. Unpublished document, 18 pp.

56. WHO. 1976. Resistance of Vectors and Reservoirs of Disease to Pesticides *WHO Tech. Rpt. Ser.* 585, 88 pp.

57. WHO. 1980. Resistance of Vectors of Disease to Pesticides. *WHO Tech. Rpt. Ser.* 655, 82 pp.

58. WHO. 1986. Resistance of Vectors and Reservoirs of Disease to Pesticides. *WHO Tech. Rpt. Ser.* 737, 87 pp.

59. Wright, J. W. 1971. The WHO Program for the Evaluation and Testing of New Insecticides. *Bull. WHO,* 44: 11–22.

60. Yates, W. E., N. B. Akesson, and D. E. Mayer. 1978. Drift of glyphosate sprays applied with aerial and ground equipment. *Weed Science* 26: 597–604.

61. Zulueta, J. de. 1959. Insecticide resistance in *Anopheles sacharovi. Bull. WHO,* 20: 797–821.

8

Pesticide Resistance in
Arthropod Natural Enemies:
Variability and Selection Responses

Marjorie A. Hoy

I. Introduction

There are several controversial issues relating to the presence, absence, or degree
of resistance to pesticides in arthropod natural enemies. Most of these issues have
been extensively reviewed. Therefore, rather than review the reviews, this chapter
will briefly review past results and present new information on variability and selection responses obtained from recent research on four natural enemy species. The
new data may alter some traditional perceptions of problems associated with detecting naturally occurring pesticide resistances and the likelihood of inducing resistance in arthropod natural enemies through artificial selection, recombinant DNA
(rDNA) techniques, or mutagenesis. As throughout this book, pesticide resistance
is defined as a genetically induced change in the ability of a population to tolerate
pesticides; no minimal level of change in tolerance need occur to be considered
resistance by this definition, as long as the measured differences are repeatable and
can be estimated in a statistically reliable manner (Chapter 2). The term *tolerance*
will be used to describe the ability of an organism to survive a specific pesticide
dose; it does not imply that a genetically determined change has occurred.

II. Recent Reviews

The effects of pesticides on arthropod natural enemies have been comprehensively, and recently, reviewed by Theiling and Croft (1988), who evaluated a
very large database containing information from original research papers as well
as reviews. They noted that early reviews of pesticide impact focused on toxicity,

I thank Frances Cave for assistance with the manuscript preparation and Elizabeth E. Grafton-
Cardwell and Jay A. Rosenheim for assistance in describing the research results. The research was
supported in part by U.S.D.A. Grant 84-CRCR-1-1452, Regional Research Project W-84, Walnut
Board of California, California Department of Agriculture, The California Agricultural Experiment
Station and two NSF Foundation Graduate Fellowships to students. Finally, I thank R. T. Roush and
Bruce Tabashnik for their assistance in the preparation of the manuscript.

selectivity, and ecological effects and were generally comprehensive (Ripper 1956, Stern et al. 1959, van den Bosch and Stern 1962, Bartlett 1964). Subsequent reviews appeared to be less comprehensive in their coverage, emphasizing more specialized aspects of natural enemy–pesticide interactions. Most recent reviews have focused on particular commodities, individual pesticide groups, specific crops, standardized testing methods, modes of pesticide uptake, physiological and ecological selectivity, resistance development, genetic improvement of natural enemies and their use in biological control, and methods of evaluation (Huffaker 1971, Croft 1972, Franz 1974, Croft and Brown 1975, Newsom et al. 1976, Brown 1977, Croft 1977, Croft 1990, Croft and Morse 1979, Hoy 1979, Roush 1979, Croft and Whalon 1982, Hoy 1982a, Croft and Strickler 1983, Croft and Mullin 1984, Hassan et al. 1983, Hoy 1985a, Hull and Beers 1985, Mullin and Croft 1985, Tabashnik and Croft 1985, Flexner et al. 1986, Fournier et al. 1986, Theiling 1987).

The topics emphasized in this paper, variability and selection responses, are of critical interest both for genetic improvement projects and for understanding of the evolution of pesticide resistance in arthropod natural enemies in the field. Arthropod natural enemies are perceived to be slow, or unlikely, to develop pesticide resistances in either the laboratory or the field because of a combination of biological, ecological, and/or biochemical reasons. The relative importance or reality of these explanations is still open to debate. Are (all or most) arthropod natural enemies biochemically incapable of developing high levels of pesticide resistances because they lack the ability to detoxify secondary plant compounds as efficiently as phytophagous arthropods (Mullin and Croft 1985, Chapter 6)? If resistances develop in arthropod natural enemies either in the laboratory or field, are the resistance levels unlikely to be sufficiently high that the strains could be used in the field with commonly applied pesticide rates? Do arthropod natural enemies have sufficient genetic variability to allow selection for resistance in either the laboratory or the field? Are there intrinsic differences between predators, parasitoids, and phytophages in their ability to develop pesticide resistances (Croft and Morse 1979, Croft and Strickler 1983, Croft and Mullin 1984)? Are ecological factors sufficient to explain observed differences in the development of resistance in pest and beneficial arthropods (Huffaker 1971, Tabashnik and Croft 1985, Chapter 6)?

This paper will not attempt to provide conclusions to these issues. Rather, new data, as well as published information, will be provided as a basis for future discussions.

III. Naturally Occurring Pesticide Resistances

A. Predators and Parasites

Compared to pest insects, few species of insect predators or parasites have been tested to determine whether they have acquired pesticide resistances under field conditions and, of these, only a few species have been demonstrated to possess acquired resistances to pesticides (Croft and Brown 1975, Croft 1977, Croft and Strickler 1983, Tabashnik and Croft 1985, Theiling 1987). Often, only a single laboratory or field population has been examined when arthropod natural enemies are evaluated in the laboratory (or field) for their responses to pesticides. The concept of intraspecific variability among laboratory colonies or field populations often seems to be ignored, with a few exceptions. The following examples, which are not exhaustive, illustrate naturally occurring resistances in arthropod natural enemies.

Atallah and Newsom evaluated the effects of DDT on the coccinellid predator *Coleomegilla maculata*; colony differences were attributed to the fact that "resistance has developed in the Boyce strain, or selection for susceptibility in the Baton Rouge laboratory population occurred during the rearing program" (Atallah and Newsom 1966). They obtained relatively flat slopes in dose–response lines with toxaphene and suggested that the "population may be in the process of developing resistance to toxaphene." Such flat slopes could also have been due to their bioassay technique; test interval, pesticide formulation, and many other nongenetic factors can yield a flat slope (Chapters 2 and 5).

Strawn (1978) found substantial variability in the tolerance of adults of California populations of the parasitoid *Aphytis melinus* (a key biological control agent of California red scale, *Aonidiella aurantii*) to four organophosphorous pesticides (parathion, methidathion, dimethoate, and malathion). However, none of the populations tested were judged to be sufficiently resistant to survive the pesticides in the field. *A. melinus* was introduced into California from India and Pakistan in a series of releases beginning in late 1957, with releases continuing through 1959. By 1961, parasites were widely established in southern California. The variability found in 1977 was therefore due to either releases of genetically diverse populations from India and Pakistan or selection for resistance had occurred in the released populations between 1957 and 1977. Strawn (1978) also tested adult *Comperiella bifasciata* but found that neither adults nor immatures of *C. bifasciata* colonies exhibited significant differences in responses to pesticides.

Schoonees and Giliomee (1982) evaluated the toxicity of methidathion to two strains of *Aphytis africanus* and *C. bifasciata*, both parasitoids of *A. aurantii* in South Africa. They found that different geographic strains varied; the Letsitele strain of *A. africanus*, which received three to four sprays of organophosphorous pesticides per year, was 5.7-fold more tolerant than the Mooinooi strain, which received only one application. *C. bifasciata* from Letsitele was 65.6 times as resistant as a laboratory culture. They concluded that these levels of resistance would not prevent most of the adults from succumbing to the commercial methidathion applications because the recommended concentration of 400–600 ppm was 17–25 and 12–18 times higher than the LC_{50} values for *A. africanus* and *C. bifasciata*. However, the relationship between the laboratory bioassay data and field survival, or lack thereof, as discussed later in this paper, should be interpreted very cautiously.

Havron (1983) collected seven field populations of *Aphytis holoxanthus* from citrus orchards in different areas in Israel. Malathion LC_{50} values were found to vary between 5.7 and 17.5 ppm. None of the populations were considered to be resistant, or developing resistance, to malathion, however.

Mansour (1984) found that colonies of clubionid spiders, *Chiracanthium mildei*, collected from a citrus grove and from a cotton field in Israel, differed in their responses to malathion by a factor of about 3.3, but he did not explicitly state whether these differences were useful for practical pest management. This spider has been shown to be a useful predator of insect pests in citrus, cotton, and apples in Israel, and Mansour noted that resistant strains of spiders "could be of great benefit in an integrated control program."

Pesticide-resistant strains of predatory mites (Acarina: Phytoseiidae) have been described, most commonly from orchard and vineyard crops, since the 1960s (Hoy 1985a). Hoyt (1969) illustrated the practical value of a strain of *Metaseiulus* (=*Typhlodromus* or *Galendromus*) *occidentalis* from apples in Washington that could survive applications of several insecticides, fungicides, and other sprays. Subsequently, other phytoseiid species, including *Amblyseius fallacis, Amblyseius potentillae, Amblyseius hibisci, Phytoseiulus persimilis, Typhlodromus pyri*, and *Amblyseius nicholsi*, have been selected for pesticide resistances either in the laboratory or in the field (Hoy 1985a).

B. Genetic Variability in the Hymenoptera

The parasitic Hymenoptera are perceived to be particularly sensitive to pesticides, and less likely to become resistant to pesticides than other insects. In general, the Hymenoptera display levels of electrophoretic variability significantly lower than those of most other insects, and a few appear to lack such variability completely (Graur 1985). These observations are thought to be due to a variety of factors, including reduced heterozygote advantage, inbreeding, eusociality, environmental stability, facilitation of exposure of deleterious genes in the haploid sex, reproduction by arrhenotokous parthenogenesis with balancing selection, and small effective population size resulting in a reduction in the amount of neutral polymorphism (Graur 1985). Assuming that there is a relationship between low levels of electrophoretic variability and genetic variability in general, this apparent lack of genetic variability suggests that parasitic hymenopterans could be limited in their responses to both natural and artificial selection for resistance to pesticides.

The responses of the honey bee, *Apis melifera*, to pesticides have been studied because honey bees are so important to agriculture. Their tolerance of or resistance to pesticides is of great potential economic value. Tahori et al. (1969) surveyed 18 colonies in Italy and found variability in tolerance to two chlorinated hydrocarbons (DDT and endrin) and two organophosphorous insecticides (trichlorfon and ronnel). The authors concluded that the differences found between these colonies "may be genetic in nature." Insecticides showing large between-colony differences did not necessarily show large within-colony differences. This is to be expected if the major portion of the bees tested from each laboratory colony were the offspring from a single drone. In this case only a small portion of the total genetic variability in the population would be found in a single colony. Thus, a breeding program with honey bees will be a time-consuming effort because of their long generation time and breeding system; breeding based on exploitation of polygenic variation is unlikely to be fruitful (Tahori et al. 1969). The efficient development of pesticide-resistant bee strains would require methods that capitalize upon major gene mutations.

Smirle and Winston (1987) found variability in polysubstrate monooxygenase activity between colonies of honey bees. They also found significant differences within the same colonies when tested in the spring, summer, and fall, suggesting that environmental factors influenced responses. Smirle and Winston concluded that differences in detoxification ability were significant factors in determining tolerance to diazinon and that "differences of this magnitude in an unselected population of colonies suggest that intercolony variation could be exploited as a means of selecting insecticide-resistant honey bees."

Atkins and Anderson (1962) found that DDT at the "standard dosage is no longer effective in killing bees from some of the colonies" in California. As early as 1952–1953, laboratory tests showed variability in susceptibility to DDT among bees from six different colonies, indicating that field selection had been effective.

Tucker (1980) increased tolerance to carbaryl in honey bees by artificial selection. By the eleventh generation, the maximum screening dose had been quadrupled for queens and doubled for drones. As selection continued, the concentration–mortality lines flattened, suggesting that a mixture of resistant and susceptible queens were present. Outcrosses and lapses in selection did not substantially reduce the gain in the last three generations, which "could suggest that one or a few major dominant genes are important for resistance to carbaryl."

These few examples of variability in responses to pesticides in the honey bee, while provocative, are far from conclusive. Where selection has been intense, selection responses have occurred through both artificial and natural selection. The frequency and level of resistances found to date do not, however, suggest that pesticide resistance is readily obtainable in honey bees.

C. Variability in Four Natural Enemy Species in California

Extensive variability in responses of California populations of *Metaseiulus occidentalis* to azinphosmethyl, diazinon, sulfur, and phosmet have been found (Croft and Jeppson 1970, Hoy and Knop 1979, Hoy and Standow 1982). Such variability suggested that this natural enemy responds to selection in local orchards and vineyards in California and that populations of *M. occidentalis* in the San Joaquin Valley are not panmictic (Hoy 1985a). Studies of the aerial dispersal of the laboratory-selected, carbaryl-resistant strain in California almond orchards supported that conclusion (Hoy 1982b, Hoy et al. 1985).

Recently, variability in responses to pesticides in three additional natural enemy species in California—a predatory insect, *Chrysoperla (=Chrysopa) carnea,* and two parasitoids, *Aphytis melinus* and *Trioxys pallidus*—was evaluated. The results of these surveys provide an expanded perspective on the types and amount of naturally occurring levels of pesticide resistance among natural enemy populations.

1. Chrysoperla carnea

Populations of common green lacewings, *Chrysoperla carnea,* were collected during 1981 and 1982 from alfalfa fields in each of four California counties separated by approximately 200–500 km (Grafton-Cardwell and Hoy 1985a). Adults, larvae, and eggs of the four colonies were then screened with six pesticides (carbaryl, methomyl, permethrin, fenvalerate, diazinon, and phosmet). Lacewing populations from the four locations responded significantly differently to all six pesticides (Fig. 8.1). In general, San Joaquin County lacewings exhibited the highest mortality, whereas Imperial County lacewings exhibited the lowest mortality; Kern and Fresno County lacewings were generally intermediate in their responses (Grafton-Cardwell and Hoy 1985a). These differences in survival generally corresponded with pesticide usage in alfalfa.

All adult lacewings tested exhibited high tolerances to both fenvalerate and permethrin, despite the fact that neither pyrethroid was registered for use in alfalfa when the colonies were collected. Pyrethroid tolerance in adult lacewings is possibly due to cross-resistance developed through earlier selection with DDT, or to a natural tolerance to pyrethroids. A natural tolerance to pyrethroids was previously reported in larvae of lacewings based on laboratory tests (Plapp and Bull 1978, Ishaaya and Casida 1981, Rajakulendran and Plapp 1982, Shour and Crowder 1980, Bigler 1984).

Significant variability in responses to permethrin and fenvalerate was found in larvae (Fig. 8.1B,C), which *suggests* that different populations had, in addition to a possible natural tolerance to permethrin, experienced some selection for enhanced permethrin resistance. Such selection could have occurred in other crops where permethrin and fenvalerate were registered for use, or could reflect past selection with DDT. It is noteworthy that the colonies with greater tolerances to pesticides were collected from areas where greater amounts of pesticides were applied (Grafton-Cardwell and Hoy 1985a). Additional evidence that selection for pyrethroid resistance has occurred in lacewings has been provided by Pree et al. (1989) from southern Ontario, Canada. In 1983, they collected a population of *C. carnea* from an orchard where insecticides had not been used. This strain appeared to be susceptible to several insecticides, including pyrethroids. Not only did Pree et al. (1989) confirm the presence of permethrin resistance in larvae of *C. carnea* from the Imperial Valley (using a strain collected from cotton in 1982), they also collected a strain

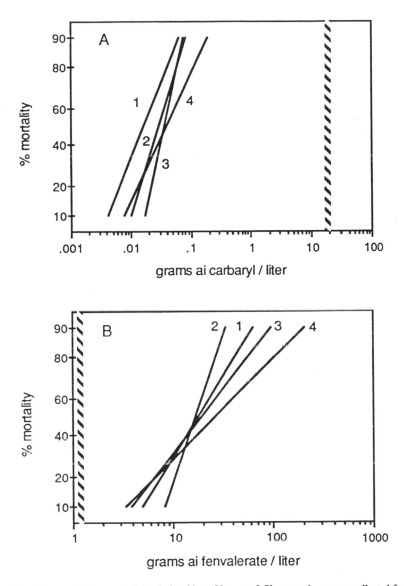

Figure 8.1. Concentration–mortality relationships of larvae of *Chrysoperla carnea* collected from San Joaquin (1), Fresno (2), Kern (3), or Imperial (4) counties in California to carbaryl (A), fenvalerate (B), permethrin (C), methomyl (D), diazinon (E), and phosmet (F). The dashed vertical line indicates a standard field application rate. (Modified, with permission, from Grafton-Cardwell and Hoy 1985.)

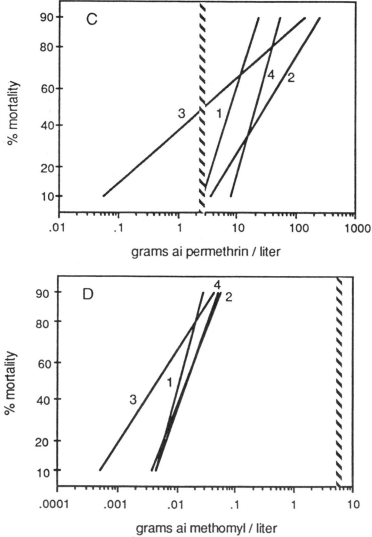

Figure 8.1. *(continued)*

from Ontario in 1981 that was 34-fold resistance to permethrin, apparently due to metabolism. Interestingly, the Canadian resistant strain was collected only about 0.5 km from where the susceptible strain was found.

Larvae of all four California colonies of *C. carnea* were also highly tolerant of phosmet, with LC_{50} values well above the field rate of 12 g Al (active ingredient) per liter water (Fig. 8.1F) (Grafton-Cardwell and Hoy 1985a). Phosmet tolerance varied among the four colonies. Although we cannot distinguish between natural

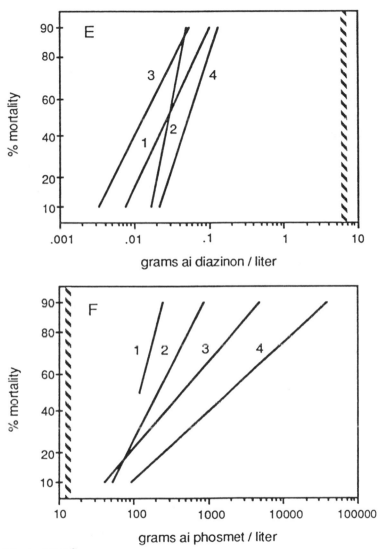

Figure 8.1 *(continued)*

tolerance or acquired resistance as the source of this variation, Pree et al. (1989) found a 62-fold difference between the resistant and susceptible Ontario strains, implying that selection may be involved. In any case, the California larvae tolerated 10 to 158 times the field rate of phosmet in laboratory bioassays (LC_{50} values ranged from 120 to 1902 g Al per liter). The relationship between bioassay LC_{50} values and the actual field dose remains somewhat speculative, however, since field trials were not conducted to document that the larvae could survive phosmet applications in the field. Grafton-Cardwell and Hoy (1985b), in evaluating the sublethal effects of permethrin and fenvalerate on oviposition by adult *C. carnea,* showed that the conclusions varied with test method. They suggested that a "single method of testing sublethal effects of pyrethroids on beneficial insects such as *C. carnea* may not be sufficient to understand the consequences of use of these insecticides in the field."

The survey for variability was a prelude to a possible selection project, and variability was found among the colonies' responses to all six pesticides (Grafton-Cardwell and Hoy 1985a). Tolerances to fenvalerate, permethrin, and phosmet were sufficiently high that selection for additional levels did not seem necessary. Because of time constraints, selection for resistance to methomyl was not attempted, but selection with both carbaryl and diazinon was carried out (Grafton-Cardwell and Hoy 1986). The variability in responses to carbaryl and diazinon appears, from Figure 8.1, to be approximately equal. However, these lines illustrate the survival of larvae between the LC_{10} and LC_{90} values only. When the complete data were examined, a small percentage (less than 5%) of Imperial larvae was able to survive on high rates of carbaryl (Fig. 8.2) (Grafton-Cardwell and Hoy 1986). Such a plateau in responses was not found with diazinon. As noted later, this plateau appeared to be a useful predictor of a selection response.

The survey data suggest that lacewing populations, like populations of *M. occidentalis* in California, respond to local pesticide selection pressures and that geographic differences in tolerances, reflecting past selection pressures, do exist (Grafton-Cardwell and Hoy 1985a, Pree et al. 1989). Furthermore, generalizations about the tolerance of *"C. carnea"* to pesticides that are based on tests with single colonies or single populations (or colonies with a narrow genetic base) may be inappropriate and misleading. Such typological thinking may be hindering our progress in understanding evolutionary responses to pesticides by arthropod natural enemies.

2. *Aphytis melinus*

Thirteen populations of *Aphytis melinus,* a parasite of California red scale, were collected from citrus-growing regions of California during 1983 (Rosenheim and Hoy 1986). Each population's history of exposure to insecticides was estimated by determining the history of insecticide use on both in-grove and county-wide geographical scales. Information on in-grove use was obtained directly from growers while county-wide use patterns were obtained from state and county records. Concentration–mortality regression lines from these populations were obtained for five insecticides widely used in citrus. For each chemical (carbaryl, chlorpyrifos, dimethoate, malathion, and methidathion), substantial variability existed in the responses of the different populations (Rosenheim and Hoy 1986). Furthermore, LC_{50} values correlated with both in-grove (Fig. 8.3A) and county-wide (Fig. 8.3B) pesticide use histories. The observed patterns of variability were best explained by

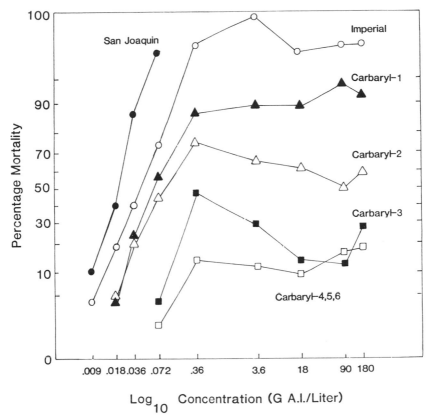

Figure 8.2. Carbaryl concentration–mortality relationships for first instars of *Chrysoperla carnea*. The strains tested include the highly susceptible San Joaquin Valley colony, the more tolerant Imperial Base colony (Imperial), and the carbaryl-resistant strains selected from the Imperial Base colony after each of six selections. The fourth through sixth selections are pooled, as they were not significantly different. (Reprinted, with permission, from Grafton-Cardwell and Hoy 1986.)

the results of a multiple regression analysis that combined the influences of both in-grove and county-wide histories (Fig. 8.3C). These data indicate that different levels of pesticide resistance are developing among different populations of *A. melinus* as they are subjected to different selection regimes.

Since *A. melinus* is an exotic species that was introduced into California citrus orchards in a classical biological control program in 1957 and presumably underwent some degree of genetic bottlenecking during quarantine and colonization, it is interesting to note that distinct populations with differing resistance levels exist. Whether this variability was present in the originally released populations or developed through mutation and natural selection after establishment in California cannot be determined.

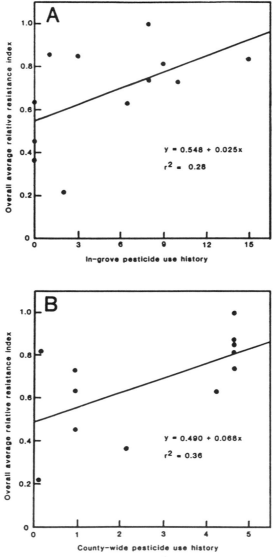

Figure 8.3. Regression analyses of 13 colonies of *Aphytis melinus'* overall average tolerance on: (A) its in-grove pesticide use history, (B) its countywide pesticide use history, and (C) both A and B simultaneously (multiple regression). (Reprinted with permission from Rosenheim and Hoy 1986.)

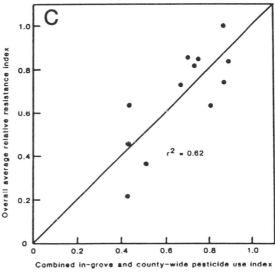

Figure 8.3. *(continued)*

$(= 0.372 + 0.024(in-grove use) + 0.066(county-wide use))$

3. Trioxys pallidus

Populations of the walnut aphid parasitoid, *Trioxys pallidus*, were collected from walnut-growing regions in California during 1985 and 1986 and tested for their tolerance to azinphosmethyl (Hoy and Cave 1988). Substantial variability was found among these populations, with the Aggregate and Los Banos colonies exhibiting the greatest tolerance for azinphosmethyl (Fig. 8.4). Azinphosmethyl is used to control the codling moth, *Carpocapsa pomonella*, and the navel orangeworm, *Ameloyis transitella*, in walnuts, and has been found to be highly disruptive to the effective biological control of the walnut aphid, *Chromaphis juglandicola*, by *T. pallidus* (Sibbett et al. 1981, van den Bosch et al. 1979). The differences in azinphosmethyl tolerance observed in laboratory bioassays among the *T. pallidus* populations have not been reflected in *observed* differences in survival or control of the walnut aphid under field conditions, but quantitative evaluations of survival of *T. pallidus* after azinphosmethyl applications in walnut orchards have not been made recently (Hoy and Cave 1988).

Again, the variation in azinphosmethyl tolerance is particularly interesting because *T. pallidus*, like *A. melinus*, is an exotic parasitoid that was introduced into California in a classical biological control project. *T. pallidus* was introduced at two different times after being collected from France (in 1959) and Iran (in 1969), respectively (van den Bosch et al. 1979, Schlinger et al. 1960, Fraser and van den Bosch 1973, van den Bosch et al. 1962). The between-population variation observed with *Trioxys* (Fig. 8.4), as well as with *Aphytis melinus*,

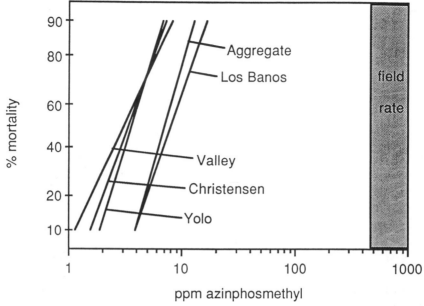

Figure 8.4. Variability in responses to azinphosmethyl for five colonies of *Trioxys pallidus* collected from California walnut orchards; the shaded portion represents the field rates (Hoy and Cave 1988).

suggests that sufficient genetic variability existed in these two exotic parasitoids to allow for a selection response to pesticides used in California walnut and citrus orchards (Rosenheim and Hoy 1986, Hoy and Cave 1988). As with *A. melinus,* the origins of this variability cannot be determined.

It could be very enlightening to evaluate pesticide tolerances of exotic natural enemies prior to their release and establishment in classical biological control programs. If subsequent monitoring indicated that pesticide resistances had developed, then we could determine whether the resistance genes existed in the colonizing population or were derived through new mutations. Careful test design will be required to ensure that the original colony is adequately sampled prior to release so that rare alleles, if present, are detected.

IV. Artificial Selection of Arthropod Natural Enemies

A. General Review

Artificial selection of arthropod natural enemies for resistance to pesticides has been proposed as a method for improving the usefulness of natural enemies in integrated pest management programs (Pielou and Glasser 1952, Robertson 1957, Adams and Cross 1967, Abdelrahman 1973, Croft and Meyer 1973, Hoy 1979,

Roush 1979, Hoy and Knop 1981, Roush and Hoy 1981, Hoy 1985a). Several selection programs with arthropod natural enemies have been attempted, but, until recently, useful levels of resistance have been demonstrated primarily in phytoseiid mite species (Croft and Meyer 1973, Hoy and Knop 1979, Hoy and Knop 1981, Roush and Hoy 1981, Strickler and Croft 1982, Hoy 1985a, Markwick 1986, Huang et al. 1987).

This apparent taxonomic bias has led to the suggestion that phytoseiid predators are uniquely preadapted to develop resistance to pesticides whereas other arthropod natural enemies lack this genetic capability. Some have speculated that the unusual parahaploid genetic system of phytoseiids could allow rapid evolution for resistance (Hoy 1985a). Others have speculated that ecological attributes of phytoseiids (i.e., low vagility and rapid generation time) predispose them to develop pesticide resistances relatively rapidly. While these factors may contribute to the evolution of pesticide resistances in phytoseiids, a counterargument would note that a relatively limited number of phytoseiid species have become resistant to pesticides, through either natural or artificial selection (Hoy 1985a). De Moraes et al. (1986) estimated that there are approximately 1,000 species of phytoseiids; of these, only a few species, such as *Amblyseius fallacis, Typhlodromus pyri, Metaseiulus occidentalis, Phytoseiulus persimilis,* and *Amblyseius nicholsi,* have been documented to have acquired resistances to pesticides through natural or artificial selection (Hoy 1985a, Fournier et al. 1986, Markwick 1986, Huang et al. 1987, Avella et al. 1985). Thus, it is unclear whether phytoseiids are more prone to develop resistances to pesticides than other arthropod natural enemies.

Few insect predators or parasitoids have become resistant to pesticides in the field (Croft and Brown 1975, Croft and Strickler 1983, Theiling and Croft 1989). Artificial selection of insect natural enemies for resistance has previously been considered a failure. However, the number of species that have been subjected to effective selection regimes is still very limited. Pielou and Glasser (1952) attempted to select *Macrocentrus ancylivorus,* a parasite of the Oriental fruit moth, *Grapholitha molesta,* for resistance to DDT. The results after nine months of selection were a modest 4.4-fold increase in LD_{50} value for females. Robertson (1957) continued the selection and increased the resistance level to a maximum of 12-fold over that of the base colony. However, when DDT selection was terminated after the F_{19} generation, the resistance level declined to nine times the initial level at the F_{29} generation. When the parasites were reared for 13 generations (F_{72-84}) without exposure to DDT, the resistance declined to nearly the initial level. The entire project lasted six years and involved more than 3 million adults. There is no evidence that the resistant strain was ever evaluated in the field.

Adams and Cross (1967) found that colonies of *Bracon mellitor,* a parasite of the boll weevil, *Anthonomus grandis,* responded to selection with DDT, carbaryl, and methyl parathion; increases were about four-fold within five generations. A population selected with a mixture of DDT and toxaphene yielded a resistance level eight-fold greater than that of the original population. Despite this moderate selection response, the LD_{50} of the resistant strains were still only in the nanogram or microgram range using a topical application test method. The authors concluded that *B. mellitor* "possesses the potential for developing a degree of resistance to some insecticides."

Abdelrahman (1973) found that *Aphytis melinus* selected for eight generations in the laboratory became 3.4 times more resistant to malathion; this modest level of resistance was achieved without depleting the variability of the colony, as suggested by the slope of the selected colony. However, since the female red scale is 707 times as tolerant of malathion as *A. melinus,* he concluded that "it is doubtful that development of a resistant strain of *A. melinus* can be of much significance for the preservation of parasites during a conventional full cover spray of malathion to control red scale."

Delorme et al. (1984) attempted to select strains of *Encarsia formosa* that were resistant to ethyl-parathion and deltamethrin by selecting for 21 successive generations. The resistance ratios obtained varied between 0.56 and 2.81 with ethyl-parathion and between 0.63 and 4.2 with deltamethrin. These authors evaluated total esterase activity in these colonies and failed to find any with an unusual activity level. They concluded that selection for resistance "seems to depend upon increasing the heterogeneity of the strains, either by searching for wild strains or by chemical or physical mutagenesis."

Havron (1983) failed to obtain a selection response to malathion with *Aphytis holoxanthus*. Lack of genetic variability was cited as the cause of the failure to achieve a response. This species had been introduced into Israel using a colony with a small founding population and thus was suspected to have limited amounts of genetic variability.

The questions raised by these apparent failures to achieve high levels of resistance to pesticides are several. Do parasitoids (and insect predators to a lesser extent) lack the ability to detoxify pesticides (Croft and Mullin 1984, Mullin and Croft 1985)? Is artificial selection unlikely to be successful because of a lack of variability upon which selection can operate? Given that variability in responses to pesticides can be identified, will the level of resistance achieved after selection ever be sufficiently high that the resistance can be useful in the field? Recent laboratory selection results with *Chrysoperla carnea, Trioxys pallidus,* and *Aphytis melinus* suggest that these issues are open to further study and evaluation.

B. *Chrysoperla carnea*

The variability in responses to pesticides observed in California populations of *C. carnea* was considered sufficiently promising to initiate an artificial selection project (Fig. 8.1) (Grafton-Cardwell and Hoy 1985a, 1986). The selection response with carbaryl was rapid, and the resistance level achieved was high (Fig. 8.2). The LC_{50} value and a resistance ratio could not be obtained, however, because concentration–mortality data did not fit a probit model (Grafton-Cardwell and Hoy 1986). The resistant strain could not be killed over a wide range of concentrations of the formulated pesticide after the third selection. The lack of a linear response after probit transformation also made a test of the mode of inheritance difficult to interpret. However, the patterns exhibited by the reciprocal F_1 and back-cross progeny suggested that carbaryl resistance in this strain is determined by one or a few major genes (Grafton-Cardwell and Hoy 1986).

When arthropods are selected in the laboratory for resistance, changes in fitness are often assumed to occur and these are thought to limit the potential of genetic improvement programs of arthropod natural enemies. The relative fitness of the resistant and base colonies of *C. carnea* were compared in the laboratory, but the results are difficult to interpret with regard to predicting field efficacy of the resistant strain. Larval and pupal stages of the base colony had higher survival rates than the carbaryl-resistant strains, but the time to develop from egg to adult was not significantly different for the two colonies. The sex ratio (female–male) was 1.6:1 for the base colony and 1.1:1 for the resistant strain. Fecundity of the resistant females was significantly higher than for the base colony and the percentage hatch of eggs was not different. The manner in which these fitness attributes would interact to influence field efficacy of the carbaryl-resistant strain of *C. carnea* is not known, as field trials were not conducted (Grafton-Cardwell and Hoy 1986).

The Imperial colony of *C. carnea* was also selected with the organophosphorous insecticide diazinon but, after three selections, no improvement in survival was discerned and the selection was discontinued (Grafton-Cardwell and Hoy 1986).

C. *Trioxys pallidus*

Trioxys pallidus colonies collected from California walnut orchards gave a slow and gradual selection response with azinphosmethyl (Hoy and Cave 1988). The concentrations of azinphosmethyl that could be used for selection, however, were very low compared to those used in the field. This led to the expectation that a dramatic change in LC_{50} value was needed before a useful level of resistance would be obtained.

Selection was achieved by exposing adult parasitoids to dried azinphosmethyl residues within plastic cups capped with treated solid plastic lids, which provided no untreated refuges for the parasitoids. Survivors after 48 hours of exposure were used to initiate the next generation. Because this selection method was so artificial, a series of bioassays was conducted during the summer of 1987 to evaluate the azinphosmethyl resistance level obtained. Assessment techniques were modified first by substituting a mesh cap for the solid plastic lid on the plastic cups and the effects of treated and untreated mesh were compared. The results were revealing. Survival of both the Selected (Fig. 8.5) and susceptible strains (data not shown) was enhanced by using a mesh cap rather than a solid cap.

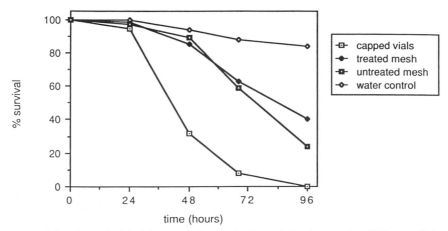

Figure 8.5. Survival of the laboratory-selected azinphosmethyl-resistant strain of *Trioxys pallidus* upon exposure to azinphosmethyl residues compared in a laboratory assay using plastic cups with plastic lids and plastic cups with untreated mesh lids (Hoy and Cave 1988).

Whether the mesh was treated with azinphosmethyl or not did not dramatically influence the results (Fig. 8.5).

Two complete concentration–mortality tests were conducted with the susceptible (Yolo), Base, and Selected colonies to confirm these results. One method used the cups capped with a solid lid (as in our selection technique); the second used cups capped with mesh (Fig. 8.6). The concentration–mortality lines of all three colonies tested with the mesh-capped cups (dotted lines) are significantly closer to the field rate than the lines obtained with the treated solid plastic caps (solid lines, Fig. 8.6). Assay technique clearly influenced the assessment of the resistance level achieved with *T. pallidus*.

In an attempt to mimic field exposure more closely, survival of the azinphosmethyl-resistant and susceptible strains was compared on potted walnut trees sprayed to drip with water, half the field rate, and the field rate of azinphosmethyl (600 ppm A.I.). Parasitoids were contained on the walnut leaves with clip cages having a mesh opening. The cages were clipped to the under surface of the sprayed walnut leaves with the treated leaf surface serving as the top of the cage. Observations indicated that the parasitoids spent most of their time on the leaf surface rather than on the cage surface. This cage design appeared to allow the parasites to behave in a more normal manner than the plastic cup assay method. Clear differences were found in the ability of the Selected and susceptible colonies to survive on the field rates of azinphosmethyl (Hoy and Cave 1988).

To further evaluate the Selected colony, we collected azinphosmethyl-treated foliage from a walnut orchard sprayed with the field rate of azinphosmethyl (1 lb 50 WP Guthion per 100 gallons water or 60 g Al per 100 liters). We collected foliage from treated and untreated tree canopies 3 days and 17 days after applica-

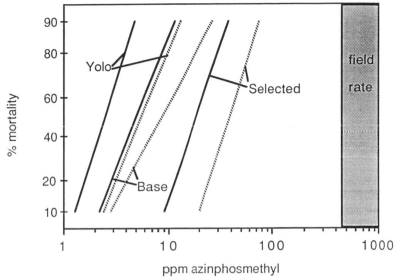

Figure 8.6. Concentration–mortality lines of azinphosmethyl-resistant (Selected) and susceptible (Yolo and Base) colonies of *Trioxys pallidus* compared in plastic cups with a mesh (dashed lines) or plastic cap (solid lines) (Hoy and Cave 1988).

tion, placed adult parasitoids in clip cages on the treated and untreated foliage, and recorded survival over a 72-hour interval. The results indicated a significant difference in the abilities of the azinphosmethyl-Selected and susceptible strains to survive on azinphosmethyl residues for up to 72 hours on three-day-old residues (Fig. 8.7) and two-week-old residues (Hoy and Cave 1988). Similar assays have shown that the Selected strain is cross-resistant to four other insecticides (endosulfan, chlorpyrifos, methidathion, and phosalone) commonly applied to walnut orchards (Hoy and Cave 1989). These results suggested that the azinphosmethyl-Selected strain was sufficiently resistant that it could be useful in walnut orchards treated with azinphosmethyl and at least some other insecticides.

These conclusions were supported by field tests at five walnut orchards during 1988. The Selected strain strain survived applications of azinphosmethyl or methidathion, parasitized aphids, and persisted in or near the release sites through the growing season in four of the five orchards (Hoy et al. 1990). Although long-term implementation will require additional information, particularly on overwintering success, the results of these first field trials suggest that it may yet be possible to genetically improve pesticide resistance in parasitic Hymenoptera.

The ability of the laboratory-selected resistant strain of *T. pallidus* to survive on pesticide-treated foliage is strongly influenced by assay methodology. The selection method (plastic-capped plastic cup with all surfaces treated) is an effective way to discriminate between resistant and susceptible phenotypes. It provides a rapid, repeatable method for comparing colonies. The plastic cup

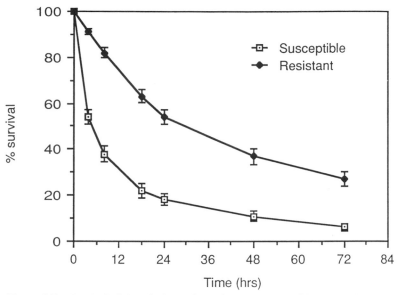

Figure 8.7. Survival of the azinphosmethyl-resistant and susceptible strains of *Trioxys pallidus* on walnut foilage treated three days previously with the field rate of azinphosmethyl (Hoy and Cave 1988).

technique, however, clearly overestimates the mortality one might expect at field rates of the pesticides, as confirmed in the field trials. The second assay technique, treated cups capped with untreated mesh, also overestimates mortality in relation to the field situation, but is somewhat more realistic about conditions the parasites might experience in the field. The sprayed potted tree assay also appeared to underestimate potential survival because of the extremely thorough coverage obtained when the potted trees were sprayed with a hand sprayer. Only when parasitoids were placed in clip cages on field-treated foliage was evidence obtained that the azinphosmethyl-selected strain of *Trioxys* would survive in the field (Hoy and Cave 1988, Hoy and Cave 1989).

D. *Aphytis melinus*

Adults of *Aphytis melinus* were selected for resistance to carbaryl, methidathion, malathion, chlorpyrifos, and dimethoate in the laboratory using a mesh-capped plastic cup technique (Rosenheim and Hoy 1988). After selection had been exerted for approximately one year, LC_{50} values of the ten selected lines were increased 1.5- to 2.6-fold over the corresponding base colonies. Three of the selected lines showed significant increases in slopes, suggesting that decreases in genetic variability had occurred (Rosenheim and Hoy 1988). Selection with carbaryl and methidathion was then continued on combined colonies during the second year, at the end of which concentration–mortality lines were obtained for the Selected and susceptible colonies. Only a modest increase in LC_{50} value was obtained for methidathion (~2-fold). The response to carbaryl was gradual but significant after 24 selections (Fig. 8.8). The concentration of carbaryl used to achieve approximately 50–80% mortality was increased four different times, from 9.6 to 14.4., to 19.2, to 24, and to 33.6 mg Al/liter (Fig. 8.8, selections at which the carbaryl concentration were increased are indicated by "I"). Interestingly, the selection response was slowed after two carbaryl-selected colonies were combined (C) after 13 selections (Fig. 8.8, "C"). If the Selected strain (after a total of 24 selections) is compared to the most carbaryl-susceptible strain (LaCouague), the resistance ratio is 20-fold (Fig. 8.9).

The concentration–mortality lines in Figure 8.9 were obtained by placing adult male and female *Aphytis* into plastic cups with dried residues; the lids were carbaryl-treated mesh. The selection concentrations of carbaryl and methidathion used were substantially lower than the concentrations used in citrus pest management and one could reasonably conclude that a selection response sufficient to provide field-usable resistance to carbaryl had not occurred. Once again, however, this conclusion might be in error.

A foliage assay was conducted to determine whether the resistant strains could

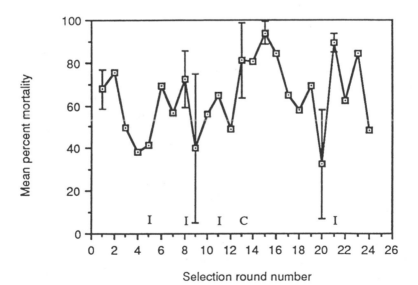

Figure 8.8. Mean percentage mortality at 24 hours of the *Aphytis melinus* colony selected with carbaryl. During selections 1–12, the Stutsman colony was selected; during selections 13–34, a combined colony including the Stutsman colony was selected. C, first selection of the combined colony. Carbaryl concentrations were increased (I) from 9.6 in selections 1–4 to 14.4 (selections 5–7) to 19.2 (selections 8–10), to 24 (selections 11–20), to 33.6. mg A.I. carbaryl/L (selections 21–24) (Rosenheim and Hoy 1988).

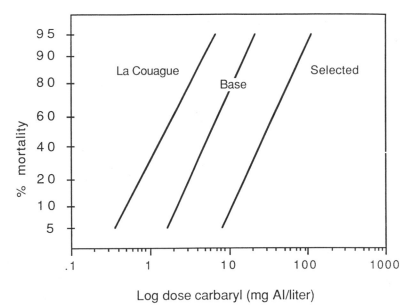

Figure 8.9. Concentration–mortality lines for the carbaryl-resistant strain of *Aphytis melinus* (Selected) compared to the base colony (Base) from which it was derived, and to a susceptible colony (La Couague) (Rosenheim and Hoy 1988).

survive on different-aged carbaryl and methidathion residues on citrus foliage collected from San Joaquin Valley citrus orchards during 1987. Foliage was collected from citrus orchards where pesticide application dates were known. For carbaryl, foliage with residues 3 to 75 days old was collected and brought into the laboratory, and two-leaf clip cages were used to contain adult parasitoids on the surfaces of the foliage. (Independent tests confirmed that *Aphytis* spends nearly 80% of its time on the undersurface of citrus leaves, if given a choice.) The results indicated that foliage with residues three weeks old allowed survival of more than 50% of the Selected adult parasitoids (Fig. 8.10); once the foliage residues were older than 21 days, mortality consistently declined to less than 50%. In contrast, the susceptible (LaCouague and Base) strains exhibited mortality rates of more than 80% on foliage with residues up to 75 days old (Rosenheim and Hoy 1988).

These bioassay data suggest that laboratory selection may have yielded a strain with a field-usable level of carbaryl resistance. Because this strain is still responding to selection, field trials are planned after a selection plateau is reached. If the carbaryl-resistant strain could reenter the citrus orchard sooner than the susceptible strain, then an economic and pest management benefit might be achieved (Rosenheim and Hoy 1988). Currently, *A. melinus* is commercially mass reared and released periodically in a number of California citrus orchards; releases immediately after carbaryl applications are, however, ineffective and

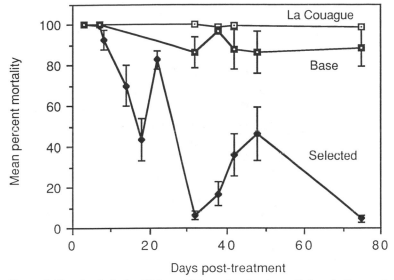

Figure 8.10. Survival after 24 hours of the carbaryl-resistant (Selected), Base, and susceptible (La Couague) colonies of *Aphytis melinus* on carbaryl-treated foilage collected from citrus orchards 3–75 days after treatment with carbaryl (Rosenheim and Hoy 1988).

typically lead to the application of sprays to control California red scale. The interval when carbaryl-treated citrus foliage is toxic is about equal to the developmental period of this parasitoid; it is possible that the carbaryl-resistant strain could survive a carbaryl spray as a protected immature stage within its scale host. If so, by the time adults emerged from the red scale host, the adults should be able to survive on the carbaryl residues (Rosenheim and Hoy 1988).

The conclusion that carbaryl residues remain highly toxic to the susceptible strain of *A. melinus* for at least 75 days postspray is dramatically different from other reports that carbaryl residues produce less than 30% mortality within three days postspray (Bellows and Morse 1988). Rosenheim and Hoy compared bioassay techniques in an effort to understand this apparent discrepancy in results, testing the efficiency of having one treated leaf (which served as the bottom of the clip cage, the top being mesh) or two treated leaves (one leaf serving as the top and one as the bottom of the clip cage) (Rosenheim and Hoy 1988). The two-leaf treatment had either a forced airflow or a passive airflow, as did the water controls. The susceptible strain of *A. melinus* had no significant differences in mortality when tested with clip cages with two treated leaves with or without forced airflow (Fig. 8.11). However, mortality was significantly reduced if the clip cage had only the bottom leaf treated rather than both the top and the bottom leaves treated (Fig. 8.11). In the clip cage with one treated leaf (on the bottom), about 26% of the cage surface was treated compared to twice that of the two-leaf clip cage. However, the refuge effect was enhanced by the behavioral preference

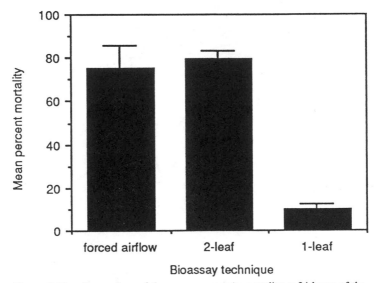

Figure 8.11. Comparison of the mean percentage mortality at 24 hours of the carbaryl-susceptible strain of *Aphytis melinus* exposed to citrus foliage with carbaryl residues 22 days old. The clip cage assays used two treated leaves (serving as the top and bottom of the cage) with either forced air flow or passive air flow, or one treated leaf with passive air flow. Clip cages with one leaf had an untreated cloth mesh top (Rosenheim and Hoy 1988).

of *A. melinus* to walk about or rest on the underside of the *top* surface of the clip cage, which was a carbaryl-treated citrus leaf in the two-leaf cage and untreated cloth mesh in the one-leaf cage (Rosenheim and Hoy 1988). Parasitoids spent about 84% of their time on treated leaf surfaces in the two-leaf cage but only 16% of their time on treated surfaces in the single leaf cage.

Once again, the difficulty of developing a laboratory bioassay technique that is relevant to the biology, behavior, and ecology of the parasite under study is evident. The bioassay technique chosen could lead to inaccurate predictions regarding the relative resistance levels of a selected strain. The clip cage test that used two treated leaves (Fig. 8.10) suggests that the carbaryl-resistant strain of *A. melinus* may be able to survive in carbaryl-treated orchards, but only field trials will confirm this.

In contrast, the methidathion-selected strain of *A. melinus* did not survive significantly better than the base strain on the methidathion-treated citrus foliage after 48 days. This colony appears to have insignificant levels of resistance and no field tests are planned with this colony (Rosenheim and Hoy 1988).

E. *Metaseiulus occidentalis*

Pesticide resistances in the phytoseiid *M. occidentalis* have been intensively selected for in the laboratory during the past 10 years (Hoy 1985a). The diversity of resistances obtained from either the field or laboratory selection probably rivals the diversity of resistances exhibited by the majority of pest arthropod species. Two resistances were obtained through identification of field-selected strains, namely, organophosphorus (Huffaker and Kennett 1953, Hoyt 1969, Hoy and Knop 1979) and sulfur resistance (Hoy and Standow 1982). Carbamate resistances were obtained (to carbaryl, propoxur, and benomyl) through laboratory selection with carbaryl (Roush and Hoy 1981, Roush and Plapp 1982). Resistance to pyrethroid insecticides has been obtained in two different laboratory selections with permethrin (Hoy and Knop 1981, Y. L. Ouyang and M. A. Hoy, in preparation).

The first permethrin resistance was obtained after screening approximately 30 field-collected colonies. The resistance obtained was relatively low, and probably polygenically determined (Hoy and Knop 1981). Recently, a new strain of *M. occidentalis* has been selected that has a higher level of resistance to permethrin (Y. L. Ouyang and M. A. Hoy, in preparation). This strain was collected from an apple orchard in Wenatchee, Washington, by S. C. Hoyt and was to serve as a susceptible native population in laboratory tests. However, it was nearly as tolerant of permethrin in that first test as our laboratory-selected strain. As far as S. C. Hoyt is aware (personal communication), this orchard population was never subjected to selection with synthetic pyrethroids. This unexpected example of permethrin resistance reinforces the belief that extensive sampling is a key to detecting variability in responses to pesticides in arthropod natural enemies.

M. occidentalis continues to surprise. Recently, this species responded to selection with abamectin, an insecticide–acaricide–nematacide derived as a fermentation product of *Streptomyces avermitilis* (Hoy and Ouyang 1989). Selection was conducted on a heterogeneous colony obtained by combining 30 females from each of 21 colonies. Twenty selections with increasing concentrations of abamectin yielded a gradual and modest shift in the concentration–mortality line. However, the modest 3.8-fold increase in the survival of adult females at the LC_{50} value was combined with an increase in the number of eggs the surviving females produced. The stability, mode of inheritance, cross-resistance(s), and resistance mechanism(s) associated with this resistance have not yet been determined. This novel acaricide is believed to affect the GABA system of arthropods, and, while the abamectin resistance mechanism in *M. occidentalis* is currently unknown, it seems likely to be different from those associated with sulfur, organophosphorus, carbamate, or pyrethroid resistances.

F. Need for Intensive Sampling

M. occidentalis has developed resistances to very different chemical groups of pesticides. The number of mechanisms is unknown but surely there are several involved. The intensive searches in the field and the intensive selections in the laboratory were probably crucial to obtaining these diverse resistances in *M. occidentalis*.

Could it be that other resistant arthropod natural enemies could be obtained if equivalent amounts of effort were expended? Literature reviews suggest that no other arthropod biological control agent has been subjected to such intense scrutiny or selection for resistances as has *M. occidentalis*. Perhaps other key natural enemy species should be studied more intensively. Substantially more effort has been invested in identifying resistant pest arthropods than resistant arthropod natural enemies.

An extreme bias in selection and sampling probably favors the evolution and detection of resistance in pests rather than in natural enemies. Because their prey or hosts often must become resistant before sufficient food is available to sustain predator or parasitoid populations, resistance development may be delayed in natural enemies compared to pests (Huffaker 1971, Rosenheim and Hoy 1986, Chapter 6). Once the pest becomes resistant, selection can begin on the natural enemy, but if use of the pesticide is then discontinued, the natural enemy can not catch up. Only in cases where hosts or prey are already resistant or tolerant to a pesticide is the natural enemy able to respond to selection for resistance to a pesticide. Furthermore, because natural enemies are often difficult to sample in the field and rear in large numbers, fewer individuals may be sampled compared to pest species. As a result, exceptional sampling efforts may be required when resistance in natural enemies is sought. Such surveys are typically time-consuming and expensive, particularly when the natural enemy is difficult to sample and rear and handle in the laboratory. Until efficacious and inexpensive artificial diets for natural enemies are widely available, such sampling biases will be difficult to overcome. Therefore, other approaches to obtaining variability in arthropod natural enemies may be necessary and justified.

V.   Potential New Sources of Variability

A.   Need for New Sources of Variability

Developing pesticide-resistant strains of natural enemies through artificial selection clearly is not always successful, but the success rate is difficult to determine. Few people publish their negative selection results. Selection of *Chrysoperla carnea* for two pesticide resistances yielded a 50% success rate: positive results with carbaryl and negative results with diazinon (Grafton-Cardwell and Hoy 1986). Likewise, selection of *Aphytis melinus* had a 50% success rate: selection for carbaryl resistance was successful while selection for methidathion resistance was not (Rosenheim and Hoy 1988).

It is not yet clear for any parasitoid or insect predator that the level of resistance achieved is sufficiently high that the strains are useful in the field. Successful field testing and implementation of pesticide-resistant *insectan* natural enemies is critically important if this approach is to continue to receive support. The positive results obtained with various phytoseiid species have encouraged research in genetic manipulation of arthropod natural enemies, but the generality of this approach will be open to question unless resistant parasitoids or insect predators can be incorporated into integrated pest management programs. Given that pesticide-resistant strains are valuable in a variety of integrated pest management programs and resistance alleles are not always readily obtained through artificial selection of naturally occurring variability, we may need to investigate ways to obtain new sources of variability to enhance our chances of obtaining pesticide-resistant strains of natural enemies.

B. Mutagenesis as a Source of Pesticide Resistance

Artificial induction of variability was not required to obtain resistances to an array of pesticides in *Metaseiulus occidentalis*. During the initial design of the genetic improvement project with *M. occidentalis,* the possibility that mutagenesis might be required to generate variability upon which we could select was considered (Hoy 1979). Mutagenesis was not pursued since it yields mutations that are generally both deleterious and random, which could result in genetically deteriorated strains that would not perform well under field conditions. Furthermore, mutations for pesticide resistance would be rare, making it crucial that very large numbers be screened to detect them.

There are a number of pros and cons to using mutagenesis as a method for inducing mutations in arthropod natural enemies. Mutagenesis has not yet been used successfully to develop pesticide resistant natural enemies, so it is not clear that it is an effective approach. A few examples provide evidence that mutagenesis can produce resistances in other arthropods: Pluthero and Threlkeld (1984) reported that mutants of *Drosophila melanogaster* with altered resistance to malathion were obtained after the flies were treated with ethyl methanesulfonate (EMS). Of 103 lines tested, three carried X-linked mutations with significantly different malathion tolerances. The authors found a significant correlation between resistance and a behavioral avoidance response to malathion. The mutations mapped to different but proximal sites, and the authors presumed them to "be evidence of a common mechanism for both the avoidance and resistance to malathion."

Buchi (1981) induced resistance to pyrethroids (deltamethrin) in the aphid *Myzus persicae* by feeding larvae an artificial diet containing EMS. The progeny were again treated with EMS and allowed to reproduce for two generations. The population was then treated with insecticides and the surviving clones were isolated. Three EMS-induced deltamethrin-resistant mutants were obtained.

Mutations can also be induced by X-irradiation. Kikkawa (1964) described the induction of a resistance gene for parathion in the susceptible Canton-S strain of *Drosophila melanogaster*. Males were treated with 2,000 or 3,000 rad, then mated to untreated virgin females; the F_1 larvae were reared on a diet with 0.5 ppm parathion. Of 12,500 larvae reared from the males treated with 2,000 rad, two progeny were obtained; likewise, two progeny were obtained from 16,000 progeny produced by males treated with 3,000 rad. These progeny were mated and lines established but only one strain derived from a male treated with 2,000 rad exhibited parathion resistance. The resistance gene was eventually mapped to the same locus as the gene that is involved in all parathion-resistant strains. This suggested that the resistance gene was actually induced from its susceptible allele. Additional tests suggested that the mutation did not take place in a single step, but occurred through an intermediate state. The intermediate type appeared to be unstable and tended to mutate to either a more resistant or more susceptible direction. Thus, the artificial induction of mutations in this case yielded a strain that was not different from or more resistant than *D. melanogaster* strains found throughout the world.

Wilson and Fabian (1987) mutagenized *D. melanogaster* with both EMS and X-irradiation (at 3,000 or 5,000 rad) and selected for methoprene resistance. A total of 2,403 F_1 progeny lines were tested following EMS mutagenesis; all but two showed less than 10-fold increases in tolerance. One of the two lines with the greater level of tolerance ($\sim$60–100-fold) was established and the resistance found to be incompletely recessive and located on the X-chromosome.

Mutagenesis, achieved either through chemical mutagens or X-irradiation, should be attempted with arthropod natural enemies. Since such mutations are random and rare, such projects are likely to be effective only with species that can be mass-reared in large numbers. Since more than one mutation may occur in a mutagenized strain, steps will have to be taken to produce a colony that does not have new deleterious genes. Mutagenesis may best be attempted with species that can be reared on artificial diets or on factitious hosts, since many parasitoids are difficult to rear in very large numbers because of the expense and difficulty of rearing their hosts. Alternatively, field collection of very large numbers of individuals for mutagenesis might be cost effective in certain situations.

C.   rDNA Techniques for Developing Pesticide-Resistant Strains

Recombinant DNA techniques (rDNA) developed for *Drosophila* could be used to develop pesticide-resistant strains of arthropod natural enemies (Beckendorf and Hoy 1985, Hoy 1987). There are several important steps involved in testing this proposal. Resistance genes must be identified, cloned, and inserted into natural enemies, be incorporated into the genome, be stable, be expressed appropriately, and be transmitted to progeny (Chapter 4). The transformed natural enemy strain must also be fit and able to perform well in the field. In addition, issues relating to the safety of releasing rDNA-manipulated arthropod natural enemies remain to be resolved. Research undertaken to fulfill all of these goals will provide important basic information on pesticide resistance in arthropod natural enemies. If we are able to insert pesticide resistance genes into beneficial species, then we can begin to compare different gene constructs with different promoters and other regulatory agents and learn more about resistance mechanisms in arthropod natural enemies.

VI. Conclusions

In *M. occidentalis, T. pallidus, A. melinus,* and *C. carnea,* the amount of intraspecific variability in tolerances to pesticides used in California crops was impressive. Artificial selection with at least one pesticide was successful in each case. The carbaryl resistance achieved with *C. carnea* was sufficiently high that the strain probably could survive field rates of carbaryl. The resistance levels achieved with *A. melinus* and *T. pallidus* appear, from preliminary evidence with field-collected treated foliage, to be sufficiently high that the strains may be proved useful in integrated pest management programs. Field trials are necessary before this conclusion can be confirmed.

The levels of resistance obtained with both *Trioxys* and *Aphytis* appeared to be inadequate when the colonies were evaluated with the bioassay techniques used for the selections. This inability to predict mortality in the field is probably due to the fact that the enclosed plastic cup techniques used to select both parasitoids are highly artificial. In contrast, the leaf spray technique that we use to select the phytoseiid, *M. occidentalis,* generally has been highly predictive of the mortality experienced under field conditions with a particular pesticide (Hoy 1985b, 1987). Clearly, caution is required when making conclusions regarding the relative toxicity of pesticides to various natural enemy species tested in the laboratory under highly artificial conditions. It is possible that these laboratory trials could lead to erroneous conclusions regarding the relative toxicity of pesticides and the level of resistance achieved in a laboratory-selected strain.

Is it possible that some laboratory-selected pesticide-resistant natural enemies developed previously were actually able to survive in the field but this was obscured by the laboratory assays employed? With both *Trioxys* and *Aphytis,* the projects were nearly abandoned without field trials because it appeared that the selection responses were insufficient for field efficacy. Clearly, we have much to learn yet about how to monitor, assay, and select for pesticide resistances in arthropod natural enemies.

The relationship between laboratory toxicity and field performance of pesticides was described as "generally unexplored" in 1966 and this statement remains essentially true more than 20 years later (Sun 1966). Yet this relationship appears to be critical in determining whether laboratory-selected natural enemy strains are sufficiently resistant that they should be field tested, a field-collected natural enemy has acquired resistance to a pesticide in the field through natural selection, or pesticides are selective to natural enemies. Developing predictive models for laboratory bioassays and field toxicity of pesticides to arthropod natural enemies remains a major challenge.

References

Abdelrahman, I. 1973. Toxicity of malathion to the natural enemies of California red scale, *Aonidiella aurantii* (Mask.) (Hemiptera: Diaspididae). Australian J. Agric. Res. 24: 119–133.

Adams, C. H., and W. H. Cross. 1967. Insecticide resistance in *Bracon mellitor*, a parasite of the boll weevil. J. Econ. Entomol. 60: 1016–1020.

Atallah, Y. H., and L. D. Newsom. 1966. Ecological and nutritional studies on *Coleomegilla maculata* DeGeer (Coleoptera: Coccinellidae). III. The effect of DDT, toxaphene, and endrin on the reproductive and survival potentials. J. Econ. Entomol. 59: 1181–1187.

Atkins, E. L., Jr., and L. D. Anderson. 1962. DDT resistance in honey bees. J. Econ. Entomol. 55: 791–792.

Avella, M., D. Fournier, M. Pralavorio, and J. P. Berge. 1985. Selection pour la resistance a la deltamethrine d'une souche de *Phytoseiulus persimilis* Athias-Henriot. Agronomie 5: 177–180.

Bartlett, B. R. 1964. Integration of chemical and biological control, pp. 489–514. *In* P. DeBach (ed.), Biological control of insect pests and weeds. Reinhold, New York.

Beckendorf, S. K., and M. A. Hoy. 1985. Genetic improvement of arthropod natural enemies through selection, hybridization or genetic engineering techniques, pp. 167–187. *In* M. A. Hoy and D. C. Herzog (eds.), Biological control in agricultural IPM systems. Academic Press, Orlando.

Bellows, T. S., Jr., and J. G. Morse. 1988. Residual toxicity following dilute or low volume applications of insecticides used for control of California red scale (Homoptera: Diaspididae) to four beneficial species in a citrus agroecosystem. J. Econ. Entomol. 81: 892–898.

Bigler, F. 1984. Biological control by chrysopids: integration with pesticides, pp. 233–245. *In* M. Canard, Y. Semeria, and T. R. New (eds.), Biology of Chrysopidae. Junk, The Hague.

Brown, A. W. A. 1977. Considerations of natural enemy susceptibility and developed resistance in the light of the general resistance problem. Z. Pflanzenkr. Pflanzensch. 84: 132–139.

Buchi, R. 1981. Evidence that resistance against pyrethroids in aphids *Myzus persicae* and *Phorodon humuli* is not correlated with high carboxylesterase activity. Z. Pflanzenkr. Pflanzensch. 88: 631–634.

Croft, B. A. 1972. Resistant natural enemies in pest management systems. Span 15: 19–21.

Croft, B. A. 1977. Resistance in arthropod predators and parasites, pp. 377–393. *In* D. L. Watson and A. W. A. Brown (eds.), Pesticide management and insecticide resistance. Academic, New York.

Croft, B. A. 1990. Arthropod biological control agents and pesticides. Wiley, New York.

Croft, B. A., and A. W. A. Brown. 1975. Responses of arthropod natural enemies to insecticides. Annu. Rev. Entomol. 20: 285–335.

Croft, B. A., and L. R. Jeppson. 1970. Comparative studies on four strains of *Typhlodromus occidentalis*. II. Laboratory toxicity of ten compounds common to apple pest control. J. Econ. Entomol. 63: 1528–1531.

Croft, B. A., and R. H. Meyer. 1973. Carbamate and organophosphorus resistance patterns in populations of *Amblyseius fallacis*. Environ. Entomol. 2: 691–695.

Croft, B. A., and J. G. Morse. 1979. Recent advances on pesticide resistance in natural enemies. Entomophaga 24: 3–11.

Croft, B. A., and C. A. Mullin. 1984. Comparison of detoxification enzyme systems in *Argyrotaenia citrana* (Lepidoptera: Tortricidae) and the ectoparasite, *Oncophanes americanus* (Hymenoptera: Braconidae). Environ. Entomol. 13: 1330–1335.

Croft, B. A., and Strickler. 1983. Natural enemy resistance to pesticides: documentation, characterization, theory and application, pp. 669–702. *In* G. P. Georghiou and T. Saito (eds.), Pest resistance to pesticides. Plenum, New York.

Croft, B. A., and M. E. Whalon. 1982. Selective toxicity of pyrethroid insecticides to arthropod natural enemies and pests of agricultural crops. Entomophaga 27: 3–21.

Delorme, R., A. Angot, and D. Auge. 1984. Variations de sensibilite d'*Encarsia formosa* Gahan

(Hym. Aphelinidae) soumis a des pressions de selection insecticide: approches biologique et biochimique. Agronomie 4: 305–309.

de Moraes, G. J., J. G. McMurtry, and H. A. A. Denmark. 1986. A catalog of the mite family Phytoseiidae. Embrapa, Brasil.

Flexner, J. L., B. Lighthart, and B. A. Croft. 1986. The effects of microbial pesticides on non-target, beneficial arthropods. Agric. Ecosyst. Environ. 16: 203–254.

Fournier, D., M. Pralavorio, J. B. Berge, and A. Cuany. 1986. Pesticide resistance in Phytoseiidae, pp. 423–432. *In* W. Helle and M. W. Sabelis (eds.), Spider mites, their biology, natural enemies and control. Vol. 1B. Elsevier, Amsterdam.

Franz, J. M. 1974. Testing of side-effects of pesticides on beneficial arthropods in laboratory—a review (in German). Z. Pflanzenkr. Pflanzensch. 81: 141–174.

Fraser, B. D., and R. van den Bosch. 1973. Biological control of the walnut aphid in California: the interrelationship of the aphid and its parasite. Environ. Entomol. 2: 561–568.

Grafton-Cardwell, E. E., and M. A. Hoy. 1985a. Intraspecific variability in response to pesticides in the common green lacewing, *Chrysoperla carnea* (Stephens) (Neuroptera: Chrysopidae). Hilgardia 53(6): 1–32.

Grafton-Cardwell, E. E., and M. A. Hoy. 1985b. Short-term effects of permethrin and fenvalerate on oviposition by *Chrysoperla carnea* (Neuroptera: Chrysopidae). J. Econ. Entomol. 78: 955–959.

Grafton-Cardwell, E. E., and M. A. Hoy. 1986. Genetic improvement of common green lacewing, *Chrysoperla carnea* (Neuroptera: Chrysopidae): selection for carbaryl resistance. Environ. Entomol. 15: 1130–1136.

Graur, D. 1985. Gene diversity in Hymenoptera. Evolution 39: 190–199.

Hassan, S. A., et al. 1983. Results of the second joint pesticide testing programme by the IOBC/WPRS Working Group "Pesticides and Beneficial Arthropods". Z. Angew. Entomol. 95: 151–158

Havron, A. 1983. Studies toward selection of *Aphytis* wasps for pesticide resistance. Ph.D. Thesis, Hebrew University of Jerusalem.

Hoy, M. A. 1979. The potential for genetic improvement of predators for pest management programs, pp. 106–115. *In* M. A. Hoy and J. M. McKelvey, Jr. (eds.), Genetics in relation to insect management. Rockefeller Foundation Press, New York.

Hoy, M. A. 1982a. Genetics and genetic improvement of the Phytoseiidae, pp. 72–89. *In* M. A. Hoy (ed.), Recent advances in knowledge of the Phytoseiidae. University of California Special Publication 3284, Division of Agricultural Sciences, Berkeley.

Hoy, M. A. 1982b. Aerial dispersal and field efficacy of a genetically improved strain of the spider mite predator *Metaseiulus occidentalis*. Entomol. Exp. Appl. 32: 205–212.

Hoy, M. A. 1985a. Recent advances in genetics and genetic improvement of the Phytoseiidae. Annu. Rev. Entomol. 30:345–370.

Hoy, M. A. 1985b. Almonds (California); integrated mite management for California almond orchards, pp. 299–310. *In* W. Helle and M. W. Sabelis (eds.), Spider mites, their biology, natural enemies and control. Vol. 1B. Elsevier, Amsterdam.

Hoy, M. A. 1987. Developing insecticide resistance in insect and mite predators and opportunities for gene transfer, pp. 125–138. *In* H. LeBaron, R. O. Mumma, R. C. Honeycutt, and J. H. Duesing (eds.), Biotechnology in Agricultural Chemistry. American Chemical Society Series No. 334, Washington, D.C.

Hoy, M. A., and F. E. Cave. 1988. Guthion-resistant strain of walnut aphid parasite. Calif. Agric. 42(4): 4–5.

Hoy, M. A., and F. E. Cave. 1989. Toxicity of pesticides used on walnuts to a wild and azinphosmethyl-resistant strain of *Trioxys pallidus* (Hymenoptera: Aphidiidae). J. Econ. Entomol. 82: 1585–1592.

Hoy, M. A., and J. Conley. 1987. Toxicity of pesticides to western predatory mite, Calif. Agric. 41(7–8), 12–14.

Hoy, M. A., and N. F. Knop. 1979. Studies on pesticide resistance in the phytoseiid *Metaseiulus occidentalis* in California, pp. 89–94. *In* J. Rodriguez (ed.), Recent advances in acarology, Vol. I. Academic, New York.

Hoy, M. A., and N. F. Knop. 1981. Selection for and genetic analysis of permethrin resistance in *Metaseiulus occidentalis:* genetic improvement of a biological control agent. Entomol. Exp. Appl. 30: 10–18.

Hoy, M. A., and Y. L. Ouyang. 1989. Selection of the western predatory mite, *Metaseiulus occidentalis* (Acari: Phytoseiidae), for resistance to abamectin. J. Econ. Entomol. 82: 35–40.

Hoy, M. A., and K. A. Standow. 1982. Inheritance of resistance to sulfur in the spider mite predator *Metaseiulus occidentalis*. Entomol. Exp. Appl. 31: 316–323.

Hoy, M. A., J. J. R. Groot, and H. E. van de Baan. 1985. Influence of aerial dispersal on persistence and spread of pesticide-resistant *Metaseiulus occidentalis* in California almond orchards. Entomol. Exp. Appl. 37: 17–31.

Hoy, M. A., F. E. Cave, R. H. Beede, J. Grant, W. H. Krueger, W. H. Olson, K. M. Spollen, W. W. Barnett, and L. C. Hendricks. 1990. Release, dispersal, and recovery of a laboratory-selected strain of the walnut aphid parasite *Trioxys pallidus* (Hymenoptera: Aphidiidae) resistant to azinphosmethyl. J. Econ. Entomol. 83: 89–96.

Hoyt, S. C. 1969. Integrated chemical control of insects and biological control of mites on apple in Washington. *J. Econ. Entomol.* 62: 74–86.

Huang, M. D., J. J. Xiong, and T. Y. Du. 1987. The selection for and genetical analysis of phosmet resistance in *Amblyseius nicholsi*. Acta Entomol. Sinica 30: 133–139.

Huffaker, C. B. 1971. The ecology of pesticide interference with insect populations, pp. 92–104. *In* J. E. Swift (ed.), Agricultural chemicals—harmony or discord for food, people and the environment. University of California Division of Agricultural Sciences, Berkeley.

Huffaker, C. B., and C. E. Kennett. 1953. Differential tolerance to parathion of two *Typhlodromus* predatory on cyclamen mite. J. Econ. Entomol. 46: 707–708.

Hull, L. A., and E. H. Beers. 1985. Ecological selectivity: modifying chemical control practices to preserve natural enemies, pp. 103–121. *In* M. A. Hoy and D. C. Herzog (eds.), Biological control in agricultural IPM systems. Academic Press, Orlando.

Ishaaya, I., and J. E. Casida. 1981. Pyrethroid esterase(s) may contribute to natural pyrethroid tolerance of larvae of the common green lacewing (*Chrysopa carnea*). Environ. Entomol. 10: 681–684.

Kikkawa, H. 1964. Genetical studies on the resistance to parathion in *Drosophila melanogaster*. II. Induction of a resistance gene from its susceptible allele. Botyu-Kagaku 29: 37–42.

Mansour, F. 1984. A malathion-tolerant strain of the spider *Chiracanthium mildei* and its response to chlorpyrifos. Phytoparasitica, 12(3–4): 163–166.

Markwick, N. P. 1986. Detecting variability and selecting for pesticide resistance in two species of phytoseiid mites. Entomophaga 31:225–236.

Mullin, C. A., and B. A. Croft. 1985. An update on development of selective pesticides favoring arthropod natural enemies, pp. 123–150. *In* M. A. Hoy and D. C. Herzog (eds.), Biological control in agricultural IPM systems. Academic Press, Orlando.

Newsom, L. D., R. F. Smith, and W. H. Whitcomb. 1976. Selective pesticides and selective use of pesticides, pp. 565–591. *In* C. B. Huffaker and P. S. Messenger (eds.), Theory and practice of biological control. Academic Press, New York.

Pielou, D. P., and R. E. Glasser. 1952. Selection for DDT resistance in a beneficial insect parasite. Science 115: 117–118.

Plapp, F. W., Jr., and D. L. Bull. 1978. Toxicity and selectivity of some insecticides to *Chrysopa carnea*, a predator of the tobacco budworm. Environ. Entomol. 7: 431–434.

Pluthero, F. G., and F. H. Threlkeld. 1984. Mutations in *Drosophila melanogaster* affecting physiological and behavioral response to malathion. Can. Entomol. 116: 411–418.

Pree, D. J., D. E. Archibold, and R. K. Morrison. 1989. Resistance to insecticides in the common green lacewing *Chrysoperla carnea* (Neuroptera: Chrysopidae) in southern Ontario. J. Econ. Entomol. 82: 29–34.

Rajakulendran, S. V., and F. W. Plapp, Jr. 1982. Comparative toxicities of five synthetic pyrethroids to the tobacco budworm (Lepidoptera: Noctuidae), an ichneumonid parasite, *Campoletis sonorensis,* and a predator, *Chrysopa carnea.* J. Econ. Entomol. 75: 769–772.

Ripper, W. E. 1956. Effect of pesticides on balance of arthropod populations. Annu. Rev. Entomol. 1: 403–438.

Robertson, J. G. 1957. Changes in resistance to DDT in *Macrocentrus ancylivorus.* Can. J. Zool. 35: 629–633.

Rosenheim, J. A., and M. A. Hoy. 1986. Intraspecific variation in levels of pesticide resistance in field populations of a parasitoid. *Aphytis melinus* (Hymenoptera: Aphelinidae): the role of past selection pressures. J. Econ. Entomol. 79: 1161–1173.

Rosenheim, J. A., and M. A. Hoy. 1988. Genetic improvement of a parasitoid biological control agent: Artificial selection for insecticide resistance in *Aphytis melinus* (Hymenoptera: Aphelinidae). J. Econ. Entomol. 81: 1539–1550.

Roush, R. T. 1979. Genetic improvement of parasitoids, pp. 97–105. *In* M. A. Hoy and J. J. McKelvey, Jr. (eds.), Genetics in relation to insect management. Rockefeller Foundation Press, New York.

Roush, R. T., and M. A. Hoy. 1981. Genetic improvement of *Metaseiulus occidentalis:* selection with methomyl, dimethoate, and carbaryl and genetic analysis of carbaryl resistance. J. Econ. Entomol. 74: 138–141.

Roush, R. T., and F. W. Plapp. 1982. Biochemical genetics of resistance to aryl carbamate insecticides in the predaceous mite, *Metaseiulus occidentalis.* J. Econ. Entomol. 75: 304–307.

Schlinger, E. I., K. S. Hagen, and R. van den Bosch. 1960. Imported French parasite of walnut aphid established in California. Calif. Agric. 14(11): 3–4.

Schoonees, J., and J. H. Giliomee. 1982. The toxicity of methidathion to parasitoids of red scale, *Aonidiella aurantii* (Hemiptera: Diaspididae). J. Entomol. Soc. South Africa 45: 261–273.

Shour, M. H., and L. A. Crowder. 1980. Effects of pyrethroid insecticides on the common green lacewing. J. Econ. Entomol. 73: 306–309.

Sibbett, G. S., L. Bettiga, and M. Bailey. 1981. Impact of summer infestation of walnut aphid on quality. Sun-Diamond Grower June–July: 8.

Smirle, M. J., and M. L. Winston. 1987. Intercolony variation in pesticide detoxification by the honey bee (Hymenoptera: Apidae). J. Econ. Entomol. 80: 5–8.

Stern, V. M., R. F. Smith, R. van den Bosch, and K. S. Hagen. 1959. The integration of chemical and biological control of the spotted alfalfa aphid. Hilgardia 29: 81–101.

Strawn, A. J. 1978. Differences in response to four organophosphates in the laboratory of strains of *Aphytis melinus* and *Comperiella bifasciata* from citrus groves with different pesticide histories, M.S. thesis, University of California, Riverside.

Strickler, K. A., and B. A. Croft. 1982. Selection for permethrin resistance in the predatory mite, *Amblyseius fallacis* Garman (Acarina: Phytoseiidae). Entomol. Exp. Appl. 31: 339–345.

Sun, Y. P. 1966. Correlation between laboratory and field data on testing insecticides. J. Econ. Entomol. 59: 1131–1134.

Tabashnik, B. E., and B. A. Croft. 1985. Evolution of pesticide resistance in apple pests and their natural enemies. Entomophaga 30: 37–49.

Tahori, A. S., Z. Sobel, and M. Soller. 1969. Variability in insecticide tolerance of eighteen honey-bee colonies. Entomol. Exp. Appl., 12: 85–98.

Theiling, K. M. 1987. SELCTV: a database management system on the effects of pesticides on arthropod natural enemies, M.S. thesis, Oregon State University, Corvallis.

Theiling, K. M., and B. A. Croft. 1988. Pesticide effects on arthropod natural enemies: a database summary. Agric. Ecosyst. Environ. 21: 191–218.

Tucker, K. W. 1980. Tolerance to carbaryl in honey bees increased by selection. Am. Bee J. January: 36–46.

van den Bosch, R., and V. M. Stern. 1962. The integration of chemical and biological control of arthropod pests, Annu. Rev. Entomol. 7:367–386.

van den Bosch, R., E. I. Schlinger, and K. S. Hagen. 1962. Initial field observations in California on *Trioxys pallidus* (Haliday), a recently introduced parasite of the walnut aphid. J. Econ. Entomol. 55:857–862.

van den Bosch, R., R. Hom, P. Matteson, B. D. Frazer, P. S. Messenger, and C. S. Davis. 1979. Biological control of the walnut aphid in California: impact of the parasite, *Trioxys pallidus*. Hilgardia 47: 1–13.

Wilson, T. G., and J. Fabian. 1987. Selection of methoprene-resistant mutants of *Drosophila melanogaster*, pp. 179–188. *In* J. Law (ed.), Molecular Entomology, UCLA Symposium on Molecular and Cell Biology, new series, no. 49.

9

Management of Pyrethroid-resistant Tobacco Budworms on Cotton in the United States

Frederick W. Plapp, Jr., Clayton Campanhola, Ralph D. Bagwell, and *Billy F. McCutchen*

I. Introduction

The bollworm, *Heliocoverpa* (= *Heliothis*) *zea,* and the tobacco budworm, *Heliothis virescens,* are important pests of cotton in the United States. Control difficulties for *Heliothis* spp. on cotton were reported in the 1950s (Ivy and Scales 1954), and resistance to DDT in the tobacco budworm was clearly demonstrated in Texas by Brazzel (1963) and in Louisiana by Graves et al. (1967). In a summary of these studies, Sparks (1981) documented DDT resistance for the bollworm in 12 states and for the tobacco budworm in eight states by 1970.

The strategy adopted for controlling *Heliothis* spp. on cotton was to shift to organophosphate insecticides. These provided satisfactory control through most of the 1960s. However, resistance to methyl parathion in the tobacco budworm appeared in Texas in 1968 (Whitten and Bull 1970, Wolfenbarger and McGarr 1970). Within a few years resistance to methyl parathion ad other organophosphates occurred throughout the Cotton Belt. No organophosphate resistance has been confirmed in the bollworm in the United States, although resistance has been reported from Central America (Wolfenbarger et al. 1981).

Whitten and Bull (1970, 1974) established that organophosphate resistance in the tobacco budworm was primarily metabolic. Bioassay tests showed that resistance extended to most dialkyl phosphates and phosphorothioates as well as aryl carbamates (Plapp 1971, 1972). In contrast, there was little resistance to enantiomeric organophosphates in the form of racemic mixtures, such as EPN, acephate, profenofos, and sulprofos, or to oxime carbamates, such as methomyl or thiodicarb. Metabolic resistance is apparently uncommon with enantiomeric

The assistance of numerous extension service and research entomologists in the five states of Arkansas, Louisiana, Mississippi, Oklahoma, and Texas who did most of the resistance monitoring and field work upon which this study was based in gratefully acknowledged. We also thank R. E. Frisbie, J. B. Graves, G. T. Payne, T. C. Sparks, and R. T. Roush for reviews of early drafts of this article.

organophosphates and oxime carbamates, although resistance to profenofos has been reported from South Carolina (Brown et al. 1982) and methomyl resistance has been reported from Mexico (Roush and Wolfenberger 1985). Indeed, insecticides of these groups are widely used as substitutes for pyrethroids when resistance is present.

During the early 1970s no useful alternative types of insecticides were available. As a consequence, cotton production technology underwent a rapid change to short season, lower input varieties. In fact, one consequence of resistance was the successful incorporation of integrated pest management strategies into cotton production.

The objectives of this chapter are to describe the collapse of the pyrethroid-only strategy for *Heliothis* control and the development of responses to the problem. These efforts have involved numerous state and federal researchers, extension service personnel, and industry cooperators. Overall, the program has been remarkably successful. We believe the experiences described here reflect the reality of the efforts made and can serve as guidelines for similar situations as they arise in the future.

II. Use of Pyrethroids and the Occurrence of Resistance

In 1978, pyrethroid insecticides became available for controlling *Heliothis* and other lepidopterous pests of cotton in the United States (Sparks 1981). As pointed out by Riley (1989), widespread resistance to organophosphates hastened the transition to pyrethroid use. Once again, production of high-input, longer-season cottons became feasible. In addition to providing superior pest control, the pyrethroids proved less phytotoxic than the alternative organophosphate and carbamate insecticides. The net result, according to local crop consultant Dr. S. J. Nemec (personal communication), was an approximately 25–33% increase in cotton yields, with the lack of phytotoxicity being at least as important as the superior insect control.

The situation continued until 1985, when resistance to pyrethroids was reported in late-season tobacco budworm populations in west Texas. Preliminary data obtained with F_1 and F_2 neonate larvae from field collections in the Garden City, Texas, area revealed an approximately 15-fold resistance to permethrin (Plapp and Campanhola 1986). Tests with a permethrin–chlordimeform combination showed a level of synergism about equal to the level of resistance.

In addition to the documented resistance in the Garden City area, widespread reports of control difficulties in other areas of Texas were reported in 1985 (Allen et al. 1987). The situation was perhaps most severe in the Wintergarden, an isolated area of high-yielding–high-input cotton production, approximately 100 miles west of San Antonio near Uvalde, Texas.

The resistance in Texas in 1985 was not the first report of control difficulties with the tobacco budworm. Earlier reports from California (Twine and Reynolds

1980, Martinez-Carrillo et al. 1983) had demonstrated up to 50-fold resistance to pyrethroids in tobacco budworm populations from the Imperial Valley, although no evidence was presented to document impact of this resistance on control (Nicholson and Miller 1985). Similarly, 1978 data indicated a low level of tolerance in central Texas populations (Plapp 1979). This early resistance may have represented cross-resistance associated with metabolic resistance to organophosphates. As pointed out by Staetz (1985), there was no consistent trend toward increased tolerance to pyrethroids in the tobacco budworm during the 1980–1985 period.

In retrospect, the earlier resistance to DDT in bollworm and tobacco budworm should have alerted us to the potential for pyrethroid resistance. Although the mechanism of DDT resistance in U.S. *Heliothis* populations was never adequately elucidated, it seems likely that a *kdr*-like (target site) resistance may have been involved. Since the same mechanism confers resistance to both DDT and pyrethroids (Plapp and Hoyer 1968, Nicholson and Miller 1985, Sparks et al.; 1988), the appearance of pyrethroid resistance was not a complete surprise (Sparks 1981).

III. Development of Resistance Monitoring Technique

The occurrence of resistance in west Texas in 1985 followed a similar experience in Australia during the 1983–1984 season (Gunning et al 1984) and prompted Texas entomologists to search for a useful resistance monitoring method. Previous tests to demonstrate resistance in *Heliothis* had usually involved collecting larvae in the field, bringing them to the laboratory, rearing them to adulthood, and testing the progeny. This process is so time-consuming, usually requiring at least a month, that it cannot be considered to provide a timely estimate of a resistance problem. That is, by the time the presence of resistance could be established, it was too late to recommend alternative strategies.

What was needed was an assay that could give an estimate of resistance within 24 hours. Therefore, we decided to develop a strategy based on monitoring for resistance in adult tobacco budworms. Fortunately, the existence of widespread population monitoring for both *H. zea* and *H. virescens* with pheromone traps provided a ready source of insects.

The system was devised in the spring of 1986 (Plapp et al. 1987). By this time we had gained considerable knowledge about the nature of resistance in tobacco budworms. Neonate larvae of field populations collected in Texas, Louisiana, and Mississippi in 1985 and 1986 were resistant to several pyrethroids but not to other types of insecticides (Luttrell et al. 1987, Campanhola and Plapp 1989a, Leonard et al. 1988a, 1988b). These data indicated resistance was of the target site type. Insects with this type of resistance are usually resistant to all insecticides with the same mode of action (Plapp 1976a). We also knew, based on previous studies with house flies and mosquitoes (Plapp and Hoyer 1968), that target site resistance to DDT and pyrethroids is usually expressed to a similar level in all life stages. That is, resistance can be determined by either adult or larval bioassays. Based on this knowledge, we assumed that measurements of resistance in tobacco budworm adults might provide a satisfactory estimate of resistance in their phytophagous progeny. We chose cypermethrin as our test pyrethroid because it was widely used and was the standard for *Heliothis* control in the United States at the time. We also elected to use a residue exposure method to test for resistance. The method involved exposing pheromone trap-collected adult males to residues of cypermethrin in 20-ml glass vials (Plapp 1979, 1987). Laboratory tests (Plapp 1987) indicated that cypermethrin residues were very persistent when prepared in this way. There was less than 10% loss of insecticide within 90 days in treated vials held in the laboratory at room temperature.

Initially, we determined the toxicity of cypermethrin to adults of a laboratory-susceptible strain of tobacco budworm. We then started monitoring the response of field-collected tobacco budworm males from the College Station area. The first tests were run at doses starting at about an LC_{75} for susceptible tobacco budworm males (2.5 μg insecticide per vial) and several higher doses (5, 10, and 25 μg per vial). As the season progressed and the response of the insects changed, the dosages used were adjusted accordingly. The highest dose used was 100 μg per vial.

In the summer of 1986, extension entomologists at numerous locations in Texas were provided with cypermethrin-treated vials to field-test the method. Late in the 1986 season, treated vials were provided to Dr. J. B. Graves at Louisiana State University and to Marvin Wall, extension entomologist at Mc-Gehee, Arkansas. These workers tested field-collected moths in Louisiana and Arkansas in August and September, 1986, and demonstrated the presence of resistance in both states (Plapp et al. 1987). Similarly, Roush and Luttrell (1989) in Mississippi prepared treated vials and compared the adult assay with several larval assays for resistance that they were evaluating. They found resistance in their state; furthermore, they demonstrated an excellent correlation between resistance as determined by the adult vial test and resistance based on several larval assays.

IV. Development of Resistance Management Strategy

The demonstration of widespread resistance to pyrethroids in the tobacco budworm throughout the mid-South in 1986 prompted state entomologists to propose resistance management strategies. In this endeavor they were greatly aided by the similar Australian experience (Gunning et al. 1984) and the management strategy adopted there (Forrester 1985, Daly and McKenzie 1986, Forrester and Cahill 1987, Chapter 5). The Australian strategy involved restricting pyrethroid use to not more than three treatments during a 42-day period against one generation of *H. armigera*. For early-season control, the use of endosulfan was recommended. For late-season control, alternate insecticides, neither pyrethroids nor endosulfan, were to be employed.

Entomologists from Arkansas, Louisiana, and Mississippi—the three major cotton-producing states of the U.S. mid-South—met late in 1986 and proposed a resistance management strategy in which pyrethroid use would be delayed until the start of the second generation of tobacco budworms, approximately the first of July (Anonymous 1986). All early-season insect control for other pests as well as *Heliothis* was to be accomplished with alternate insecticides, such as organophosphates and carbamates. Pyrethroid use was recommended midseason and late season; alternate insecticides were to be used if pyrethroid resistance became a problem late in the season. The pyrethroid use period was timed to start when bollworm populations moved into cotton from corn, sorghum, and other alternate hosts. Endosulfan, specifically recommended as an early-season alternative in Australia, was excluded from the U.S. recommendations based on lack of efficacy and availability.

A very similar resistance management program was proposed for Texas (Plapp 1987, Frisbie and Plapp 1987). Pyrethroid use was to be avoided early season and restricted to the critical midseason period. The use of chlordimeform in combination with pyrethroids was recommended, particularly after resistance was first suspected.

The management programs were implemented for the 1987 season and continued in 1988 and 1989. From all reports, compliance levels have been high (80–90% in most areas where resistance has been a problem), and as described later, the programs seem to be working.

An important component in the success of the U.S. programs has been the backing and cooperation of interested components of the pesticide industry. Pyrethroid producers, working through "PEG–US" (Pyrethroid Efficacy Group– United States), have supported the restricted pyrethroid use program and also participated in and partially funded the extensive national monitoring program instituted in 1987. A report on activities of this group has recently been prepared (Riley 1989).

V. Resistance Monitoring and Success of the Management Strategy

The management strategy described earlier was adopted essentially in the absence of data indicating its possible effectiveness. Neither adequate monitoring data nor an understanding of the basis of field resistance in the tobacco budworm was available. Since the inception of the program, more than 50,000 pheromone trap–collected tobacco budworm males have been tested for resistance in the five states of Arkansas, Louisiana, Mississippi, Oklahoma, and Texas. Smaller numbers have been tested in other cotton-producing states, coast to coast. In 1986, tests were run at doses of 5, 10, 25, and 50 μg cypermethrin per vial. In 1987, the 50-μg dose was dropped, and in 1988, the 25-μg dose was dropped. With experience we gained confidence in the use of the two lower doses only. As described earlier (Plapp 1987), the dose of 5 μg cypermethrin per vial approximated the LC_{50} for heterozygous resistant males and the LC_{98} for susceptible males collected in the Brazos Valley near College Station in 1986. At 10 μg cypermethrin per vial, all susceptibles and more than 90% of heterozygotes died, whereas all resistant homozygotes lived.

Based on the assumption that resistance is due to a single gene, we propose that the Hardy–Weinberg formula $(p + q)^2$ for a single incompletely recessive resistance gene is applicable to the observed field data. In this formula, where p represents the frequency of the susceptible allele of the gene for resistance and q is the resistance allele, measurements of resistance gene frequency can easily be derived. Thus, percentage of survival at 5 is q, which is $q^2 + 50\%$ of 2 pq, whereas percentage of survival at 10 gives q^2 and the square root of this value provides a second measurement of resistance gene frequency.

If this model is correct and resistance is due primarily to a single incompletely recessive gene, estimates of resistance gene frequencies at 5 and 10 μg per vial will yield similar results. If, on the other hand, resistance is due to more than one gene, the estimates of resistance at these two doses will not yield comparable results. The results of multiple measurements from 1986 through 1988 (Plapp et al. 1989) indicate the former hypothesis is usually close to observed responses. Based on these results, the hypothesis that pyrethroid resistance is primarily due to a single incompletely recessive gene seems valid.

Therefore, the two doses of 5 and 10 μg per vial represent useful discriminating doses for pyrethroid resistance monitoring with the tobacco budworm. As discussed by Roush and Miller (1986), the use of discriminating doses is a highly efficient way to monitor for resistance in a mixed population and has numerous advantages over trying to establish complete dose–mortality lines (Chapter 2). The use of discriminating doses has also been endorsed by Sawicki and Denholm (1987), who considered it to be one of the vital techniques needed for satisfactory resistance management.

Resistance monitoring data for the five-state region of Texas, Oklahoma, Mississippi, Louisiana, and Arkansas for 1986–1988 (Plapp et al. 1989) showed that resistance is present in all states of the area. They also show that the management strategy appears to be working (Table 9.1). That is, for the most part, resistance is not increasing from year to year.

Characteristically, resistance in an area seems to follow a fairly consistent pattern. A summary of data for southeast Arkansas (Table 9.2) shows a high level of resistance late in the 1986 and 1987 seasons. However, the resistance level was low early in the season (1987) or declined in June (1988) and did not reach levels of more than 30% until August of either year. This pattern clearly shows the advantages associated with nonuse of pyrethroids early in the season and attests to the validity of a resistance management strategy based on restricting early season use of pyrethroid insecticides. The decline in resistance from values

Table 9.1. Resistance to Cypermethrin in Tobacco Budworm Males in the Southern U.S.*

| State | (% Survival at 10 μg Cypermethrin/Vial | | |
	1986	1987	1988
Arkansas	39	23	16
Louisiana	37	15	16
Mississippi	—	27	16
Oklahoma	—	25	41
Texas	25	18	23

*Data from Plapp et al. (1989).

seen at the start of the season until the midsummer time when pyrethroid use is recommended is evidence for this. These declines are probably due to decreased fitness of resistant insects as well as to the incompletely recessive nature of the major resistance gene. Data on fitness effects are described in Section VI of this chapter.

At the same time states were monitoring for resistance with two doses of cypermethrin, PEG–US was conducting similar tests throughout the United States, but with a wider range of doses. The doses used in PEG–US assays ranged from 1 to 100 μg cypermethrin per vial. Results of 1988 tests (Riley; 1989, Riley and Staetz 1989) showed almost no resistance in the southeastern states of Georgia, Florida, Tennessee, and South Carolina and limited resistance in Alabama (presumably because of less intense cotton production and greater availability of alternative host crops). Resistance was shown in the mid-South, Oklahoma, and Texas, and results were similar to those obtained by state entomologists (Plapp et al. 1989). Indeed, some of the data in both reports overlap. The PEG measurements showed no resistance in New Mexico and high resistance in Arizona and the Imperial Valley of California.

Overall, the data on adult resistance monitoring have provided baseline data adequate to measure the success or failure of future tobacco budworm resistance management efforts. The present authors favor use of discriminating dose tech-

Table 9.2. Percent Resistance in Tobacco Budworms by Month in Arkansas*

| Year | % Survival of 10 μg Cypermethrin/Vial | | | | |
	May	June	July	Aug.	Sept.
1986					56
1987	7	18	33	30	61
1988	36	9	28	38	28

*Data from Plapp et al. (1989).

niques (see Roush and Miller 1986) on the basis that (1) they are highly efficient and (2) they give information of more immediate usefulness to producers than do dose–mortality assays. The question that needs answering is how many of the insects are resistant, not how resistant are all the insects.

VI. Nature of Resistance: Extent, Biochemistry, and Genetics

Major factors that may confer resistance to pyrethroids in the tobacco budworm include changes at the target site, a mechanism similar to the *kdr* gene for knockdown resistance to DDT and pyrethroids in the house fly (Chapter 4), and an increased ability to detoxify insecticides (metabolic resistance). In this chapter we shall, for the sake of brevity, refer to target site resistance as *kdr*-like. In addition, a decreased rate of insecticide uptake may contribute to resistance, as may changes in insect behavior. Based on studies published to date, both target site and metabolic resistance are present in the tobacco budworm, as are changes in insecticide uptake. Presently, there are no data on behavioral resistance in this insect, although resistance of this type has recently been proposed for the tobacco budworm (Sparks et al. 1988).

The first study of resistance mechanisms to pyrethroids in tobacco budworms was that of Nicholson and Miller (1985). They measured the fate of *trans*-permethrin in larvae of a susceptible strain of tobacco budworm from Arizona and several resistant populations from the Imperial Valley of southern California. They concluded that increased detoxification of the insecticide was the major resistance mechanism in their populations. In addition, they reported evidence for target site insensitivity in at least a portion of their field strains. The authors found, as have other researchers since, that resistance in their field strains tended to be easily lost when the insects were reared in culture, even when regular selection with insecticides took place.

Dowd et al. (1987) also studied resistance in tobacco budworms from the Imperial Valley collections. They reported evidence for higher levels of pyrethroid hydrolases in resistance as compared with susceptible tobacco budworms and concluded that increased hydrolytic detoxification was an important mechanism of resistance in the Imperial Valley strain.

Most studies on the biochemistry of resistance have utilized the resistant strain maintained by ICI Americas at Goldsboro, North Carolina, and designated as PEG–87, PEG–88, and so on, depending on the time it was sampled. The strain was established from multiple field collections of resistant insects. In addition, it has been subjected to regular selection with pyrethroids (topical application to larvae) in the laboratory. A difficulty in using the strain has been that resistance tends to decline rapidly in the absence of continued selection pressure. Thus, Campanhola and Plapp (1989c) reported that the 1,000-fold resistance to cypermethrin present in third instars of a sample of the PEG–87 strain declined to 100-fold within four generations. Payne et al. (1988) made similar observations. In addition, resistance in first-instar larvae in different samples of the PEG strain has been quite variable in our own experience, with resistance ratios for cypermethrin ranging from 10- to 77-fold in different samples (Campanhola and Plapp 1989b, Bagwell 1989).

In a recent study, third instars of the PEG strain were compared with third instars of a susceptible strain of tobacco budworm for mixed function oxidase (MFO) activity (Little et al. 1989). Data obtained indicated a several-fold increase in enzyme activity in the resistant as compared to the susceptible strain. Based on these results, the authors concluded that increased oxidative activity is a major mechanism of resistance in the PEG strain. In a second study, Lee at al. (1989) compared the metabolism of cypermethrin in susceptible strains of *H. virescens* and *H. armigera*. The latter is an Australian species where resistance was found to depend primarily on MFOs (Forrester 1988, Chapter 5). *H. virescens* had more MFO activity than *H. armigera*, additional evidence for the importance of mixed-function oxidase activity in the tobacco budworm.

Campanhola and Plapp (1989b,c) measured the effect of piperonyl butoxide and chlordimeform as synergists for cypermethrin against first and third instar larvae of susceptible and PEG strains of tobacco budworms. Chlordimeform may be a target site synergist for pyrethroids (Chang and Plapp 1983b), whereas piperonyl butoxide is known to block oxidative detoxification of insecticides (Chapters 3 and 4). With first instar larvae of both strains, chlordimeform was a more effective synergist than piperonyl butoxide and synergism was greater against susceptible insects (Fig. 9.1). In contrast, when a three-way combination of cypermethrin plus both synergists was tested, more synergism (34-fold vs. 10-fold) was seen with resistant neonates. In tests with third-instar larvae, there was only slight synergism by both chlordimeform and piperonyl butoxide against susceptible larvae. In contrast, both synergists produced major increases in toxicity in resistant larvae. The three-way combination produced greater than 250-fold synergism, reducing the LC_{50} from more than 1,000 μg cypermethrin per vial to less than 4 μg. Based on these findings Campanhola and Plapp concluded that both target site and metabolic resistance are present in the PEG strain. In first-instar larvae and in adults, target site resistance seems to be the more important factor, whereas in third instars, metabolic resistance makes a major contribution.

The real question is which is the more important resistance mechanism in the field? Studies with the PEG strain cannot really address this question because the strain has been intensively selected in the laboratory to much higher levels of resistance than any found in the field, presumably by the addition of resistance mechanisms not prevalent in the field. In tests with third-instar larvae in Mississippi and Louisiana (Luttrell et al. 1987, Graves et al. 1988, Leonard et al. 1988a,1988b), only low levels of resistance (50-fold or less) were observed with most pyrethroids. These levels, far less than those observed for third-instar ICI larvae, are about those to be expected of target site resistance combined with occasional heterozygosity for metabolic resistance.

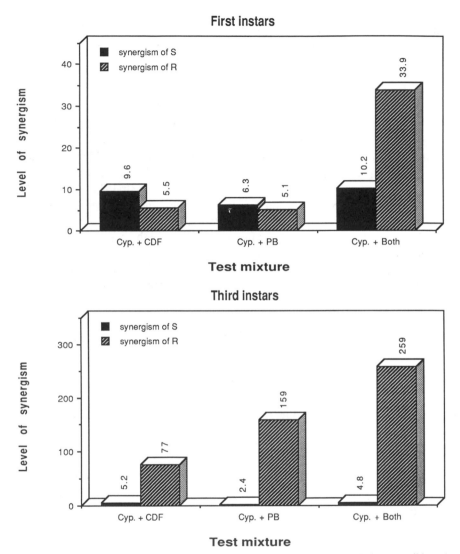

Figure 9.1. Cypermethrin synergism by chlordimeform and piperonyl butoxide in susceptible and resistant tobacco budworms. (Data from Campanhola and Plapp [1989b, 1989c]).

Several lines of evidence suggest that target site resistance is the prime resistance factor. One is the excellent correlation between resistance in adults as measured by the monitoring technique and control difficulties in the field (Roush and Luttrell 1989). Since the monitoring of adults measures primarily target site resistance (the major mechanism expressed in adults), the correlation suggests target site resistance is the mechanism usually present. Second, measurements of resistance in F_1 and F_2 generations of populations collected in the field and in

Texas have provided evidence that target site resistance to pyrethroids is the major mechanism present (Campanhola and Plapp 1989a). Additional evidence for the primacy of target site resistance includes the demonstrated instability of a major factor associated with piperonyl butoxide suppressible resistance in the PEG strain (Campanhola and Plapp 1989c) and evidence for a fitness deficit in field populations that seems to be associated with *kdr* (McCutchen et al. 1989). Finally, control measures are usually aimed at first- and second-instar larvae. This is because these larvae are usually wandering on the plant and feeding on terminal growth. In contrast, third instars and larger larvae are usually feeding in cotton fruit and thus are harder to control with insecticide residues applied to plant surfaces (Roush and Luttrell 1989). Based on the preceding discussion, resistance studies with third instars may be irrelevant to understanding resistance unless it can be shown that resistance mechanisms demonstrated in third instars are also important to resistance expression in first and second instars.

Extensive bioassays of first and third larvae performed in College Station (Campanhola and Plapp 1989b,c) support the hypothesis that although two resistance mechanisms are present, target site insensitivity is the more prevalent. All tests of Texas field-collected strains have shown broad-spectrum resistance to pyrethroids in neonates, and there was no resistance to alternate types of insecticides in this stage. In addition, limited tests with third instars of field-collected Texas populations failed to reveal high levels of resistance such as those present in the PEG strain. As stated earlier, similar results were obtained in Mississippi and Louisiana, where 50-fold or less resistance was seen in third instars. Thus, the data indicate that low levels of resistance characteristic of target site insensitivity are always present, whereas high levels of resistance are present only in strains maintained under high selection pressure in the laboratory.

Only one major study of the genetics of pyrethroid resistance in tobacco budworms has been reported (Payne et al. 1988). In this investigation, which utilized the highly resistant PEG strain, the authors concluded that resistance was inherited as a single, major, incompletely recessive gene. This is analogous to the situation with the *kdr* gene for pyrethroid and DDT resistance in the house fly (Plapp and Hoyer 1968). However, a careful analysis of the data presented in the Payne et al. study (see particularly Figure 2, their paper) suggests that an additional major resistance gene (possibly metabolic resistance) was present as a heterozygote in the resistant strain. Overall, the evidence for a single major incompletely recessive *kdr*-like gene for target site resistance is similar to that seen recently in other studies of resistance (see citations in Payne et al. 1988, Argentine et al. 1989, Chapter 4).

A possible difficulty in accepting the primacy of target site resistance may lie in the difficulties in measuring it as opposed to metabolic resistance. The latter is readily measured in studies on the in vivo and in vitro fate of radiolabeled pyrethroids with techniques familiar to most toxicologists. In contrast, measurements of *kdr*-like resistance require electrophysiological techniques that are less readily available in many laboratories (Chapter 3). An alternative approach to measuring *kdr*-like resistance in tobacco budworms was described by Sparks et al. (1988), who utilized the "hot-probe" technique of Bloomquist and Miller (1985). Widespread adoption of this procedure might ease acceptance of the importance of *kdr*-like resistance in *Heliothis*.

VII. Biological Costs of Resistance to the Tobacco Budworm

For a resistance management plan to work, there must be biological costs to the insect that are associated with resistance. If not, the best that could be done would be to delay the loss of insecticide effectiveness. On the other hand, if the costs of resistance to the pest are high enough, successful long-term management is possible.

The monitoring data accumulated to data suggest that the tobacco budworm pyrethroid resistance management program is succeeding. The reasons are not immediately apparent and have been variously ascribed to the weather, low levels of pests, or just plain luck. These viewpoints represent a failure to appreciate the value of the main reason for success, the abolition of early-season pyrethroid use on cotton.

In our laboratory we have performed several studies that have provided data on the costs of resistance to the tobacco budworm. In the first of these, Campanhola (1988) measured reproductive success in a population of a laboratory susceptible strain in comparison with the PEG–US strain. Several parameters were measured, including developmental time, egg production, and fertility. Campanhola found a slight, but statistically significant, increase in developmental time for insects of the resistant strain. In contrast, differences in fecundity were very large. Susceptible females produced an average of more than 2,500 eggs per female, whereas resistant females produced about 1,200 eggs each. In addition, fertile eggs were produced by 94% of susceptible females compared with only 63% of resistant females. These two deficits, 50% in egg production and 33% for number of females laying eggs, represented a total fitness deficit of approximately 67% for the resistant as compared to the susceptible strain. These studies have been criticized (Roush, personal communication) on the basis that the strains were of different origins and are essentially uncomparable. However, similar data were obtained for house flies with metabolic resistance to insecticides (Roush and Plapp 1982), and in both cases, the data suggest that reduced fecundity is a consequence of metabolic resistance to insecticides.

Monitoring data indicated that the main loss of resistance each season occurs during the first generation, when most tobacco budworms are feeding on wild hosts rather than on cotton. McCutchen et al. (1989b) investigated reasons for this early-season loss of resistance. In one test, numbers of males attracted by live susceptible or resistant females from laboratory strains caged in pheromone traps were measured. With one female per trap, susceptibles captured on average

4.6 times as many males as resistant females. With three females per trap, susceptibles attracted three times as many males as resistant females. However, at six females per trap, the numbers of males attracted were approximately identical for the two strains. Also, comparisons were made of pheromone production between susceptible and resistant females. GLC experiments indicated nearly twice as much pheromone was produced for each susceptible female as for each resistant female.

In other studies, the response of resistant and susceptible males was compared by measurements of resistance in adult males collected in pheromone traps as compared with resistance in adult males collected by hand and sweep net in adjacent fields. Data from six comparisons indicated greater susceptibility to pyrethroids (42.5%, $n = 296$) in pheromone trap–collected males than in hand and sweep net–collected males (20.25%, $n = 198$). That is, resistant males were less responsive to pheromone than susceptible males. If this is true, it means that estimations of resistance based on tests of pheromone trap–collected males may underestimate the actual extent of resistance in field populations of the tobacco budworm.

The experiments showed differences between strains that may explain the success of the management programs. The decreased mate attractiveness of resistant females and the decreased responsiveness of resistant males can explain the early-season decline in resistance. Since the experiments with both field males and laboratory females yielded differences, the differences appear to relate to resistance and not to origin of the strains. Again, the critical factor for resistance management seems to be avoidance of selection with pyrethroids early in the season when pest numbers are low.

The question arises, "Why has a mating deficiency as described here not been reported previously?" We have, for example, maintained *kdr* strains of house flies in this laboratory for years without observing any of the phenomena described here. The answer seems to be that the tobacco budworm experiments were largely field tests, and measurements made under these conditions allowed the expression of mating deficiency patterns.

Another question is, "Why should changes in mating success, pheromone production, and pheromone responsiveness relate to target site resistance to pyrethroids?" The answer is we don't know and probably won't until the nature of target site resistance to pyrethroids is better understood at the molecular level.

VIII. Field Test of Insecticide Efficacy in Relation to the Resistance Management Strategy

The proposed management plans for pyrethroid resistance are based on limiting pyrethroid use to one generation per year in Australia and one of two generations per year in the United States. We conducted a field test of this plan in the Brazos Valley area near College Station, Texas (Bagwell et al. 1989). This is an area where resistance problems have been severe. In the experiments, we tested potential alternative insecticides early in the season, pyrethroids midseason, and potential alternative insecticides late in the season. The purpose was to evaluate efficacy of the proposed alternative insecticides under actual field conditions when the management strategy was followed.

The results (Table 9.3) revealed that several alternative insecticides worked as well as a cypermethrin standard early in the season. Outstanding were an acephate–amitraz combination, endosulfan–amitraz, and a mixture of acepthate, endosulfan, and amitraz. Midseason, six pyrethroids combined with the formamidine amitraz were compared for efficacy. All produced satisfactory control until resistance levels as monitored by both adult and larval testing reached high levels. A cypermethrin–amitraz mixture was more effective than cypermethrin only. Late season, several alternative insecticides and mixtures were tested. Effective control was obtained with several organophosphates, a carbamate, and a mixture of cypermethrin with amitraz and piperonyl butoxide. Overall, the results supported the management plans by demonstrating that pyrethroids are effective for only part of the growing season in areas where resistance is a problem and showing that satisfactory alternatives are available as substitutes for pyrethroids.

IX. The Role of Formamidines in Pyrethroid Resistance Management

The pyrethroid insecticides used on cotton are generally halogenated and resistant to biodegradation, particularly in comparison with organophosphate insecticides. Therefore, antimetabolic synergists such as piperonyl butoxide have not been useful in combination with pyrethroids for *Heliothis* control, at least as long as no resistance was present. As reported earlier in this chapter, combinations of cypermethrin and piperonyl butoxide never showed more than marginal synergism against first or third instars of a susceptible strain or first instars of the PEG-resistant strain.

Table 9.3. Effective *Heliothis* Insecticides in Relation to Proposed Resistance Strategy.*

Time	Effective Insecticide and Doses (kg/ha)
Early season (22 June–11 July)	Acephate + amitraz (1.12 + 0.14), endosulfan + amitraz (0.84 + 0.14), acephate + endosulfan + amitraz (0.56 + 0.25 + 0.14)
Midseason (16–21 July)	Cypermethrin (0.07), esfenvalerate (0.04), biphenthrin (0.07), cyhalothrin (0.04), cyfluthrin (0.037), tralomethrin (0.02), all in combination with amitraz (0.14)
Late season (15–19 August)	Sulprofos (1.12), chlorpyrifos + profenofos (0.56 + 0.56), methomyl (0.67), Cypermethrin + amitraz + piperonyl butoxide (0.056 + 0.14 + 0.28)

*Data from Bagwell et al. (1989).

The formamidine compound chlordimeform has long been used on cotton, where it has served several functions. Initially, it was used on the basis of its ovicidal activity. However, it is also an effective synergist for pyrethroids and other insecticides against both first and third instars of susceptible and resistant tobacco budworms (Plapp 1976b, 1979; Campanhola and Plapp 1989b,c). Synergism was often greater with first instars than with third instars.

The use of formamidines may serve to delay the onset of resistance. Selection with a chlordimeform–permethrin combination failed to result in resistance in a tobacco budworm population, even though selection with permethrin alone resulted in the rapid development of resistance (Jensen et al. 1984). In a second study, a substrain of the population selected with the combination was selected with permethrin only. Again, resistance development was rapid (Bohmann et al. 1988). The reasons for the lack of resistance development in tobacco budworms selected with the chlordimeform–pyrethroid combination are unknown, and no rationale is immediately apparent. Nevertheless, the data suggest that use of formamidines in combination with pyrethroids may be an effective way to prevent, or at least slow down, resistance development.

Concern over carcinogenicity resulted in removal of chlordimeform from the market in the rest of the world by 1987 and in the U.S. Cotton Belt at the end of the 1989 season. The loss of this material has prompted a search for both alternate ovicides and alternate pyrethroid synergists. Bagwell (1989) evaluated the formamidine miticide and ovicide amitraz as a substitute for chlordimeform as a pyrethroid synergist against tobacco budworm in both field and laboratory experiments. Results of laboratory tests indicated amitraz synergism of pyrethroids against first instars of both susceptible and pyrethroid-resistant strains of the tobacco budworm. Generally, the level of synergism was less than that with chlordimeform. These results confirmed earlier findings of Plapp (1979), who showed amitraz synergism against third instars of a susceptible tobacco budworm strain, but less synergism than was observed with chlordimeform. Bagwell also showed that amitraz itself was strongly synergized against resistant, but not susceptible neonates by piperonyl butoxide. In addition, he found synergism of cypermethrin by amitraz was increased upon addition of piperonyl butoxide. These results were similar to those obtained by Campanhola (1988) for combinations of cypermethrin with chlordimeform and piperonyl butoxide in laboratory tests with resistant insects (Fig. 9.1).

Amitraz is a dimeric formamidine and probable pro-insecticide that apparently must be activated by cleavage to the monomeric form (Knowles and Hamad 1989). The formamidine chemical BTS–27271 (half amitraz) is the activation product of amitraz. Bagwell showed this material to be more active as a synergist than amitraz in laboratory tests. In addition, synergism was observed with 1:1 molar and 1:1 weight mixtures of amitraz and half amitraz. Generally, half amitraz produced maximum synergism when insects were exposed to fresh residues while amitraz synergism persisted longer.

Bagwell et al. (1989) also evaluated amitraz synergism of pyrethroids and other insecticides in field tests. In a comparison between cypermethrin and cypermethrin + amitraz, they found that the combination yielded better control of midseason *Heliothis* on cotton, particularly on the upper region of treated plants. The data suggested that activation of amitraz in the field may have been responsible for the observed synergism. In a separate test, evidence was obtained that a cypermethrin–amitraz–piperonyl butoxide combination was effective against a mixed resistant and susceptible late season *Heliothis* population. The results indicate that amitraz has the potential to replace chlordimeform as a pyrethroid synergist on cotton. Additional laboratory and field studies are needed to evaluate this possibility. Amitraz is not yet labeled for use on cotton in the United States, but registration is possible before the 1991 season.

X. Future Research Needs

The research reviewed in this paper has described the development and implementation of a successful resistance management strategy for tobacco budworm on cotton. The strategy was born out of necessity; producers had to manage resistance or quit growing cotton. Additional research needs must be addressed if this or any similar strategy is to succeed. The following section briefly addresses some of these.

A. Formamidine Synergism in Relation to Target Site Effects of Pyrethroids

The work reviewed in this chapter has shown the usefulness of formamidines when combined with synthetic pyrethroids for control of the tobacco budworm. Two effects have been demonstrated. These include the synergism of pyrethroids by formamidines and the lack of resistance development in a tobacco budworm population selected with a pyrethroid–formamidine combination as compared to selection with the pyrethroid only.

The mechanisms by which formamidines produce synergism and block resistance development remain unknown. In view of the usefulness of formamidines in resistance management, an elucidation of these mechanisms is a high-priority research need.

B. Pyrethroid Resistance in the Bollworm

A potentially very serious problem in pyrethroid resistance management on cotton concerns the bollworm, *H. zea*. Numbers of bollworms are frequently greater in cotton than numbers of the tobacco budworm. If resistance were to develop in the bollworm, it would represent a serious threat to successful *Heliothis* resistance management on cotton.

Entomologists in the United States have done limited monitoring for pyrethroid resistance in pheromone trap–collected bollworm males since 1986. Initial tests were performed with 5 and 10 μg cypermethrin per vial, the same doses used to test for resistance in tobacco budworms. Results (unpublished data, this laboratory) revealed almost no survival at these doses.

Tests performed in this laboratory (unpublished results) revealed that susceptible strain bollworm males were about 10 times more sensitive to cypermethrin that budworm males with an LC_{50} of between 0.1 and 0.25 μg cypermethrin per vial and an LC_{98} of about 1μg per vial. Accordingly, in 1988 and 1989 monitoring of bollworms has been performed at cypermethrin doses of 0.5, 1, and 2.5 μg cypermethrin per vial. Responses in May and June, 1989, from southeast Arkansas and the Corpus Christi, Texas, areas show (Table 9.4) much higher survival in the Texas samples than in the Arkansas samples and suggest that resistance may be present in Texas. Certainly, the data suggest that monitoring for pyrethroid resistance in the bollworm should be continued.

C. Biochemical Characterization of Target Site Resistance

Insecticide resistance may depend on decreased uptake of toxicants, increased breakdown, or changes at target sites. As described earlier, much bioassay work in the United States has suggested the presence of broad-spectrum, low-level (less than 50-fold) resistance to pyrethroids in the field in U.S. tobacco budworm populations. Such resistance is characteristic of target site resistance associated with *kdr*-like mechanisms. However, proof of this hypothesis has been slow to appear, mainly because of difficulties in making biochemical measurements of *kdr*-like resistance genes.

In Australia, Gunning (1988) described electrophysiological techniques for measuring *kdr* in resistant *H. armigera* populations. In the United States, Sparks et al. (1989) utilized a hot probe technique to demonstrate the presence of *kdr*-like resistance in *Heliothis* larvae. However, use of the hot-probe technique has not been widely adopted for demonstration of *kdr*, mainly because the relationship between response to the hot probe and *kdr* is not immediately obvious. Alterna-

Table 9.4. Responses of Adult Male Bollworms to Cypermethrin*

Dose of Cypermethrin (μg/vial)	% Survival of Adult Male Bollworms in Cypermethrin			
	May 1989		June 1989	
	AR	TX	AR	TX
0.5	45	85	25	84
1.0	16	75	19	68
2.5	3	18	5	46

*Data from McCutchen et al. (1989b).

tively, measurements of sodium channels, the probable pyrethroid target site, have so far failed to resolve the problem. Several reports (Chang and Plapp 1983, Kasbekar and Hall 1988, Rossignol 1988) have indicated a decline in numbers of target sites may be the mechanism of *kdr*-like resistance in insects. Other research (Grubs et al. 1988, Pauron et al. 1989, Chapter 4) has indicated that changes in the affinity of resistant-strain sodium channels are more likely to be the mechanism of resistance. Resolution of this problem seems critical to future efforts to develop adequate biochemical methodologies to characterize target site pyrethroid resistance in insects.

D. Monitoring for Resistance in Larvae

Adult monitoring of pheromone trap–collected tobacco budworm males has proved valuable in determining the distribution and extent of pyrethroid resistance. As reported recently (Riley 1989, Riley and Staetz 1989, Plapp et al. 1989), more than 50,000 tobacco budworm adult males have been tested for resistance. The technique is useful because the type of resistance present seems to be expressed to a similar level in adults and neonate larvae, the stage at which control is normally directed. However, dispersal of adults may mean that resistance estimates based on adult collections may not accurately reflect the situation in larvae, the actual pest stage, at a particular location.

Tests on field-collected neonate *Heliothis* larvae would have considerable utility in making treatment decisions for individual cotton fields. A technique to do this has been developed in our laboratory (McCutchen and Plapp 1988) and tested in the field in Texas (McCutchen et al. 1989b). In this procedure neonate larvae reared from field-collected eggs are exposed to a discriminating dose of 0.5 μg cypermethrin in glass vials to determine if resistance is present. The information gained is timely in that it is available when larvae hatch and a treatment decision must be made. Extensive field tests in 1988 (McCutchen et al. 1989b, Bagwell et al 1989) revealed that larval resistance estimates based on the vial technique yielded more accurate results than simultaneous adult monitoring in determining the extent of resistance present in field populations.

The larval procedure measures resistance in the total *Heliothis* population, not just in the tobacco budworm. This is advantageous to the producer who, after all, needs to control all *Heliothis,* not just the tobacco budworm. In addition, because it involves collections of eggs, parasitism can be determined. Tests in 1989 (unpublished data) indicated that in several instances there were no control problems associated with high levels of egg infestation in cotton, since parasitism levels were in excess of 90%. The data obtained to date suggest the technique has great promise for future resistance management in the field and can be used not only to decide whether or not pyrethroids will yield control, but also to determine whether or not any insecticide control is needed. The net effect should be to extend pyrethroid usefulness by eliminating unnecessary insecticide treatments and thereby reduce selection pressure on pest populations.

E. How to Use Insecticides: Mixtures vs. Rotation

The problem of managing insecticide use in relation to resistance has been discussed extensively in recent years (Sawicki and Denholm 1987, Curtis 1987, Roush 1989). If no resistance were present, the use of mixtures could be useful, simply because it is highly unlikely that an individual insect would gain resistance to different types of insecticide at the same time. Even so, there are problems with the use of mixtures, due in part to differences in persistence (Roush 1989). In the real world, insecticide resistance management is not invoked as a strategy until after a control failure occurs (i.e., until after resistance is common in the pest population). In such situations the use of mixtures seems inappropriate, since if resistance genes are present at a measurable frequency, the use of mixtures will act to concentrate these genes rapidly in the population.

Our experience with the tobacco budworm suggests that rotational use of insecticides, rather than use of mixtures, may represent an optimum strategy. In this regard our experience parallels that in Australia and supports the Australians' initial development of a rotational strategy. The real strength of this approach lies in the fact that development of resistance may impose costs on the insect, and these costs are expressed as reduced reproductive success. As reviewed previously in this chapter, both reduced fecundity and reduced mating success seem to accompany pyrethroid resistance in the tobacco budworm, and in the absence of pyrethroid use, resistant insects are rapidly replaced by their more reproductively successful susceptible compatriots.

Similar costs may be associated with target site resistance to insecticides other than pyrethroids in *Heliothis* and no doubt in other pest species. Tests of field-collected tobacco budworms in Texas in 1986 and 1987 revealed that earlier resistance to cyclodiene insecticides had disappeared and there was no resistance to endrin. This represents a dramatic loss of cyclodiene resistance since the demise of toxaphene as a cotton insecticide about 1980 and implies the presence of reproductive costs associated with cyclodiene resistance. Evidence for costs in other cases of cyclodiene resistance is reviewed in Chapter 5. Also, resistance to the carbamate insecticide thiodicarb (probably target site) was seen in one Texas field population in 1986 (Campanhola and Plapp 1989a), but it disappeared rapidly and was not seen in succeeding years.

The data obtained with both cyclodienes and cholinesterase inhibitors suggest that target site resistance may occur to these insecticides, and that it is unstable. Tests are needed on field populations to verify these findings and to attempt to establish mechanisms of resistance and biological costs responsible for the observed rapid decline in resistance.

Although the use of mixtures of insecticides may be disadvantageous, the same limitations may not apply to use of mixtures of insecticides with synergists. That is, use of mixtures in which only one component is inherently toxic seems acceptable. Based on this hypothesis, combinations of pyrethroids with formamidines or piperonyl butoxide—or even three-way mixtures of pyrethroids, formamidines, and piperonyl butoxide—may be useful.

To summarize, the use of mixtures of insecticides seems unwise since the effect is to negate the advantages associated with limiting selection with a particular insecticide type to brief time periods during the year. An alternative strategy, rotation by time of insecticides with different modes of action, has proved effective. Finally, use of insecticides plus synergists represents another strategy that may prove useful. The latter strategy seems particularly attractive based on the lack of resistance development to a combination of permethrin and chlordimeform shown previously (Jensen et al. 1984, Bohmann et al. 1988).

XI. Conclusions

The research reviewed for this study has taken place since the fall of 1985, when pyrethroid resistance was first reported in Texas. As such, the data cannot be said to have withstood the passage of time. Therefore, it is well to realize that future experience is likely to modify some of the conclusions reached and reported in this paper.

Nevertheless, the experience of the past three years has resulted in the development of a successful resistance management plan and its very rapid implementation. A major reason the scheme was implemented so rapidly was the availability of test insects based on a population monitoring scheme already in place. The critical need to manage resistance was also responsible for the rapid adoption of the strategy. Finally, although there may be differences in interpretation and emphasis in the U.S. experience as compared to that in Australia, the plan of action adopted in the United States owes a major debt to the prior experience of the Australians and their successful development of the first *Heliothis* resistance management strategy.

References

Allen, C. T., W. L. Multer, R. R. Minzenmayer, and J. S. Armstrong. 1987. Development of pyrethroid resistance in *Heliothis* populations in cotton in Texas, pp. 332–335. *In* Proceedings Beltwide Cotton Production Research Conferences, Dallas, Tex., Jan. 4–9, 1987. National Cotton Council of America, Memphis, Tenn.

Anonymous. 1986. Cotton entomologists seek to delay pyrethroid resistance in insects. MAFES Res. Highlights 49 (12): 8.

Argentine, J. A., J. M. Clark, and D. N. Ferro. 1989. Genetics and synergism of resistance to azinphosmethyl and permethrin in the Colorado Potato Beetle (Coleoptera: Chrysomelidae). J. Econ. Entomol. 82: 698–705.

Bagwell, R. D. 1989. Evaluation of a pyrethroid resistance management plan and the use of amitraz as an insecticide synergist for resistant tobacco budworms (Lepidoptera: Noctuidae) in cotton. M.S. thesis, Texas A&M University, College Station.

Bagwell, R. D., F. W. Plapp, Jr., and S. J. Nemec. 1989. Field evaluation of management plans using alternate insecticides for pyrethroid resistant tobacco budworms, pp. 359–364. *In* Proceedings Beltwide Cotton Production Research Conferences, Jan. 4–7, 1989. National Cotton Council of America, Memphis, Tenn.

Bloomquist, J. R., and T. A. Miller. 1985. A simple bioassay for detecting and characterizing insecticide resistance. Pestic. Sci. 16: 611–614.

Bohmann, D. J., T. F. Watson, L. A. Crowder, and M. P. Jensen. 1988. Repression of permethrin resistance by chlordimeform in the tobacco budworm (Lepidoptera: Noctuidae). J. Econ. Entomol. 81: 1536–1538.

Brazzel, J. R. 1963. Resistance to DDT in *Heliothis virescens*. J. Econ. Entomol. 56: 571–574.

Brown, T. M., K. Bryson, and G. T. Payne. 1982. Pyrethroid susceptibility in methyl parathion-resistant tobacco budworm in South Carolina. J. Econ. Entomol. 75: 301–303.

Campanhola, C. 1988. Resistance to pyrethroid insecticides in the tobacco budworm (Lepidoptera: Noctuidae). PhD. diss., Texas A&M University, College Station.

Campanhola, C., and F. W. Plapp, Jr. 1989a. Pyrethroid resistance in the tobacco budworm (Lepidoptera: Noctuidae): insecticide bioassays and field monitoring. J. Econ. Entomol. 82: 22–28.

Campanhola, C., and F. W. Plapp, Jr. 1989b. Toxicity and synergism of insecticides against susceptible and pyrethroid resistant neonate larvae and adults of the tobacco budworm (Lepidoptera: Noctuidae). J. Econ. Entomol. 82 1527–1533.

Campanhola, C., and F. W. Plapp, Jr. 1989c. Toxicity and synergism of insecticides against susceptible and pyrethroid resistant third instars of the tobacco budworm (Lepidoptera: Noctuidae). J. Econ. Entomol. 82 1495–1501.

Chang, C. P., and F. W. Plapp, Jr. 1983a. DDT and pyrethroids: receptor binding and mechanism of knockdown resistance (kdr) in the house fly. Pestic. Biochem. Physiol. 20: 86–91.

Chang, C. P., and F. W. Plapp, Jr. 1983b. DDT and synthetic pyrethroids: mode of action, selectivity, and mechanism of synergism in the tobacco budworm, *Heliothis virescens* F., and a predator, *Chrysopa carnea* Stephens. J. Econ. Entomol. 76: 1206–1210.

Curtis, C. F. 1987. Genetic aspects of selection for resistance, pp. 150–161. *In* M. G. Ford, D. W. Holloman, B. P. S. Khambay, and R. M. Sawicki (eds), Combatting resistance to xenobiotics. Ellis Horwood, Chichester, U.K.

Daly, J. C., and J. A. McKenzie. 1986. Resistance management strategies in Australia: The Heliothis and "wormkill" programmes, pp. 951–959. *In* British Crop Protection Conference, 1986, Brighton. British Crop Protection Council, Croydon, U.K.

Dowd, P. F., C. C. Gagne, and T. C. Sparks. 1987. Enhanced pyrethroid hydrolysis in pyrethroid-resistant larvae of the tobacco budworm, *Heliothis virescens* (F.). Pestic. Biochem. Physiol. 28: 9–16.

Forrester, N. W. 1985. Pyrethroid resistance strategy—retrospect and prospect. Australian Cotton-grower 6(3): 5–7.

Forrester, N. W. 1988. Field selection for pyrethroid resistance genes. Australian Cottongrower. 9 (3): 48–51.

Forrester, N. W., and M. Cahill. 1987. Management of insecticide resistance in *Heliothis armigera* (Hubner) in Australia, pp. 127–137. *In* M. G. Ford, B. P. S. Khambay, D. W. Holloman, and R. M. Sawicki (eds.), Combatting resistance to xenobiotics. Ellis Horwood, Chichester, U. K.

Frisbie, R. E., and F. W. Plapp, Jr. 1987. Managing insecticide resistant tobacco budworm in Texas cotton. Texas Agri. Ext. Serv., The Texas A&M Univ. System. 500–5–87, 2 pp.

Graves, J. B., B. R. Leonard, A. M. Pavloff, G. Burris, K. Ratchford, and S. Micinski. 1988. Monitoring pyrethroid resistance in tobacco budworm in Louisiana during 1987: Resistance management implications. J. Agric. Entomol. 5: 109–115.

Graves, J. B., D. R. Clower, and J. R. Bradley, Jr. 1967. Resistance of the tobacco budworm to several insecticides in Louisiana. J. Econ. Entomol. 60: 887–888.

Grubs, R. E., P. M. Adamson, and D. M. Soderlund, 1988. Binding of (^{3}H) saxitoxin to head membrane preparations from susceptible and knockdown-resistant house flies. Pestic. Biochem. Physiol. 32: 217–223.

Gunning, R. 1988. The pyrethroids-how they work and why they fail, pp. 45–53. *In* Proceedings Australian Cotton Conference, Queensland.

Gunning, R. V., L. R. Easton, L. R. Greenup, and V. E. Edge. 1984. Pyrethroid resistance in *Heliothis armigera*. (Hübner) (Lepidoptera: Noctuidae) in Australia. J. Econ. Entomol. 77: 1283–1287.

Ivy, E. E., and A. L. Scales. 1954. Are cotton insects becoming resistant to insecticides? J. Econ. Entomol. 47: 981–984.

Jensen, M. P., L. A. Crowder, and T. F. Watson. 1984. Selection for permethrin resistance in the tobacco budworm (Lepidoptera: Noctuidae). J. Econ. Entomol. 77: 1409–1411.

Kasbekar, D. P., and L. M. Hall. 1988. A *Drosophila* mutation that reduces sodium channel number confers resistance to pyrethroid insecticides. Pestic. Biochem. Physiol. 32: 132–145.

Knowles, C. O., and M. S. Hamed. 1989. Comparative fate of amitraz and N'(2,4-dimethylphenyl)-N-methylformamidine (BTS–27271) in bollworm and tobacco budworm larvae (Lepidoptera: Noctuidae). J. Econ. Entomol. 82: 1328–1334.

Lee, K.-S., C. H. Walker, A. McCaffery, M. Ahmad, and E. Little. 1989. Metabolism of *trans*-cypermethrin by *Heliothis armigera* and *H. virescens*. Pestic. Biochem. Physiol. 34: 49–57.

Leonard, B. R., J. B. Graves, T. C. Sparks, and A. M. Pavloff. 1988a. Variation in resistant of field populations of tobacco budworm and bollworm (Lepidoptera: Noctuidae) to selected insecticides. J. Econ. Entomol. 81: 1522–1528.

Leonard, B. R., T. C. Sparks, and J. B. Graves. 1988b. Insecticide cross-resistance in pyrethroid-resistance strains of tobacco budworm (Lepidoptera: Noctuidae). J. Econ. Entomol. 81: 1529–1535.

Little, E. J., A. R. McCaffery, C. H. Walker, and T. Parker. 1989. Evidence for an enhanced metabolism of cypermethrin by a monooxygenase in a pyrethroid-resistant strain of the tobacco budworm (*Heliothis virescens* F.). Pestic. Biochem. Physiol. 34: 58–68.

Luttrell, R. G., R. T. Roush, A. Ali, J. S. Mink, M. R. Reid, and G. L. Snodgrass. 1987. Pyrethroid resistance in field populations of *Heliothis virescens* (Lepidoptera: Noctuidae) in Mississippi in 1986. J. Econ. Entomol. 80: 985–989.

Martinez-Carrillo, J. L., and H. T. Reynolds. 1983. Dosage-mortality studies with pyrethroids and other insecticides on the tobacco budworm (Lepidoptera: Noctuidae) from the Imperial Valley, California. J. Econ. Entomol. 76: 983–986.

McCutchen, B. F., and F. W. Plapp, Jr. 1988. Monitoring procedure for resistance to synthetic pyrethroids in tobacco budworm larvae, pp. 356–358. *In* Proceedings Beltwide Cotton Production Research Conferences, New Orleans, La., Jan. 5–8, 1988. National Cotton Council of America, Memphis, Tenn.

McCutchen, B. F., F. W. Plapp, Jr., S. J. Nemec, and L. Nemec. 1989a. The use of larval and adult monitoring techniques for the detection and determination of the critical frequency for pyrethroid resistance in *Heliothis* spp. on cotton, pp. 348–352. *In* Proceedings Beltwide Cotton Production Research Conferences. Nashville Tenn., Jan. 4–7, 1989. National Cotton Council of America, Memphis, Tenn.

McCutchen, B. F., F. W. Plapp, Jr., H. J. Williams, and D. A. Kostroun. 1989b. Reproductive deficiencies associated with pyrethroid resistance in the tobacco budworm, pp. 364–366. *In* Proceedings Beltwide Cotton Production Research Conferences. Nashville, Tenn., Jan. 4–7, 1989. National Cotton Council of America, Memphis, Tenn.

Nicholson, R. A., and T. C. Miller. 1985. Multifactorial resistance to trans-permethrin in field-collected strains of the tobacco budworm, *Heliothis virescens*, F. Pestic. Sci. 16: 561–570.

Pauron, D., J. Barhanin, M. Amichot, M. Pralavorio, J.-B. Berge, and M. Lazdunski. 1989. Pyrethroid receptor in the insect Na' channel: Alteration of its properties in pyrethroid-resistance flies. Biochemistry 28: 1673–1677.

Payne, G. T., R. G. Blenk, and T. M. Brown. 1988. Inheritance of permethrin resistance in the tobacco budworm, *Heliothis virescens* (Lepidoptera: Noctuidae). J. Econ. Entomol. 81: 65–73.

Plapp, F. W., Jr. 1971. Insecticide resistance in *Heliothis:* tolerance in larvae of *H. virescens* as compared with *H. zea* to organophosphate insecticides. J. Econ. Entomol. 64: 999–1000.

Plapp, F. W., Jr. 1972. Laboratory tests of alternate insecticides for the control of methyl parathion-resistant tobacco budworm larvae. J. Econ. Entomol. 65: 903–904.

Plapp, F. W., Jr. 1976a. Biochemical genetics of insecticide resistance. Annu. Rev. Entomol. 21: 179–197.

Plapp, F. W., Jr. 1976b. Chlordimeform as a synergist for insecticides against the tobacco budworm. J. Econ. Entomol. 69: 91–92.

Plapp, F. W., Jr. 1979. Synergism of pyrethroid insecticides by formamidines against *Heliothis* pests of cotton. J. Econ. Entomol. 72: 667–670.

Plapp, F. W., Jr. 1987. Managing resistance to synthetic pyrethroids in the tobacco budworm, pp. 224–226. *In* Proceedings Beltwide Cotton Production Research Conferences. Dallas, Tex., Jan. 4–8, 1987. National Cotton Council of America, Memphis, Tenn.

Plapp, F. W., Jr., and R. F. Hoyer. 1968. Possible pleiotropism of a gene conferring resistance to DDT, DDT analogs and pyrethrins in *Musca domestica* and *Culex tarsalis*. J. Econ. Entomol. 61: 761–765.

Plapp, F. W., Jr., and C. Campanhola. 1986. Synergism of pyrethroids by chlordimeform against susceptible and resistant *Heliothis*, pp. 167–169. *In* Proceedings Beltwide Cotton Production Research Conferences, Las Vegas, Nev., Jan. 4–9, 1986. National Cotton Council of America, Memphis, Tenn.

Plapp, F. W., Jr., G. M. McWhorter, and W. H. Vance. 1987. Monitoring for pyrethroid resistance in the tobacco budworms in Texas - 1986, pp. 324–326. *In* Proceedings Beltwide Cotton Production Research Conferences. Dallas, Tex., Jan. 4–8, 1987. National Cotton Council of America, Memphis, Tenn.

Plapp, F. W., Jr., J. A. Jackman, C. Campanhola, R. E. Frisbie, J. B. Graves, R. G. Luttrell, W. F. Kitten, and M. Wall. 1989. Monitoring and management of pyrethroid resistance in the tobacco budworm (Lepidoptera: Noctuidae) in Texas, Mississippi, Louisiana, Arkansas, and Oklahoma. J. Econ. Entomol. 82: 335–341.

Riley, S. L. 1989. Pyrethroid resistance in *Heliothis virescens:* Current U.S. management programs. Pestic. Sci. 411–421.

Riley, S. L. and C. A. Staetz. 1989. PEG-U.S. pyrethroid resistance monitoring data for 1988, in press. *In* Proceedings Beltwide Cotton Conferences, Nashville, Tenn., Jan. 4–7, 1989. National Cotton Council of America, Memphis, Tenn.

Rossignol, D. P. 1988. Reduction in number of nerve membrane sodium channels in pyrethroid resistant house flies. Pestic. Biochem. Physiol. 32: 146–152.

Roush, R. I. 1989. Designing resistance management programs: How can you choose? Pestic. Sci. 26: 423–441.

Roush, R. T., and F. W. Plapp, Jr. 1982. The effects of insecticide resistance on the biotic potential of house flies. J. Econ. Entomol. 75: 708–713.

Roush, R. T., and D. A. Wolfenbarger. 1985. Inheritance of methomyl resistance in the tobacco budworm (Lepidoptera: Noctuidae). J. Econ. Entomol. 78: 1020–1022.

Roush, R. T., and G. L. Miller. 1986. Considerations for design of insecticide resistance monitoring programs. J. Econ. Entomol. 79: 293–298.

Roush, R. T., and R. G. Luttrell. 1989. Expression of resistance to pyrethroid insecticides in adults and larvae of tobacco budworm (Lepidoptera: Noctuidae): implications for resistance monitoring. J. Econ. Entomol. 82: 1305–1310.

Sawicki, R. M., and I. Denholm. 1987. Management of resistance to pesticides in cotton pests. Tropical Pest Management. 33: 262–272.

Sparks, T. C. 1981. Development of insecticide resistance in *Heliothis zea* and *Heliothis virescens* in North America. Bull. Entomol. Soc. Amer. 27: 186–192.

Sparks, T. C., B. R. Leonard, and J. B. Graves. 1988. Pyrethroid resistance and the tobacco budworm: Interactions with chlordimeform and mechanisms of resistance, pp. 366–370. *In* Proceedings Beltwide Cotton Conferences, New Orleans, La., Jan. 5–8, 1988. National Cotton Council of America, Memphis, Tenn.

Sparks, T. C., J. A. Lockwood, R. L. Byford, J. B. Graves, and B. R. Leonard. 1989. The role of behavior in insecticide resistance. Pestic. Sci. 26: 383–399.

Staetz, C. A. 1985. Susceptibility of *Heliothis virescens* (F.) (Lepidoptera: Noctuidae) to permethrin from across the cotton belt: a five year study. J. Econ. Entomol. 78: 505–510.

Twine, P. H., and H. T. Reynolds. 1980. Relative susceptibility and resistance of the tobacco budworm to methyl parathion and synthetic pyrethroids in Southern California. J. Econ. Entomol. 73: 239–242.

Whitten, C. J., and D. L. Bull. 1970. Resistance to organophosphorus insecticides in tobacco budworm. J. Econ. Entomol. 63: 1492–1495.

Whitten, C. J., and D. L. Bull. 1974. Comparative toxicity, absorption, and metabolism of chlorpyrifos and its dimethyl homologue in methyl parathion-resistant and - susceptible tobacco budworms. Pestic. Biochem. Physiol. 4: 266–274.

Whitten, C. J., and D. L. Bull. 1978. Metabolism and absorption of methyl parathion by tobacco budworms resistant or susceptible to organophosphorus insecticides. Pestic. Biochem. Physiol. 9: 196–202.

Wolfenbarger, D. A., and R. L. McGarr. 1970. Toxicity of methyl parathion, parathion, and monocrotophos applied topically to populations of lepidopteran pests of cotton. J. Econ. Entomol. 63: 1762–1764.

Wolfenbarger, D. A., P. R. Bodegas, and R. Flores. 1981. Development of resistance in *Heliothis* spp. in the Americas, Africa, and Asia. Bull Entomol. Soc. Am. 27: 181–185.

10

Resistance Management in Multiple-pest Apple Orchard Ecosystems in Eastern North America

David J. Pree

I. Introduction

One of the great challenges in the development of strategies to manage pesticide resistance in multipest ecosystems is that any change made in control procedures for one pest may interfere with systems for other pests or may exacerbate the development of new pests. Tree fruit ecosystems, especially apples, provide several classic examples of this. When the spotted tentiform leafminer, *Phyllonorycter blancardella,* developed resistance to azinphosmethyl in the mid-1970s, growers n the northeastern United States and Canada were forced to shift to pyrethroids and methomyl or oxamyl for control. This in turn disrupted an integrated system of control for European red mite, *Panonychus ulmi,* and two-spotted spider mite, *Tetranychus urticae*, using the phytoseiid predator *Amblyseius fallacis*. The predator had become resistant to azinphosmethyl but remained highly sensitive to pyrethroids. Consequently, control of phytophagous mites in this system reverted to a purely chemical program with predictable acaricide resistance problems.

This chapter reviews ongoing research efforts to devise resistance management strategies for these species and to describe the difficulties associated with current strategies. The review is restricted to apple orchard ecosystems in the northeastern United States and Canada, which are largely similar in pest and predator complexes. However, these systems illustrate the problems faced in many other multiple-pest agroecosystems.

Over 100 arthropod species are potential pests of apples in eastern North America (Oatman et al. 1964). Over 50 of these are relatively common in unsprayed orchards in southern Ontario (Hagley and Hikichi 1973), but only 10–15 are currently pests in commercial orchards. Six or seven of these are classified as direct pests, that is, they damage the fruit directly. The others are termed indirect pests and are feeders on foliage, wood, or roots. In Ontario and the neighboring U.S. states of New York and Michigan, somewhat similar systems of pest control for apple pests have evolved. Codling moth, *Cydia pomonella,* apple maggot *Rhagoletis pomonella,* and plum curculio *Conotrachelus nenuphar* are the most important direct pests and require regular insecticidal controls. These may be applied on a scheduled program or, preferably, timed by the use of pheromone trap catches and/or predictive phenological data. Resistance has not been a problem with these pests, although codling moth populations have, in the past, developed resistance to lead arsenate and to DDT (Glass 1960). As with many other crop systems, most of the resistance problems have occurred with the indirect pests, especially the foliage feeders (see also Chapter 11).

II. Resistance to Acaricides

The European red mite, *P. ulmi,* is the most important mite pest in all of eastern North America. Problems associated with control were increased with the introduction of DDT and later by some of the organophosphorous (OP) insecticides, at least in part because of their toxic effects on predaceous mites. In fact, it was these increasing difficulties with European red mite and induced outbreaks of several insect pests that led Pickett and his coworkers (see Pickett et al. 1946) to begin their efforts in Nova Scotia on integrated control, the forerunner of today's orchard IPM programs. Resistance to acaricides rapidly followed the first outbreaks of European red mite and since then *P. ulmi* and *T. urticae* have become resistant to a wide range of acaricides. This topic has been reviewed several times, most recently by Cranham and Helle (1985). With the evolution of native populations of the predaceous mite *A. fallacis* that were resistant to phosmet and azinphosmethyl (but not to other OPs such as diazinon or phosalone) (Croft and Nelson 1972, Croft and Stewart 1973, Watve and Lienk 1976, Herne unpublished data), coupled with the judicious use of selective acaricides such as dicofol or cyhexatin, an integrated system of mite control developed. Chemical programs, consisting mainly of sprays of azinphosmethyl or phosmet were used to control other pests and were compatible with the OP-resistant populations of *A. fallacis.* Acaricide use declined to at most one application per season (Pree, unpublished data).

With development of resistance to OP insecticides in the spotted tentiform leafminer (STLM), *P. blanchardella,* in 1975–1977, many growers began to include broader-spectrum pyrethroids or methomyl in these programs. As had been widely predicted (see, for example, Croft and Hoyt 1978), the use of these insecticides has induced outbreaks of both European red mite and two-spotted spider mite (Penman and Chapman 1988, Gerson and Cohen 1989, Pree, unpublished data). Certainly, part of this was due to the loss of effective predation by *A. fallacis.* This has forced a reliance on the repeated use of acaricides to control phytophagous mites with the predictable development of resistance.

Dicofol, which has been available for about 25 years, was not heavily used through most of the period from 1970 to 1980. Resistance had become widespread in the 1960s prior to the establishment of IPM programs for mite control (Herne

1971), and many growers had lost confidence in this product. However, after over a decade of minimal use, resistance to dicofol had declined. Pree and Wagner (1987), in a 1984 survey, detected resistance in European red mite populations at only two of 45 sites surveyed. Dennehy et al. (1988) surveyed both *P. ulmi* and *T. urticae* populations in New York in 1985–1988 and found wide variations in the susceptibility of various populations to dicofol. Both of these studies suggested that resistance to dicofol was unstable (i.e., declined when selection pressure was reduced) and might be managed at low levels. Resistance occurred less frequently with *T. urticae* populations than with *P. ulmi*. Dicofol resistance has been shown to be inherited as a single recessive or incompletely recessive gene in both *T. urticae* (Rizzieri et al. 1988) and *P. ulmi* (Pree 1987). Cross-resistance to organotin acaricides did not occur. Differences in frequencies of resistance between the two species might well be related to differences in host range and dispersal abilities (Dennehy et al. 1988). *P. ulmi* is found almost exclusively on fruit trees, whereas *T. urticae* is noted for its wide host range (Lienk et al. 1980). Thus, the potential for migration of susceptible *T. urticae* into orchard environments is much higher. Because dicofol resistance declined in *P. ulmi* during periods of light use of dicofol and because migration of susceptible individuals from outside is unlikely, it follows that there may be fitness differences between resistant and susceptible populations. However, Pree (1987) found no large differences in the various parameters measured and in further population cage tests found only a slight deviation from the phenotypes predicted by the Hardy-Weinberg equation. This suggested that if fitness differences occurred, they were probably not large and that resistance would be lost slowly over many generations.

The results of the 1984 survey by Pree and Wagner (1987), which indicated that most populations were sensitive, prompted a renewed interest in dicofol by advisory service personnel in Ontario. Dicofol was recommended to growers to be used a maximum of once per season. However, when populations were resurveyed in 1988 (Pree et al. 1989), resistance to dicofol was detected in 11 of 22 orchards surveyed. The frequency of resistant individuals ranged from 15% to 94%. Dicofol failures were reported by growers in 1987 or 1988 in at least two of the sites. Resistant individuals ranged from 80% to 94% in these populations. Based on grower's spray records, resistance was always detected where more than one application of dicofol had been made in a single year (during the years 1985–1987) and was likely to be detectable where single applications had been made in two consecutive years. The net effect was to increase the numbers of populations in which resistance was detected from 4% to 50%, similar to those indicated in Herne's (1971) survey.

These data could indicate the inadequacy of the restricted use program in current strategies and also raise questions about the adequacy of the various surveys. It is probable that the 1984 survey failed to detect small numbers of resistant individuals. In those tests, 100–200 females were tested against dicofol and cyhexatin at single discriminating concentrations. However, increases in the numbers of mites tested per sample would likely reduce the numbers of orchards surveyed. This is a major concern if such data are to be used as a basis for the selection of the acaricide to be used in a resistance management strategy. The earlier surveys (Herne 1971, Pree and Wagner 1987) used a sprayed leaf disc as a test surface. This technique requires a Potter spray tower and a supply of uniform (peach) leaves and also results in high levels of control mortality when mites collected in the field are tested directly. This also prevents calculation or even estimation of the percentage of resistant individuals in field populations.

Dennehy et al. (1987a) have devised a technique that utilizes a plastic petri dish with a tight-fitting lid as the medium. This assay requires little equipment to prepare and might allow for larger samples to be handled more readily. When compared with the leaf disc technique (Pree et al. 1989), both indicated similar resistance levels, but in field samples the petri dish assay identified resistance in more populations. The petri dish technique had lower control mortality and allowed an estimate of the frequency of dicofol resistance in the various populations. This will likely be a major advantage in future resistance management programs for dicofol. In areas where dicofol resistance is common it may be feasible to use this method to test populations prior to usage. If the criteria for usage of dicofol is whether mites will be controlled, sample sizes of 50–100 may be adequate to estimate the frequency of resistant phenotypes and the potential for control when coupled with estimates of population size. Before this can be reliably accomplished, it will be necessary to determine the frequency of resistant individuals at which control failures occur (see Chapter 2). Dennehy et al. (1988) have suggested that control failures for both *P. ulmi* and *T. urticae* are likely when resistant phenotypes are about 20% or more of the population.

Because of widespread dicofol resistance, most integrated systems developed from about 1968–1975 relied on cyhexatin or propargite for the chemical control phase. In Ontario, where an IPM system was well established, one to one and one half acaricides per season was the maximum expected use level. Similar patterns of use also developed in New York (Tette et al. 1979). This rose to three to four within three seasons of pyrethroid use in Ontario for control of STLM. Control failures with cyhexatin were first noted in 1981 in both New York and Ontario (Welty et al. 1987; Pree 1987). Resistance in both areas was widespread but did not occur in all populations. In a survey conducted in Ontario in 1984, Pree and Wagner (1987) detected (but could not quantify) resistance in 29% of the populations they surveyed. Welty et al. (1987, 1988) also found that susceptibility to cyhexatin varied widely in New York populations of ERM and that susceptibility of some of these populations decreased through the season as cyhexatin was applied. Pree (1987) showed that resistance to cyhexatin was polygenic in *P. ulmi* and suggested that resistance might decline rapidly in field situations. In laboratory populations, beginning with 1:1 ratios of R:S *P. ulmi*, cyhexatin resistance was not detectable after four to six generations. Since there are eight to 10 generations of European red mites in most production areas of Ontario, New York, and Michigan, management of cyhexatin resistance might be accomplished by restricting use to one or at most two widely spaced applications per year (Pree 1987). Edge and James (1986) report that, for *T. urticae* in Australia, resistance to organotins was also not stable but that reversion was slow and that resistance remained detectable at low levels (5%) for 160 generations.

Inheritance of resistance to cyhexatin has also been shown to be polyfactorial in the two-spotted spider mite (Croft et al. 1984; Hoyt et al. 1985) and in the Kanzawa spider mite *Tetranychus kanzawai* (Mizutami et al. 1988). However, resistance may not be genetically similar in all Tetranychidae, since Hoy et al. (1988) have recently shown that resistance to the organotins cyhexatin and fenbutatin-oxide in *Tetranychus pacificus* appeared to be due to a single, almost recessive, gene, and further, that resistance was stable under glasshouse conditions. Whether resistance selected by cyhexatin confers cross-resistance to fenbutatin oxide has not been shown for European red mite but has been demonstrated in *T. urticae* by Edge and James (1986). Flexner et al. (1988) (see also Chapter 11) have shown in the field that relaxation of selection pressures on organotin-resistant *T. urticae* for two seasons has resulted in a significant increase in susceptibility. Similar tests have not been published to show the efficacy of such strategies for European red mite. Because cyhexatin has been withdrawn from legal registration, such studies are now less critical. Fenbutatin oxide is considered less effective than cyhexatin (Herne, unpublished data) has not been extensively used on fruit trees in Ontario.

Another IPM-compatible acaricide, propargite, has not been extensively used in mite control programs on apples in Ontario but is the preferred compound in IPM programs for peach. While resistance to propargite in *T. urticae* and *T. pacificus* has been described from the western United States (Keena and Granett 1987, Hoy and Conley 1989) and New Zealand in both *T. urticae* and *P. ulmi* (Chapman and Penman 1984), it has not been reported for the eastern U.S. or Canada. In addition, Dennehy et al. (1987b) found that resistance (42-fold) in *T. urticae* did not reduce the effectiveness of propargite in field trials (see also Chapter 2). This compound, if it remains available, will likely become more important in growers' programs. The new acaricides clofentezine and hexythiazox have also been shown to be nontoxic to some species of beneficial mites (Hoy and Ouyang 1986) and have potential for use in pest management programs for mites (see also Chapter 11). However, resistance to clofentezine and cross-resistance to hexythiazox have been shown in colonies of *T. urticae* from Australia (Edge et al. 1987). In addition, unconfirmed reports of control failures with one or both of these products have come from Europe, Japan, and the eastern United States. Therefore, any integrated mite management system that is adopted must have management of acaricide resistance as a major objective. Other acaricides, such as formetanate, which are toxic to beneficial mite species (e.g., *A. fallacis*) (Watve and Lienk 1976), could be utilized along with petroleum oils in resistance management schemes but are difficult to use in integrated mite management systems.

Resistance management programs for *T. urticae* and *P. ulmi* on fruit trees are likely to consist of rotations of acaricides. The logical sequence of such rotations is still to be determined. Combinations (i.e., mixtures) seem less than acceptable because their effectiveness depends on low frequencies of resistance to each of the pesticides used (see Chapters 5 and 6). However, detectable frequencies of resistance to dicofol and organotins are already common in many populations. Still to be determined are reliable estimates of the rate of decline or increase of resistance to dicofol, which will greatly affect the acceptable frequency of applications. Obviously, the longer-term solution is to reintroduce a biological control component into mite management systems (see also Chapter 6). This component is vital in IPM systems but is destroyed when broad-spectrum pesticides are added to the system for control of other pests or where resistance to a selective pesticide has forced a change. This is precisely the situation in eastern North America, where resistance has forced changes in control strategies for the spotted tentiform leafminer (see also Chapter 11).

III. Insecticide Resistance in the Spotted Tentiform Leafminer

A. Resistance to OP's and Pyrethroids

The spotted tentiform leafminer, *P. blancardella*, has become the most important of several species of Gracillariidae that mine apple foliage in the northeastern United States (Pottinger and LeRoux 1971). Initial infestations were considered minor, but damaging populations became recurrent in the late 1960s. Severe infestations cause premature leaf drop, production of undersized fruit, and premature fruit drop, especially on the cultivars McIntosh, Greening, or Ida Red. Control recommendations (azinphosmethyl 1.0 kg a.i./ha) first appeared in Ontario spray calendars in 1971. By 1976 an outbreak of STLM occurred over most of the host range of *P. blancardella* in eastern North America. Control failures with OP insecticides were also common with *Phyllonorycter crataegella* at the same time (Weires 1977).

Populations of *P. blancardella* from southern Ontario were shown to be highly resistant (about 160-fold at the LC_{50}) to azinphosmethyl (Pree et al. 1980). Resistance has now been shown to extend to most types of OP insecticides (Pree et al. 1989b). Resistance was not overcome by addition of any of several synergists, likely precluding enhanced metabolism by esterases, oxidases, or glutathione-S-transferases in resistant populations (Chapter 3). Resistance appears due, at least in part, to selection for populations with acetylcholinesterases (AChE) less sensitive to inhibition by OP insecticides. Whether OP resistance is associated with reduced fitness and whether resistance would decline over time if selection pressures were replaced have not been studied.

There are three generations of STLM annually in southern Ontario and New York; controls have been recommended preferentially for the first generation (Fisher 1988). However, OP insecticides continue to be used several times (up to 10 are possible) for other insects in this multiple-pest system. These include codling moth, *C. pomonella*, plum curculio, *C. nenuphar;* tarnished plant bug, *Lygus lineolaris;* apple maggot, *R. pomonella;* and two or three species of aphids. This makes it difficult to develop resistance management strategies which would require removal of selection pressures for reversion of resistant populations. At this time, resistance to OP insecticides has become general in commercial orchards in Ontario (Table 10.1.)

When resistance to azinphosmethyl (OP resistance) was described in 1977 (Weires 1977, Pree et al. 1980), pyrethroids (initially permethrin and fenvalerate), methomyl, or oxamyl were developed as alternatives. Pyrethroids were the preferred strategy and control was largely directed toward the overwintering or spring population. Applications were initially timed for emergence of moths or when eggs were first detected in the spring. In 1982, resistance to pyrethroids (4- to 10-fold at the LC_{50}) was discovered after only about five years of annual use (Pree et al. 1986). Although resistance was at much lower levels than that for OP insecticides, control failures with pyrethroids occurred. Although most usage prior to 1982 was permethrin or fenvalerate, resistance extended to all other types of pyrethroids. Resistance to DDT in laboratory bioassays was 39-fold, higher

Table 10.1. Occurrence of Insecticide Resistance in Populations of Spotted Tentiform Leafminers from Apple Production Areas in Southern Ontario—1987*

		No. of Populations Rated Resistant		
Production Area	No. of Orchards Surveyed	OP	Pyrethroid	Methomyl†
London	5	5	0	0
Simcoe	9	9	9	3 (1M)
Georgian Bay	4	4	2 (M)	0
Milton	2	2	2	0
Niagara	1	1	0	0
Smithfield	4	4	4	0
TOTAL	25	25	17	3

*Ratings are based on response to discriminating concentrations. Populations resistant to azinphosmethyl at 2,000 ppm suffered <20% mortality compared to 100% mortality of susceptible populations. Moths were exposed to two discriminating concentrations of permethrin. At 7 ppm about 65% of a susceptible population was affected and about 5% of resistant populations; at 50 ppm, 100% of susceptibles was killed versus 65–70% of a resistant population. Responses of 25–40% morality at 7 ppm were rated as mixed (M). For methomyl resistance, 700 ppm was used. This killed 65–90% of susceptible populations versus <30% of resistant populations. Populations whose responses were between 40% and 60% mortality were rated as mixed (M).

†Cf. Pree et al. (1990a).

than that for any of the pyrethroids. Lack of synergistic effects by known pyrethroid synergists such as piperonyl butoxide or DEF indicated the occurrence of a nonmetabolic type of resistance. Later tests indicate that differences in rate of penetration of DDT into resistant and susceptible populations were not involved (Pree unpublished). All these data suggest that resistance is most likely kdr-type DDT resistance (Chapter 3). It has been speculated (Pree et al. 1986) that pyrethroid resistance might be better described as DDT resistance, reselected by pyrethroid use. DDT was last widely used in Ontario about 1970 for tarnished plant bug control. Applications were just prior to bloom, also coinciding with STLM activity, but whether resistance to DDT occurred in the STLM at that time is not known. This rapid development of resistance to pyrethroids in five years or less suggests that a significant portion of the populations might have been resistant to DDT before initial selection with pyrethroids. Whether this resistance also occurs in other areas has not been tested.

B. Management of Pyrethroid Resistance

When pyrethroid resistance was discovered, some growers switched to methomyl or oxamyl timed for a predominance of early larval stages. However, laboratory tests with pyrethroids against eggs and larval stages indicated that although resistance also occurred in these stages, differences between populations could be largely overcome by concentrations of permethrin not greater than those used in practice (Table 10.2) (Marshall and Pree 1986). It is also clear that although insecticides used against resistant populations are less effective, considerable mortality of the resistant populations can still be achieved. Hayden and Howitt (1986) also showed the toxicity of permethrin and fenvalerate to egg stages. Field trials have confirmed that timing applications to coincide for first hatch of eggs is a viable approach to overcoming the low levels of pyrethroid resistance in STLM (Marshall and Pree 1986). It might be argued that modification of spray timing to coincide with the most sensitive life stage is similar to an increase in application rate (see Chapter 5). The use of pyrethroids against all STLM populations, whether resistant or susceptible, has become the preferred strategy for leafminer control in Ontario (Fisher 1988). Although the numbers of orchards rated as pyrethroid resistant may have increased since this strategy was adopted, the most serious concern is whether higher levels of pyrethroid resistance will be selected. This has not occurred in one orchard monitored from 1984 through 1987 (Table 10.3), but the continued success of this technique (i.e., treatment of the most sensitive life stage) will require regular monitoring.

Whether removal of selection pressure by replacement of pyrethroid sprays with alternative strategies for extended periods would result in a decline in resistance levels has not been reported. Observations (Pree unpublished) indicate that the pyrethroid resistance persists without large changes in resistance frequencies at least one season after pyrethroid use ceased and studies to document

Table 10.2. Toxicity of Permethrin to Eggs and Larvae of Resistant and Susceptible Populations of Spotted Tentiform Leafminer in the Laboratory*

	% Mortality		
Concentration	Resistant	Susceptible	Calculated *t*-value†
0–2-Day-Old Eggs‡			
6.3	44.8	73.0	3.41
30.0	98.0	98.7	0.05
63.0	93.0	99.7	31.29
4–6-Day-Old Eggs			
6.3	17.6	57.3	3.25
30.0	69.8	99.2	5.94
63.0	73.0	96.1	3.84
First- and Second-Instar Larvae			
6.3	19.8	59.4	12.16
30.0	66.1	88.2	16.88
63.0	73.4	90.2	2.63
Third-Instar Larvae			
63.0	43.5	83.9	2.16

*Adapted from Marshall and Pree, 1986.

†Critical $t_{.05} = 3.182$.

‡Seedling apple trees with eggs or larvae on leaves were dipped five seconds in permethrin suspensions. See Marshall and Pree (1986) for full details.

persistence are under way. Marshall and Pree (1986) reported that three times more eggs were laid by S than by R populations but were unable to state whether this trend was related to fitness differences associated with resistance or to other factors.

In the Ontario leafminer control program, pyrethroid applications are made in the early season, largely prebloom, against the first of the three generations. Details of timing and thresholds in current use are described by Fisher (1988). Advantages associated with early-season pyrethroid treatments are that populations are usually synchronous and effects on beneficial insects and mites, many of which have not yet moved onto the trees, could be minimized. (The validity of this last point is discussed later.) There are also some advantages in this early-season usage with regard to the selection of resistance. Spring is a period of rapid leaf growth on apple, and toxic residues are rapidly diluted. Later-season applications of pyrethroids, when leaves are fully formed (e.g., July) have been shown to exert selection pressure for up to three months (Marshall et al. 1988). Persistent pesticides are more likely to select resistance than those whose toxic residue life is short (Taylor and Georghiou 1982, Chapters 5 and 6).

Table 10.3. Toxicity of Permethrin and Methomyl to Resistant Populations of STLM, 1984–1987[*,†]

Year Tested	Population	LC$_{50}$ (95% CL)	Slope ± S.E.	Resistance Level
Permethrin Resistance				
1984	S‡	3.6(2.4–4.6)	2.3 ± .34	—
	R	22.0(15.9–28.3)	2.6 ± .30	6.1
1987	S	4.5(4.0–5.0)	2.7 ± .21	—
	R	27.3(24.6–29.7)	3.0 ± .32	6.1
Methomyl Resistance†				
1984	S	285.1(216.7–352.2)	1.5 ± .15	—
	R	2562.2(1921.1–3509.2)	2.1 ± .25	9.0
1987	S	209.3(154.3–277.6)	1.3 ± .16	—
	R	964.3(771.0–1159.7)	2.3 ± .28	4.6

*Data from Pree et al. (1989a).

†Insecticides were applied to moths with a Potter spray tower.

‡Susceptible populations were from the same orchard in each sample. Resistant populations (both permethrin and methomyl) were from the same orchard in Norfolk County in each year.

C. Resistance to Methomyl and Oxamyl

Resistance to methomyl and oxamyl in moths was first detected in 1984 (Pree et al. 1990a) at low levels (four- to nine-fold at the LC$_{50}$). The origin of this resistance is not clear. Examination of growers' spray records in these areas does not reveal extensive evidence of repeated use in the same season. Methomyl residues are not active for extended periods (Marshall et al. 1988), and repeated applications are discouraged (Fisher 1988), which should have precluded selection of resistant populations. In the latest surveys (see Table 10.1), resistance to methomyl was restricted to a few orchards in one production area (Simcoe in Norfolk County). This may also reflect current use patterns; most growers prefer to use pyrethroids because of their lower mammalian toxicity and have been discouraged (Fisher 1988) from making annual applications of methomyl.

Resistance to methomyl and oxamyl is likely caused by a combination of altered target acetylcholinesterase (AChE) and enhanced metabolism by esterases (Pree et al. 1990b). The data on inhibition of AChE (Table 10.4) indicate that there is a difference in the sensitivity of AChE between methomyl-resistant populations also resistant to OP insecticides and those resistant only to OP types. This suggests that methomyl resistance has been selected separately from OP resistance.

D. Management of Resistance to Methomyl

Two approaches to the management of methomyl resistance were described by Pree et al. (1990a). The low levels of resistance found in adults also occurred in larval stages but could be overcome at the higher methomyl concentrations tested.

Table 10.4. Inhibition of Acetylcholinesterases in Populations of Spotted Tentiform Leafminer*

		Percent Inhibition	
		Azinphosmethyl Oxon	Methomyl
Population	AChE Hydrolytic Rate† (pmoles/min/mg protein)	$10^{-7}M$	$10^{-5}M$
Susceptible	49,655 ± 14,630‡	100	100
OP-resistant, methomyl-susceptible	38,290 ± 10,800	0	100
OP-resistant, methomyl-resistant	47,469 ± 8,530	0	34.9 ± 14.4

*Cf. Pree et al. (1990b).

†As measured by hydrolysis of acetylcholine chloride. Methodology described in Pree et al. (1987).

‡Mean ± S.D. of three separate assays.

Field tests indicated the success of this approach when applications were timed for early larval instars, which are the most sensitive life stage. This approach is analogous to the present strategy for overcoming resistance to pyrethroids. An alternative is to stop using methomyl until resistance is no longer detected. Where this has been tested, resistance declined from nine-fold to 4.6-fold in three years. Whether this decline will continue requires further assessment, but the data indicate that resistance is lost slowly over time (Table 10.3). Once resistance has been selected, annual or even biannual use of methomyl could ensure the maintenance of resistance.

E. The Future of Resistance Management for STLM

The STLM has developed resistance to all three groups of insecticides used for control over the past 10 years. The benzoyl-phenyl urea (BPU) group of insect growth regulators (IGRS) is extremely toxic to STLM eggs and hatching larvae (Marshall et al. 1988). One of these, diflubenzuron, is currently pending for commercial use against STLM in Canada. These compounds are also attractive because of their low toxicity to predaceous mites (Anderson and Elliot 1982), which are of critical importance in the reestablishment of IPM systems for phytophagous mites. However, the potential for selection of resistance to BPU-type insecticides in the STLM must also be considered. There are reports of BPU resistance in a blotch leafminer of apples, *Leucoptera scitella* (Lepidoptera:Lyonetiidae), in northern Italy (Oberhofer 1986), and resistance may be easily selected in some North American populations. Naturally occurring resistance (tolerance) to diflubenzuron has been recently shown for codling moth in an isolated situation in Oregon (Moffitt et al. 1988).

Mixtures of insecticides have been suggested as the preferred way to manage resistance (Curtis 1985, Mani 1985), provided certain criteria are met. Resistance genes (for both compounds) need to be extremely rare, and cross-resistance between the two compounds must not occur. In addition, residues of both materials should be active for the same length of time. Mixtures of pyrethroids and methomyl are also not useful because the preferred target stage (i.e., the most sensitive life stage) is different for each pesticide and occurs up to two weeks apart. Optimal timing of IGRs and that for pyrethroids are similar, and mixtures of an IGR such as diflubenzuron and pyrethroids are theoretically feasible. However, the high frequency of pyrethroid resistance that already occurs likely means the pyrethroid resistance would not be much affected by such a system.

Roush (1989) has examined the value of mixtures versus rotations and concluded that the severe limitations associated with the success of mixtures (i.e., equal persistence of residue and high efficacy levels) made alternation (or rotation) the preferred approach. Limits to both approaches are discussed in Chapters 5 and 6. For STLM, rotation or alternation of pyrethroids (or methomyl) with an IGR might be considered. The chief difficulty with any rotation or combination system (beyond whether it is successful in reducing resistance) is that the inclusion of broad-spectrum pesticides such as pyrethroids in a resistance management system is extremely disruptive to IPM programs. This is a critical issue in a multipest ecosystem such as apple, where IPM systems make extensive use of beneficial insects and mites.

Naturally occurring or field-selected populations of beneficials that are multiresistant (i.e., resistant to several types of insecticides) are rare and may only be produced by extensive genetic manipulation (Chapter 8). The development of IPM programs for STLM, which have management of resistance as one of their major goals, should be a major objective (see also Chapter 6). Incorporation of action thresholds, differences in varietal tolerances (Fisher 1988), and biological controls (Hagley 1985) may reduce the need for annual applications of insecticides. Other procedures that are potentially useful include mass trapping with pheromones to reduce populations of STLM (Trimble and Hagley 1988). These could minimize the selection pressure on STLM populations and reduce the risks associated with continued use of a single type of (selective) insecticide.

The current approach to management of resistance in STLM is to overcome the low levels of pyrethroid resistance by modification of pesticide application to coincide with a predominance of the most sensitive life stages (i.e., eggs and hatching larvae). After four years of this approach, higher levels of resistance have not been selected. This does not indicate the success or failure of this strategy, since no other strategy has been comparatively tested.

Unfortunately, this approach has precluded the reestablishment of management systems for mites that include reliable biological components. Predaceous mites resistant to pyrethroids have been shown to occur (Croft 1983), but the resistance was polygenic and unstable. Effective populations of A. fallacis have not developed under the current resistance management scheme for STLM. However, it is not likely that all the difficulties associated with current mite outbreaks are due to destruction of *A. fallacis* populations by pyrethroids. Penman and Chapman (1988) cite enhanced dispersal, increased reproductive rates, and shorter development times as possible indirect effects of pyrethroid usage that exacerbate phytophagous mite populations. It is probable that the redevelopment of an integrated system of mite controls for either *T. urticae* or *P. ulmi* will be difficult to achieve until modifications of the system used to manage resistance in STLM are made.

It is obvious from studies with the STLM that insecticides used against resistant populations still have a considerable impact. This effect can be increased by treatment of the most sensitive life stages. For indirect pests, where a tolerance threshold can be established, this approach may provide adequate controls. Certainly, there is a continued risk that higher levels of resistance may be selected. Whether this approach might also be feasible with acaricides has not been clearly demonstrated. Roush (1989) has pointed out that levels of resistance tend to be lower for certain pesticides within a group or chemistry (discussed in Chapters 2 and 5). For STLM, resistance to DDT was higher than that to pyrethroids, 39-fold versus four- to 10-fold for permethrin or fenvalerate (Pree et al. 1986). However, resistance levels for other pyrethroids have (largely) not been determined.

IV. Conclusions

Neither the proposed rotational systems of acaricide use nor the adjustment in timing of controls for STLM (to treatment of the most sensitive life stage) has been demonstrated as an effective resistance management strategy in the apple ecosystem. Considerable effort will be required to develop these proofs, but it should be possible to design suitable experiments to test resistance management strategies. Initial studies (see Table 10.3) are under way, but these will need to be expanded. Assays of gene frequency changes in spider mites, which probably disperse less extensively than other arthropods, may be especially valuable in developing general principles of the dynamics of resistance development and loss. Some of the required data will likely be based on case histories and will require long-term field studies.

Multipest complexes increase the difficulty of establishing and maintaining pest management programs. Nonetheless, pest management in such systems must incorporate resistance management options such as pesticide rotation and the targeting of the more sensitive life stages. The introduction of new selective pesticides such as diflubenzuron will also be important in further modifications of pest management programs for apples.

References

Anderson, D. W., and R. H. Elliott. 1982. Efficacy of diflubenzuron against the codling moth, *Laspeyresia pomonella* (Lepidoptera:Olethreutidae) and impact on orchard mites. Can. Entomol. 114: 733–737.

Chapman, R. B., and D. R. Penman. 1984. Resistance to propargite in European red mite and two spotted mite. N.Z. J. Agric. Res. 27: 103–105.

Cranham, J. E., and W. Helle. 1985. Pesticide resistance in Tetranychidae, pp. 405–421. In W. Helle and M. W. Sabelis (eds.), World crop pests Vol. B, Spider mites. Their biology, natural enemies and control. Elsevier, Amsterdam.

Croft, B. A. 1983. Status and management of pyrethroid resistance in the predatory mite, Amblyseius fallacis. Great Lakes Entomol. 16: 17–32.

Croft, B. A., and S. C. Hoyt. 1978. Considerations for the use of pyrethroid insecticides for deciduous fruit pest control in the U.S.A. Environ. Entomol. 7: 627–630.

Croft, B. A., and E. E. Nelson. 1972. Toxicity of apple orchard pesticides to Michigan populations of Amblyseius fallacis. Environ. Entomol. 1: 576–579.

Croft, B. A., and D. G. Stewart. 1973. Toxicity of one carbamate and six organophosphorus insecticides to OP resistant strains of Typhlodromus occidentalis and Ambylseius fallacis. Environ. Entomol. 2: 486–488.

Croft, B. A., R. W. Miller, D. R. Nelson, and P. H. Westigard. 1984. Inheritance of early-stage resistance to formetanate and cyhexatin in Tetranychus urticae Koch. J. Econ. Entomol. 77: 574–578.

Curtis, C. F. 1985. Theoretical models of the use of insecticide mixtures for the management of resistance. Bull. Ent. Res. 75: 259–265.

Dennehy, T. J., E. E. Grafton-Cardwell, J. Granett, and K. Barbour. 1987a. Practitioner-assessable bioassay for detection of dicofol resistance in spider mites (Acari: Tetranychidae). J. Econ. Entomol. 80: 998–1003.

Dennehy, T. J., J. Granett, T. F. Leigh, and A. Colvin. 1987b. Laboratory and field investigations of spider mite resistance to the selective acaricide propargite. J. Econ. Entomol. 80: 565–574.

Dennehy, T. J., J. P. Nyrop, W. H. Reissig, and R. W. Weires. 1988. Characterization of resistance to dicofol in spider mites (Acari:Tetranychidae) from New York apple orchards. J. Econ. Entomol. 81: 1551–1561.

Edge, V. E., and D. G. James. 1986. Organo-tin resistance in Tetranychus urticae in Australia. J. Econ. Entomol. 79: 1477–1483.

Edge, V. E., J. Rophail, and D. G. James. 1987. Acaricide resistance in two spotted mite, Tetranychus urticae in Australian horticultural crops, pp. 87–90. In W. G. Thwaite (ed.), Proc. Symp. of Mite Control in Hort. Crops. Orange, N.S.W. Australia. 1987. Agdex 200/622.

Fisher, P. 1988. Spotted tentiform leafminer. OMAF Factsheet 88–067. 4 pp.

Flexner, J. L., P. H. Westigard, and B. A. Croft. 1988. Field reversion of organotin resistance in the two-spotted spider mite following relaxation of selection pressure. J. Econ. Entomol. 81: 1516–1520.

Gerson, U., and E. Cohen. 1989. Resurgences of spider mites (Acari:Tetranychidae) induced by synthetic pyrethroids. Exp. Appl. Acarol. 6: 29–46.

Glass, E. H. 1960. Current status of pesticide resistance in insects and mites attacking deciduous orchard crops. Misc. Publi. Entomol. Soc. Am. 2: 17–25.

Hagley, E. A. C. 1985. Parasites recovered from the overwintering generation of the spotted tentiform leafminer, Phyllonorycter blancardella (Lepidoptera:Gracillariidae), in pest management apple orchards in southern Ontario. Can. Entomol. 117: 371–374.

Hagley, E. A. C., and A. Hikichi. 1973. The arthropod fauna in unsprayed apple orchards in Ontario. I. Major pest species. Proc. Entomol. Soc. Ont. 103: 60–64.

Hayden, P. J., and A. J. Howitt. 1986. Ovicidal activity of insecticides on the spotted tentiform leafminer. J. Econ. Entomol. 79: 258–260.

Herne, D. H. C. 1971. Methodology for assessing resistance in the European red mite, pp. 663–667. In Proc. Third International Congress of Acarol, Prague, Czechoslovakia.

Hoy, M. A., and J. Conley. 1989. Propargite resistance in Pacific spider mite (Acari:Tetranychidae): Stability and mode of inheritance. J. Econ. Entomol. 82: 11–16.

Hoy, M. A., and Y. Ouyang. 1986. Selectivity of the acaricides clofentezine and hexythiazox to the predator Metaseiulus occidentalis. J. Econ. Entomol. 79: 1377–1380.

Hoy, M. A., J. Conley, and W. Robinson. 1988. Cyhexatin and fenbutatin-oxide resistance in Pacific spider mite: Stability and mode of inheritance. J. Econ. Entomol. 81: 57–64.

Hoyt, S. C., P. H. Westigard, and B. A. Croft. 1985. Cyhexatin resistance in Oregon populations of *Tetranychus urticae* Koch. J. Econ. Entomol. 78: 656–659.

Keena, M. A., and J. Granett. 1987. Cyhexation and propargite resistance in populations of spider mites from California almonds. J. Econ. Entomol. 80: 560–564.

Lienk, S. E., C. M. Watve, and R. W. Weires. 1980. Phytophagous and predacious mites on apple in New York. Search:Agriculture 6, 14 pp.

Mani, G. S. 1985. Evolution of resistance in the presence of two insecticides. Genetics 109: 761–783.

Marshall, D. B., and D. J. Pree. 1986. Effects of pyrethroid insecticides on eggs and larvae of resistant and susceptible populations of spotted tentiform leafminer. Can. Entomol. 118: 1123–1130.

Marshall, D. B., D. J. Pree, and B. D. McGarvey. 1988. Effects of benzoylphenylurea insect growth regulators on eggs and larvae of the spotted tentiform leafminer *Phyllonorycter blancardella*. Can. Entomol. 120: 49–62.

Mizutani, A., F. Kumayama, K. Ohba, T. Ishiguro, and Y. Hayashi. 1988. Inheritance of resistance to cyhexatin in the Kanzawa spider mite, *Tetranychus kanzawai* Kishida. Appl. Ent. Zool. 23: 215–255.

Moffit, H. R., P. W. Westigard, K. D. Mantey, and H. E. Van De Baan. 1988. Resistance to diflubenzuron in the codling moth. J. Econ. Entomol. 81: 1511–1515.

Oatman, E. R., E. F. Legner, and R. F. Brooks. 1964. An ecological study of arthropod populations on apple in Northeastern Wisconsin: Insect species present. J. Econ. Entomol. 57: 979–983.

Oberhofer, H. 1986. Diflubenzuron-ein wirkstoff seiner zeit voraus. Ostbau Weinbau, July–August 1986, p. 173.

Penman, D. R., and R. B. Chapman. 1988. Pesticide-induced mite outbreaks: pyrethroids and spider mites. Exp. Appl. Acarol. 4: 265–276.

Pickett, A. D., N. A. Patterson, H. T. Stultz, and F. T. Lord. 1946. The influence of spray programs on the fauna of apple orchards in Nova Scotia. I. An appraisal of the problem and a method of approach. Sci. Agr. 26: 590–600.

Pottinger, R. P., and E. J. Leroux. 1971. The biology and dynamics of *Lithocolletis blancardella* (Lepidoptera:Gracillariidae) on apple in Quebec. Mem. Entomol. Soc. Canada 77. 437 pp.

Pree, D. J. 1987. Inheritance and management of cyhexatin and dicofol resistance in the European red mite. J. Econ. Entomol. 80: 1106–1112.

Pree, D. J., and H. W. Wagner. 1987. Occurrence of cyhextain and dicofol resistance in the European red mite, *Panonychus ulmi* (Koch) in southern Ontario. Can. Entomol. 119: 287–290.

Pree, D. J., E. A. C. Hagley, C. M. Simpson, and A. Hikichi. 1980. Resistance of the spotted tentiform leafminer, *Phyllonorycter blancardella*, to organophosphorous insecticides in southern Ontario. Can. Entomol. 112: 469–474.

Pree, D. J., D. B. Marshall, and D. E. Archibald. 1986. Resistance to pyrethroid insecticides in the spotted tentiform leafminer, *Phyllonorycter blancardella* in southern Ontario. J. Econ. Entomol. 79: 318–322.

Pree, D. J., J. L. Townshend, and D. E. Archibald. 1987. Sensitivity of acetylcholinesterases from *Aphelenchus avenae* to organophosphorous and carbamate pesticides. J. Nematol. 19: 188–193.

Pree, D. J., K. J. Cole, and P. A. Fisher. 1989. Comparisons of leaf disc and petri dish assays for assessment of dicofol resistance in populations of European red mite from southern Ontario. Can. Entomol. 121: 771–776.

Pree, D. J., D. B. Marshall, and D. E. Archibald. 1990a. Resistance to methomyl in populations of the spotted tentiform leafminer *(Lepidoptera:Gracillaridae)* from southern Ontario. J. Econ. Entomol. 83: 320–324.

Pree, D. J., D. E. Archibald, and K. J. Cole. 1990b. Insecticide resistance in the spotted tentiform leafminer: mechanisms and management. J. Econ. Entomol. 83: 678–685.

Rizzieri, D. A., T. J. Dennehy, and T. J. Glover. 1988. Genetic analysis of dicofol resistance in two populations of twospotted spider mite from New York apple orchards. J. Econ. Entomol. 81: 1271–1276.

Roush, R. T. 1989. Designing resistance management programs: How can you choose? Pestic. Sci. 26: 423–441.

Taylor, C. E., and G. P. Georghiou. 1982. Influence of pesticide persistence in evolution of resistance. Environ. Entomol. 11: 746–750.

Tette, J. P., E. H. Glass, D. Bruno, and D. Way. 1979. The New York tree-fruit pest management project 1973–1978. New York Food and Life Sci. Bull. 81: 1–10.

Trimble, R. M., and E. A. C. Hagley. 1988. Evaluation of mass trapping for controlling the spotted tentiform leafminer, *Phyllonorycter blancardella*. Can. Entomol. 120: 101–107.

Watve, C. M., and S. E. Lienk. 1976. Toxicity of carbaryl and six organophosphorous insecticides to *Amblyseius fallacis* and *Typhlodromus pyri* from New York apple orchards. Environ. Entomol. 5: 368–370.

Weires, R. W. 1977. Control of *Phyllonorycter crataegella* in eastern New York. J. Econ. Entomol. 70: 521–523.

Welty, C., W. H. Reissig, T. J. Dennehy, and R. W. Weires. 1987. Cyhexatin resistance in New York populations of European red mite. J. Econ. Entomol. 80: 230–236.

Welty, C., W. H. Reissig, T. J. Dennehy, and R. W. Weires. 1988. Comparison of residual bioassay methods and criteria for assessing mortality of cyhexatin-resistant European red mite. J. Econ. Entomol. 81: 442–448.

11

Developing a Philosophy and
Program of Pesticide Resistance Management

Brian A. Croft

I. Goals, Tools, Implementation, and Constraints to Pesticide Resistance Management

The primary goals of Pesticide Resistance Management (PRM) are (1) avoiding resistance development in pest populations, (2) slowing the rate of resistance development, and (3) causing resistant populations to "revert" to more susceptible levels and thereafter keeping resistance below some threshold.

Pesticide resistance also can be managed in some beneficial arthropod populations, such as natural enemies of pests (Croft and Strickler 1983, Tabashnik 1986a). With natural enemies the objectives and tactics of PRM are often the opposite of those used for pests. The evolution of resistance in predators and parasitoids to achieve selective use of broad-spectrum pesticides for pest control has been encouraged in a number of crops, including greenhouse crops and deciduous tree fruits (Hoy 1985, Croft and van de Baan 1988, Chapter 10) Selection of arthropod natural enemies in the laboratory or the field by artificial means to develop resistant strains is another aspect of PRM (Croft and Strickler 1983, Hoy 1985, Chapter 8).

In this chapter some of the factors influencing PRM are identified and discussed. PRM, as a subcomponent of Integrated Pest Management (IPM), is still at an early stage of development (Carlson et al. 1986, NAS 1986). Ideally, both PRM and IPM can be viewed as components of crop or other resource production systems (Croft 1986). Efforts to implement PRM use a variety of tactics (NAS 1986), but many biological, technical, economic, and sociological constraints limit the success of PRM.

With both pests and natural enemies, PRM is greatly facilitated by an understanding of the genetic and biochemical mechanisms of resistance (Chapters 3 and 4). At another level, understanding the life history and, especially, ecology and population dynamics of pests and natural enemies is essential (Croft and van de Baan 1988, Roush and Croft 1986, NAS 1986, Chapter 6). These levels of information must then be applied to the specific "pesticide resistance situation" before appropriate decisions can be made and resistance development curtailed or encouraged in the field.

Tactics for PRM of pests usually involve changing pesticide use patterns, or even better, using nonchemical control measures (e.g., biological or cultural controls) (Dover and Croft 1984, 1986). Operational tactics for pests include varying dose or frequency of pesticide application; using local rather than area-wide applications; applying treatments only when economic levels of pests are present; using less persistent pesticides; treating only certain life stages of pests; using pesticide mixtures; using alternations, rotations, or sequences; using improved pesticide formulation; using synergists; exploiting unstable resistance; and identifying new toxicants with alternate sites of activity (NAS 1986, Chapters 5 and 6).

Implementation of PRM can involve many different spatial and temporal scales and levels of management complexity. Consideration of these elements becomes important in designing and implementing a system of PRM. (Specific examples of these levels will be discussed later.) PRM involves personnel, monitoring systems, communications, and decision making. These must be integrated into an overall delivery system. The design of a PRM implementation system is usually specific to a particular species and an ecological or management environment.

Resistance management can be facilitated by enlightened policies governing pesticide development, registration, marketing, regulation, and education. Understanding the overall technical, implementation, and policy issues of a given resistance situation is critical in developing an effective PRM program.

This chapter focuses on some of the basic principles useful in developing a philosophy and program of PRM. Some recent case histories of resistance management are reviewed to illustrate where the principles of PRM are being applied effectively. These case histories also indicate some of the constraints that are encountered when seeking to implement PRM. The scope of presentation will include the temporal and spatial scales of PRM and the components of PRM, including detection and monitoring of resistance, integration of tactics, PRM delivery and implementation, environmental monitoring, and economic and policy aspects.

II. A Philosophy of Resistance Management

A. The need for a Philosophy of Resistance Management

By raising the question of the need for a philosophy of PRM, I imply that this area of science is still developing and that the factors most important for PRM decision making have not been fully agreed upon. PRM has long been a goal of pest control practitioners, but not always explicitly or with a focus on prevention as well as remedial action. The recent emphasis on PRM as part of IPM has been more directed in scope and specific in action than in earlier periods. For example, many of the concepts of resistance management have been examined in detail through simulation and experimental approaches. However, the validation and testing of some of these principles in the field are still limited (Chapters 5, 6, and 10). Part of the reason for the lack of validation of PRM is that it is very difficult to do resistance management research with large populations of "treated" and "untreated" pests or natural enemies. Few experimental plot areas are large enough to mimic treatment conditions and the spatial scales of resistance conditions that occur in conventional agriculture. Furthermore, few growers will allow scientists to produce populations of pesticide-resistant pests for experimental study in their fields (Taylor 1983). For these and other reasons, some tactics of PRM and their integration have yet to be fully tested in the field. However, with a few pest species (e.g., less mobile pests), actual testing of PRM alternatives is possible. Some examples of experimental testing of PRM alternatives will be presented later.

A PRM philosophy provides a perspective in choosing management options that may be applied in individual PRM situations. Despite our limited knowledge of some of the principles and tactics of PRM, most scientists believe that there is much more that can be done to minimize pesticide resistance development in pests. Furthermore, one's perspective can be updated as new research information is obtained and greater experience is achieved. Although one is not always sure which tactic will bring about the best PRM, one can often experiment with different approaches. Most attempts to lessen selection pressure will be better than not doing anything about resistance problems (other than finding a new chemical to use). In rare cases—for example, in using high doses to curtail resistance (see Georghiou and Taylor 1976, Tabashnik and Croft 1982, Chapters 5 and 6)—one might inadvertently increase the likelihood of resistance through PRM attempts, but this would be the exception rather than the rule.

My current philosophy of PRM is discussed here as an example of a philosophy of PRM. I acknowledge that there are other useful perspectives that can be taken as the basis of minimizing pesticide resistance in pests. Also, one's philosophy can and should change over time as new research information and experience are obtained.

B. A Philosophy of Paradigm of Resistance Research and PRM

Figure 11.1 illustrates the spectrum of disciplines that contribute to PRM. They range from basic studies at the cellular or suborganismal levels—including the genetics, biochemistry, and physiology of resistance—to the other extreme of population dynamics and community ecology. In between are elements of organismal biology that include many aspects of life history. Traditionally, physiologists, toxicologists, geneticists, and those examining life history have studied resistance development more than have population and community ecologists. It is my

Discipline Needs in Pesticide Resistance Management

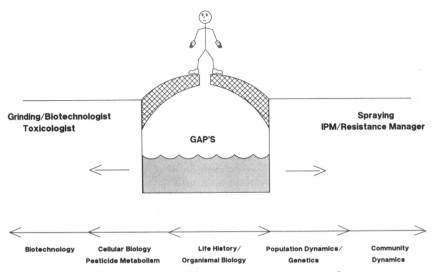

Figure 11.1 Discipline needs in a pesticide resistance management research program.

contention, however, that information from the entire spectrum is often necessary to manage resistance development effectively.

Before the mid-1970s, bridges between the cellular–physiological and ecological perspectives on resistance development were limited and large gaps in the PRM information base were present (Fig. 11.1). Few researchers could bridge these gaps. Teams of scientists that included individuals having perspectives across the entire range of disciplines were seldom formed. Mostly because of recent modeling and some experimental research studies (Georghiou and Taylor 1976, 1977a, 1977b, Comins 1977, Tabashnik and Croft 1982), a broadened philosophy and appreciation of the factors affecting PRM have emerged.

Out of the broadened perspective of resistance evolution has arisen an overall philosophy of PRM. It is the consensus of many scientists that although genetics, biochemistry, and physiology determine whether resistance develops, these are usually not the primary determinants influencing the rate of resistance development. Instead it is the ecological genetics of resistance (including gene flow and population dynamics of selection) that largely determines the status of resistance development over time (NAS 1986, especially pp. 143–270; Roush and McKenzie 1987).

Features of population dispersal, survivorship, and residency in treated habitats become critical elements in understanding the broader features of resistance development or lack thereof in populations. As we proceed in our discussion of resistance management and the case histories cited later, the specifics of this philosophy of resistance management will become more obvious.

III. Temporal–Spatial Scales and Components of PRM

A. Overview

Next to having a comprehensive and broadly integrated philosophy of resistance management, a second important need is to develop a temporal and spatial perspective of resistance and PRM in the target pest(s) and key natural enemies with which one is concerned. For example, what is the effective area in which the genetic factors influencing resistance are maintained through immigration–emigration of susceptible or resistant populations? What is the size of the resident population in which resistance is maintained as a whole? Is it at a field or orchard level, or does it extend throughout a regional environment? A temporal and spatial perspective can be gained from consideration of the organism itself. For example, is one dealing with a wingless pest that moves only from one plant to another over a distance of a few meters in a single season or with a highly mobile species that can move intercontinentally during a single generation or during the course of several generations? Management-related factors also define the temporal–spatial scales of a PRM system as well. Such differences in temporal–spatial perspectives can be seen in the contrasting examples discussed later (Section V) for relatively sedentary species such as spider mites and predatory mites versus highly mobile species such as the diamondback moth and pear psylla (Croft and van de Baan 1988, Tabashnik et al. 1987, Follett et al. 1985, Croft et al. 1988). These extremes of PRM perspective commonly exist in other species and are important considerations in designing a PRM program.

The temporal–spatial scale of pesticide resistance should determine the appropriate scale of operation for the four PRM components (discussed later). Restated another way, the biological dimensions should dictate the features of PRM rather than the human-related features of the management system. Of course, in practice, there must be a compromise between these factors, but it is important to point out that the biological features of the system should not be ignored.

The components of a PRM program for a general case are illustrated in Figure 11.2. The major elements may include (1) a detection and monitoring system for assessing the early incidence and progress of a resistance in a population or several pest–natural enemy populations; (2) a model of the PRM system, including the effects of different biological and operational factors; (3) a delivery system for implementation of tactics; (4) an environmental monitoring system necessary to provide data to drive models of pest development, tactics, and effects; and (5) an economic–policy component that monitors or predicts the "sociological environment" (and constraints) under which a particular PRM system will operate. Additionally, the design of a particular PRM program should consider the spatial and temporal scales of biological and management resolution, as noted earlier. Each of these subcomponents of PRM design is discussed later.

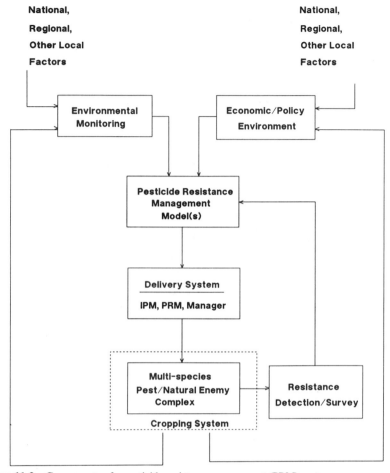

Figure 11.2 Components of a pesticide resistance management (PRM) system.

B. PRM Monitoring and Detection (Tools, Timeliness, Spatial Scales)

Implementation of a resistance detection and monitoring system for PRM was considered in detail in Chapter 2 and in a recent National Academy of Sciences report (NAS 1986). The role of detection and monitoring in the overall context of PRM is briefly considered here.

Resistance monitoring research has shown considerable progress in recent years, largely paralleling the reemphasis on dealing with pesticide resistance problems in IPM programs (Brown and Brogdon 1987). More refined classical methods of resistance monitoring have been developed as well as a number of new techniques for bioassay, including biochemical enzyme test methods, immunological techniques (e.g., ELISA), and biotechnology probes using DNA or RNA segments (NAS 1986, Brown and Brogdon 1987). Improvements in the statistical methods and sampling requirements for resistance monitoring at specified levels are beginning to emerge (e.g., Roush and Miller 1986). These tools help to estimate the onset of resistance monitoring in the early stages, when it is still possible to do something to limit resistance development.

Other developments in the design of resistance detection and monitoring systems are coming in the form of databases containing baseline susceptibility for key pest and natural enemy species. In the United States, national and statewide monitoring information networks are being established to help PRM practitioners find up-to-date information on resistance development.

Factors not emphasized enough in these developments are the timeliness of monitoring and the cost–benefits of such systems. In designing such monitoring system for PRM, these factors must be carefully evaluated before a full-scale system is put in place. A preplanning exercise or feasibility analysis of these aspects is critical to the widespread adoption, implementation, and performance of a PRM program once it is established.

C. Control Models: Dynamics–Tactics–Integration

The term *control models* is used broadly here to mean the understanding of the pest and natural enemy biology and associated IPM system and ultimately the entire complex of species and factors involved in PRM. These models should include the influences of management and environmental variables on system performance. This knowledge can be in the form of a mental model of experience and logic or it can be as formal as a simulation model of these processes and associated decision rules. Whatever the form of PRM model, it is a critical core element for decision making by the IPM–PRM practitioner.

As noted, PRM uses a variety of tactics. The range of factors that can be managed in a PRM program should be viewed broadly and include tactics other than pesticides (although altered or new pesticide products and changing pesticide use patterns are probably the most commonly employed operational measures, Dover and Croft 1984). Biological controls, for example, are certainly alternate tactics in PRM. Also, although immigration of susceptible pests into a treated habitat is most often thought of as a factor that cannot be influenced appreciably, in some cases it can be modified to enhance PRM (see later example of this type of manipulation).

As intimated earlier, the decision of which PRM tactic to select is not always apparent. For example, there is still much debate over whether it is better to use mixtures or rotation of pesticides to ameliorate resistance problems (NAS 1986, Tabashnik 1989). Experimental studies to answer these questions in each specific case are not available. Until they are, the PRM practitioner will have to make the best possible judgment considering the factors operating in a specific case.

As might be expected, the biggest challenge of tactic selection is in estimating the integrated effects of multiple tactics on an entire complex of species. Very few formal means to estimate the "optimal solution" to PRM have been developed. Even in IPM there are difficulties in making these types of complex decisions (Frisbie and Adkisson 1986).

D. PRM Implementation Delivery Systems

As mentioned earlier, the delivery system is the assemblage of persons, equipment, supplies, and communications necessary actually to put a PRM system in place. However, it should be recognized that this usually is not just a one-time decision, but rather a dynamic system of evaluation, response, and feedback evaluation (Fig. 11.2). These resources should be optimally allocated in both time and space. Depending on the temporal scales of operation, decision making and a PRM response may need to take place very rapidly, over a period of hours for an organism such as rapidly developing plant disease.

As noted earlier, since PRM programs are often imbedded as parts of IPM systems in many crops, the components of an effective PRM system may already be in place. It is then just a matter of training personnel in the details of the system and its implementation. Appropriate monitoring protocols must be established and timely information distributed to decision makers in ways and in time frames that meet management needs. Again, because different species may require different spatial and temporal scales of management action, an overall schedule and optimal scheme for management of entire complexes of pest–natural enemy species will be necessary on heavily sprayed crops, such as cotton, apple, citrus, and corn.

E. PRM Environmental Monitoring Systems

Environmental monitoring data can be useful inputs into PRM models. They reflect the influence weather or climate or more specific variables such as temperature, relative humidity, and wind has on the biologies of the species involved in a PRM program. These abiotic variables also influence the effectiveness of control tactics. An example of such effects might be from using synthetic pyrethroid insecticides on a particular target arthropod population. These pesticides are known to exhibit negatively correlated temperature coefficients of activity on insects (Vijverberg et al. 1983). Under different environmental conditions the selection of pests and natural enemies by pesticides could vary appreciably and therefore influence the time course of a resistance episode among these species. (See Waddill [1978] for an example of how pyrethroids influence physiological selectivity differences among species in a vegetable cropping system.)

IV. The Economic–Policy Environment and Constraints to PRM

The benefits of PRM depend ultimately on production economics, which therefore must always be considered in decision making (Fig. 11.2). While local economics is usually of greatest concern to the PRM practitioner, regional and even global economic factors can influence production outcomes. In fact, resistance development often takes on a global scale, so that looking only at local conditions of PRM and production economics is often unwise. Beyond the economic considerations, the policy environment may provide many opportunities to improve PRM systems or reduce constraints to their operation.

Various policy-related issues and constraints have been raised by academic, government, industry, and public groups interested in facilitating greater implementation of PRM in the United States (Dover and Croft 1984, 1986; NAS 1986). A comparative list of issue areas identified by two study groups is given in Table 11.1. Policies associated with federal pesticide regulation, state and local regulation and management of pesticide use, regulation and management of pesticide marketing, antitrust concerns by industry, education of a wide diversity of personnel, funding of resistance management programs, and international considerations are included. Generally, these issues are not addressed by scientific groups because of their lack of experience in dealing with policy aspects of science. Each of these topics cannot be discussed here, but several will be referred to in the next section on sample case histories describing the practice of PRM in

Table 11.1.   Policy and Implementation Issues Limiting Resistance Management

| | Policy Study Group/Importance Rating* | |
Policy Issue†	Dover and Croft (1984)	NAS (1986)
Information storage, retrieval, and dissemination	X X X	X X X
Resistance risk assessment	X X X	X
International coordination	X	X X X
International policy development	ND	X X X
Regulatory reform	X X X	X
Federal pesticide regulation	X X X	X X
State and local pesticide regulation	X X X	X X
Industry self-regulation	X X X	X X X
Industry marketing tactics	X X X	X X
Antitrust limitations	X	X X
Educational needs	X X X	X X X
Implementation infrastructure	X X	X X X
Research support	X X	ND

*X X X = of great importance, X X = of moderate importance, X = of little importance, ND = not discussed.

†See Dover and Croft (1985) and NAS (1986) for details.

the field. Both Dover and Croft (1984, 1986) and the NAS report (NAS 1986) treat these important issues in considerable detail. Policy considerations are inevitably involved in implementation of resistance management programs. In fact, they are often more important constraints to PRM than are technical limitations. Many policies must be altered if we are to improve greatly the environment for achieving greater use of PRM for pest control in the future.

Three examples of policy trends that may soon affect PRM implementation at the local, state, or federal levels are discussed next.

1. Recently, the U.S. Environmental Protection Agency has begun to consider a policy of identifying panels with resistance expertise, developing databases, and recommending test protocols that include resistance information for use in support of registration and reregistration of pesticides. Support for development of appropriate resistance monitoring systems to evaluate product efficacy over time is being considered. Such inputs in the registration process will do two important things. First, when products are no longer effective because of resistance problems, this factor will be included as part of the risk–benefit equation used to evaluate whether reregistration is appropriate. Currently, such changing risk–benefit aspects are not explicitly considered. Second, the registration process might be hastened for new chemical products that fit well into PRM programs and contribute to improved systems of IPM. These developments will greatly improve the methodology of resistance monitoring as well as increase the number of personnel involved in this type of product evaluation.

2. Policies are being developed at federal, state, and local levels to identify mechanisms to generate funds for research on pesticide side-effects in areas not adequately covered by current fiscal appropriations (e.g., environmental research on groundwater, endangered species, selective pesticide development, PRM, IPM, etc.). Proposals for resource generation include such diverse means as fees for state product registration, taxes on producers and users of pesticides, and user fees. Probably a combination of different mechanisms should be used to promote equitable funding of this type of environmental research. The generation of resources must be conducted in ways fair to the benefitted parties of pesticide sales and use. Policies are needed to determine the priorities for use of funds generated for environmental research.

3. Recently, pesticide industry personnel have begun to consider PRM research as essential step in new pesticide development, especially for some of the pest species that are highly prone to resistance development (Croft 1990). A case in point is in the development, registration, and use of new acaricidal pesticides. With new acaricides, research funds are more commonly being invested in PRM research before a product is registered for use to evaluate how the product might be most effectively used in the field. Conservation of the product from a resistance perspective is being considered in labeling and marketing plans as well. How any new product fits into the whole range of acaricides available for spider mite control from a PRM perspective is increasingly a critical element in research

planning. In general, an industry-wide perspective of PRM has become established among research teams and increasingly so among producers and users of acaricidal products.

V.   Case Histories of PRM

Some recent cases where resistance management is being applied in the field with good success are described in this section. Also, examples are presented where considerable efforts have been made but where less progress in stemming the onset of resistance has occurred (Table 11.2). In some instances, the same species or closely related ones are discussed in both categories of classification. This illustrates that success in achieving resistance management can be very site specific. It can vary greatly from area to area because of a wide variety of biological, management, and policy-related factors. In making such a general comparison, our goal is to highlight some of the key factors that seem to determine relative success or lack thereof in a given set of circumstances. As might be expected, no single factor is totally responsible, but some general trends can be discussed. Determining the reasons that resistance management does or does not work is a tremendous challenge, given the wide variety of factors that can be involved. Also, it is probably inappropriate to characterize certain case histories cited here as failures, just because progress has been slow. In fact, we are learning much about what does and does not work in a given situation from these examples.

A.   More Successful Examples

In Australia, resistance to synthetic pyrethroid (SP) insecticides was confirmed in laboratory tests of the cotton bollworm, *Heliothis armigera,* after field control failures in cotton and soybean occurred in the Emerald Irrigation District of Queensland in 1983. Within a few months a strategy to manage the resistance problem was proposed by state authorities after consultation with federal CSIRO research authorities, pesticide companies, and growers. The program involved restricting SP use on all crops in the area to a 42-day period during the middle of the growing season (Daly and McKenzie 1987, Chapter 5). Adherence to the program is voluntary in New South Wales but is potentially enforceable in Queensland. Even though these restrictions in pesticide use are a financial burden in some cases, there has been no effort to enforce use of the program in either state, and compliance appears to be almost complete (J. C. Daly, personal communication). The program has been judged a success inasmuch as there have not been further major field control failures resulting from resistance development in any of the areas where the program has been attempted. The key to success in this case has been the unusual, voluntary cooperation that has been achieved in implementing the resistance management strategy.

House flies in Danish farms have become resistant to almost every new insecti-

Table 11.2. Examples of Recent Trends in Practical Application of Resistance Management Practices in the Field

Class Species/ Prod. System	Site	Recommended Tactics/ Limiting Factors	References
More Successful			
Heliothis armigera/ cotton	Australia	Monitoring, thresholds, areawide compliance	Daly and McKenzie (1987)
Musca domestica/ animals	Denmark	Monitoring, short-residue compounds, regulations/ environmental limitations	Keiding (1986)
Psylla pyricola/pear	Western U.S.	Monitoring, rotations, regulation, industry compliance/difficult biology, chemical alternatives	Riedl et al. (1981), Dover and Croft (1984), Follett et al. (1985)
Tetranychus urticae/ pear–apple	Western U.S., Australia	Monitoring, unstable-R, selective Cpd, rotation formulation, biological control/grower compliance	Croft et al. (1987), Edge and James (1986), Flexner (1987)
Pest/natural enemy complex/apple	U.S.	Monitoring, lack of resistance in key pests, resistant N.E., selective pesticides, biological control/grower compliance, improved monitoring methods	Croft (1982), Croft et al. (1984), Croft et al. (1987)
Less Successful			
Heliothis virescens/ cotton	Southern USA	Monitoring, thresholds, mixtures, synergists/grower apathy, limited compliance	Plapp and Campanhola (1986), Luttrell and Roush (1987), Roush, pers. comm.
Boophilus microplus/ cattle	Australia	Monitoring, strategic dipping, grazing mgmt./grower compliance	Sutherst and Comins (1979)
Plutella xylostella/ vegetables	Tropical areas	Monitoring, cultural and biological controls/biological constraints, low thresholds.	Tabashnik (1986b), Tabashnik et al. (1987), Cheng (1988)
Psylla pryi/*Psylla pyricola*/pear	Italy, Eastern U.S.	Monitoring, synergists/no rotation, no RM program per se	E.C. Burts (pers. comm.), Enduro (pers. comm.) Follett et al. (1985), Croft et al. (1988)

cide introduced for their control since the 1950s (Keiding 1986). In the mid-1970s, researchers showed that resistance to DDT and the new synthetic pyrethroid insecticides was due to a common resistance factor (the *kdr* and super-*kdr* genes), and they predicted that rapid resistance development to the more persistent SP compounds would rapidly evolve. In 1978–1979, surveys for SP resistance indicated that resistance to long-residual compounds was beginning to develop in the field. Overall, the survey data indicated that resistance would soon be widespread. Several steps were immediately taken to evaluate the resistance potential of short-residual SP's and additional monitoring and regulatory actions were implemented. A decision was made in collaboration among scientists, industry, and regulatory personnel against registering long-residual compounds for use. Only short-residue SP's were made available for use. Strict regulation and monitoring of SP resistance was employed. So far, SP resistance in Danish house flies has been manageable. In contrast, resistance has evolved in house flies against long-residual SP's in Britain (Chapter 6) and the United States (Scott et al. 1989). Through appropriate registrations, continued use of SP compounds has been sustained in a very difficult resistance management environment. The keys to success in this case have been the effectiveness of the resistance monitoring effort and the critical regulatory actions taken to ensure that certain pesticidal products were not used.

Resistance management in the pear psylla (*Psylla pyricola*) has involved conservative pesticide use programs and good cooperation among industry and grower groups. This species has previously developed resistance to virtually all synthetic pesticides registered for its control, often over large areas (Riedl et al. 1981; Follett et al. 1985). In the late 1970s, experiments demonstrated conclusively that resistance development to the SP's would occur if these products were used unilaterally in the early season before bloom and during midseason for summer control of this pest (Riedl et al. 1981). It was therefore recommended that SP's be used only against overwintering psylla before bloom and that the formamidine amitraz be used during the summer. In many areas, growers readily complied in restricting SP use in this manner. Furthermore, SP chemical producers did not put summer use of SP's on the label, thus making this use of the product less likely. In some pest management districts of California, state officials with grower support made summer use of SP's illegal, thus going one step further in ensuring that these highly valuable products were not misused. To date, resistance to the SP's has not developed in psylla populations in most areas of the western United States, where strict limits on SP's have been promoted (Croft et al. 1989). In contrast, in areas where summer use of SP's has been common or prebloom use intensive, resistance to the SP's in *P. pyricola* and a related species (*Psylla pyri*) has developed rapidly (see further discussion later). The key to success in this case has been the far-sighted resistance risk assessment research that was initially done and the excellent industry cooperation in fostering the program.

The case history of resistance management of organotin resistance in the spider mite *Tetranychus urticae* is an interesting one because it involves the optimal use of physiologically selective acaricides and biological control of this pest by insecticide-resistant strains of predatory mites (Croft et al. 1987). Organotin resistance plateaus in the field and rapidly reverts toward susceptibility in the absence of selection by the organotins (OT) or other closely related acaricides (Edge and James 1986, Flexner 1987, Flexner et al. 1989, Chapter 10). Also, a number of tactics to limit OT resistance are available, including formulations that effectively enhance toxicity to resistant mites (Edge and James 1986, Flexner 1987), critical timing of organotin applications to control more susceptible immature mite stages (Edge and James 1986, Croft et al. 1987, Flexner 1987), alternation of organotins with unrelated acaricides (Croft et al. 1987, Flexner 1987), and increased use of predators in biological control (Croft et al. 1987). Implementation of these measures in an integrated way allows for continued use of the organotins in areas where previous use was limited because of presence of highly resistant mites. Furthermore, with the introduction of new acaricides (e.g., hexythiazox, clofentezine), PRM programs are being promoted before the products are first registered and used, rather than waiting until resistance problems begin to appear. (However, there are already reports of some resistance to these compounds; see Chapter 10.) The key to success in the case of spider mites has been the instability of OT resistance, the wide variety of alternative tools available to combat resistance development, the extensive research effort made to integrate these methods, and the emerging industrywide attitude that PRM is an essential part of any IPM program for this group of pests.

One feature of progress with spider mites is that researchers are beginning to evaluate PRM alternatives experimentally in the field (Flexner et al. unpubl.). Replicated, small blocks of orchard trees with appropriate buffer rows between treatments are being used to test acaricide rotations, alternations, mixtures, and various levels of application. Because of the limited movement of mites between treatment plots, evaluation of selection effects over three to seven years is feasible. Similar studies for more mobile pests or natural enemies are less feasible, but are greatly needed to test the efficacy of PRM.

A final success story of resistance management involves not just a single key pest species, but a whole complex of phytophagous arthropod pests of apple in the United States (Croft 1982). It also focuses on resistance management of a number of beneficial arthropod predators and parasitoids that provide significant biological control of a variety of secondary pests of this crop (Tabashnik and Croft 1985, Tabashnik 1986a). In this case, a rather serendipitous pattern of resistance has developed among species to certain organophosphate (OP) insecticides. This resistance or lack thereof in some species has been exploited and provided a good example of IPM, as well as resistance management.

To summarize the resistance situation in broad terms, no key pest, such as the codling moth, apple maggot, or plum curculio, has developed resistance to the OP azinphosmethyl, whereas various secondary pests, such as mites, aphids, leafhoppers, leafminers, and their natural enemies have developed resistant strains, thus rendering the compound more selective. In fact, the example cited earlier of management of cyhexatin resistance management in the spider mite *T. urticae* using biological control by insecticide-resistant predatory mites is just a subsystem of this larger resistance management program.

Essential maintenance components of the program include (1) carefully monitoring OP resistance in key pests such as the codling moth, apple maggot, plum curculio, and several leafroller species; (2) seeking to minimize further resistance development in secondary pests, such as mites, aphids, leafminers, leafhoppers and scales, through minimum use of selective pesticides and maximum exploitation of biological control agents; and (3) exploiting resistance development for key natural enemy species by monitoring for natural development of resistance or by genetic improvement of resistant strains using hybridization or artificial selection techniques.

The success of the resistance management program for the complex of pests and natural enemies of apple in the United States, although generally good, has been mixed. In some areas where OP resistance has developed in several key leafroller species (e.g., the eastern United States), increased use of nonselective pesticides, such as the synthetic pyrethroids, has occurred. This change has brought resistance problems in species like the spotted tentiform leafminer (Pree et al. 1986, Chapter 10). However, in the western United States, OP pesticides still provide adequate control of key pests, while not upsetting biological control of certain secondary pests by their resistant predators and parasites (Croft 1982). In those areas where the program has been successful, the keys to success have been good monitoring of resistance (e.g., among leafrollers, Croft and Hull 1990), rapid responses to early signs of resistance (e.g., Croft et al. 1984), maximum use of IPM and alternative control tactics other than pesticides (Croft et al. 1987), and a certain amount of luck (i.e., the lack of resistance development in key pests such as the codling moth).

B. Less Successful Examples

Pyrethroid resistance in North American *Heliothis* spp. has also been of great concern to scientists since these pesticides were first introduced in the late 1970s. However, many cotton growers have been unconcerned, apparently out of a faith that pesticide manufacturers will develop replacement compounds. Despite efforts of cotton entomologists to promote judicious use of the SP's, there has been a tendency to use these chemicals intensively. Not surprisingly, and with striking parallels to the Australian situation, control failures occurred in *Heliothis virescens* in certain areas of western Texas in 1985 (Plapp and Campanhola 1986, Chapter 9). Resistance in Texas was confirmed by Roush and Luttrell (1987), who also documented the independent evolution of resistance in Mississippi in 1986. A resistance management program has been developed by state entomologists for Mississippi, Louisiana, and Arkansas growers (Luttrell and Roush 1987), but at present it is only a recommendation for cotton growers to use. Adoption of the program ranges from about 70% in some areas of Mississippi to about 90% in some areas of Texas (Roush, personal communication; Chapter 9) and support appears to be wavering in some areas because of the absence of recent control failures. In the absence of some enforcement capability (which is not currently legal), it appears unlikely that the program will be more widely adopted unless a larger consensus of industry, public, and private groups joins in supporting such a program.

Another case showing lack of progress despite extensive research on resistance management is resistance in the cattle tick, *Boophilus microplus*, in Australia. This species has been a critical threat to the cattle industry, particularly in southeastern Queensland, primarily because of its development of resistance to almost every pesticide introduced for control (Sutherst and Comins 1979). Considerable effort has been devoted to resistance studies, and Australian tick resistance authorities now believe that they have some very clear ideas on how resistance can be delayed (J. Nolan, remarks at a joint CSIRO/DSIR Workshop on Insecticide Resistance Management held July, 1986, in Canberra, Australia, Chapter 5). Many of these concepts were previously described and analyzed by Sutherst and Comins (1979). However, Australian cattle producers are not as cohesive as a group as are Australian cotton growers. Although the Australian cotton industry is relatively young (about 20 years), cattle production has been a major enterprise for more than 100 years, and many production practices are strongly influenced by traditional approaches to management. Whereas the cotton growers, perhaps rather uniquely in Australian agriculture, tend to band together, cattle producers tend to be very independent and are very slow to adopt new resistance management methods.

The diamondback moth, *Plutella xylostella* is a major pest of cruciferous vegetables in more than 80 countries worldwide. It has developed resistance to all major classes of insecticides (Georghiou 1981, Cheng 1988). Three factors promote its rapid evolution of resistance to pesticides: its biology, the relatively low level of acceptable damage, and the intensive use of insecticides for its control (Liu et al. 1981). In tropical regions this pest can complete more than 15 generations yearly (Liu et al. 1981), which accelerates resistance development (Tabashnik and Croft 1982, Chapter 6). Because of the direct use of the crop for human consumption, the economic threshold (pest density at which control must be applied) is only one or two larvae per plant. Insecticides have been used heavily for diamondback moth control. For example, many farmers in Taiwan spray weekly. Consequently, more than 2,000-fold resistance to the pyrethroid insecticides has developed there in less than four years after initial use of these products (Cheng 1981, Liu et al. 1981). Simulation studies suggest that under tropical conditions, insecticide use must be reduced to two or fewer sprays per crop cycle to slow resistance development substantially (Tabashnik 1986b, Chapter 6). Although this moth can disperse great distances, some local variation in resistance development among individual farms does occur (Tabashnik et al. 1987). This suggests that individual growers could retard resistance development in their own fields by reducing insecticide use. Integration of insecticides with biological, cultural, and microbial methods for diamondback moth control is the most promising way to retard resistance development. However, implementation of these methods is complex and difficult to achieve with growers (B. E. Tabashnik, personal communication).

A final example of where little progress has been made in achieving resistance management is with the pear psylla in parts of the midwestern United States and Europe. (see Table 11.2). In these cases, efforts at developing and implementing a resistance management strategy have been less extensive than in western North America (see earlier). In areas where both prebloom and summer applications of synthetic pyrethroid insecticides have been used unilaterally without alternation with other pesticides, resistance to the SP has developed widely. SP resistance now extends to all registered compounds over widespread areas of production. Only the addition of a synergist (piperonyl butoxide, Chapter 3) can provide reasonable control of this pest in the resistance-affected regions. It is expected that the use of synergists to reinstate effective control with the SP's will only be a short-term solution to the continuing problems of resistance in pear psylla (Burts et al. 1989).

VI. Synthesis of Principles

Several generalizations can be made from the selected case histories cited earlier. They apply as well to many other cases of resistance management described in the entomological literature. Discussion here focuses on gaps in our research base and on the factors limiting implementation of PRM.

1. Improved resistance monitoring is an essential component if resistance management is to be practiced effectively. Too often in the past, resistance monitoring was implemented only after resistance developed. More frequently now, monitoring is being done to anticipate resistance management needs. New methods of resistance monitoring are being developed that are sensitive to small changes in gene frequencies when resistance is first beginning to develop (Chapter 2). Resistance monitoring usually comes at considerable cost and often can be justified only on certain high-value production systems. Resistance monitoring methods are also necessary for verifying the effectiveness of the resistance management programs and to anticipate any changes in the resistance response of various species over time (Chapter 5).

2. Expertise, training, and methodology are needed in the ecology and population genetics of pesticide resistance. This is a critical gap in our spectrum of disciplines necessary to solve resistance problems and to maintain resistance management programs. Many avenues to solve these problems could be pursued, ranging from better training of applied scientists, better cooperation between applied and basic scientists, and increased use of resistance models for evolutionary biology studies. Cross-disciplinary funding for research is also a key element in bridging this gap.

3. New institutional policies of cooperation and change are needed to better facilitate the implementation of resistance management tactics across broad

boundaries of societal groups, including producers, industry personnel, regulators, and pesticide users. For example, resistance management districts could be organized to operate much like mosquito abatement districts (or similar units of management) that are established, at least temporarily, to solve a persistent problem. IPM districts operate in many regions where producers see a common benefit from organizing themselves. Most cases of resistance management are unique in terms of what an appropriate organizational structure might be. Each resistance episode requires a very specific, tailored response. Little effort has been made to study how to improve the implementation of resistance management in the field. Research, extension, private groups, and other interested personnel must put their collective heads together and develop some innovative ideas on how these problems can be dealt with in the future.

Finally, a utopian goal for resistance management would be to stem the increasing tide of resistance to most of our valuable chemical pest control products. Instead of experiencing an effective life of 5–15 years until a compound is rendered ineffective because of resistance problems, we might see these products last for longer periods—say up to 50 years. This would allow time to develop other, more safe and effective products and control methods. Too often in the past the obituary of a particular product has not been written by us, but by our competitors, the pests, as they have continued to find new ways to evolve resistant strains and circumvent our best efforts to contain them.

References

Brown, T. M., and W. G. Brogdon. 1987. Improved detection of insecticide resistance through conventional and molecular techniques. Ann. Rev. Entomol. 32: 145–162.

Burts, E. C., H. E. van de Baan, and B. A. Croft. 1989. Pyrethroid resistance in pear psylla, *Psylla pyricola* Foerster (Homoptera:Psyllidae) and synergism of pyrethroids with piperonyl butoxide. Can. Entomol. 121: 219–223.

Carlson, G. A. (Leader). 1986. Implementing management of resistance to pesticides, pp. 371–387. *In* Pesticide resistance: strategies and tactics for management. National Academy of Sciences, Washington, D.C.

Cheng. E. Y. 1981. Insecticide resistance study in *Plutella xylostella* L. I. Developing a sampling method for surveying resistance. J. Arg. Res. China. 30: 277–284.

Cheng, E. Y. 1988. Problems of control of insecticide-resistant *Plutella xylostella*. Pestic. Sci. 23: 177–188.

Comins, H. N. 1977. The development of insecticide resistance in the presence of immigration. J. Theor. Biol. 64: 177–197.

Croft, B. A. 1982. Developed resistance to insecticides in apple arthropods: A key to pest control failures and successes in North America. Entomol. Exper. and Appl. 31: 88–110.

Croft, B. A. 1986. Integrated pest management: the Agricultural and Environmental Rationale, pp. 712–728. *In* R. F. Frisbie and P. L. Adkisson (eds.), Integrated pest management of major agricultural systems. CIPM-IPM Project Rept. Texas Agric. Exper. Sta., Pub. MP–1616.

Croft, B. A. 1990. Technical and policy issues in pesticide resistance management. *In* W. K. Moberg and H. M. LeBaron (eds.), Fundamental and Practical Approaches to Combating Resistance. (ACS Symposium Series No. 421, American Chemical Society, Washington, D.C.)

Croft, B. A., and L. A. Hull. 1990. Chemical control and resistance in tortricid pests of pome and stone fruits, Ch. 2.3.4. *In* L. P. S. van der Geest and H. H. Evenhius (eds.), Torticid pests. Elsevier, Amsterdam, The Netherlands.

Croft, B. A., and K. Strickler. 1983. Natural enemy resistance to pesticides: Documentation, characterization, theory and application. pp. 669–702. *In* G. P. Georghiou and T. Saito (eds.), Pest Resistance to Pesticides. Plenum, New York.

Croft, B. A., and H. E. van de Baan. 1988. Ecological and genetic factors influencing evolution of pesticide resistance in tetranychid and phytoseiid mites. Exper. and Appl. Acarol. 4: 277–300.

Croft, B. A., R. W. Miller, R. D. Nelson, and P. H. Westigard. 1984. Inheritance of early stage resistance to cyhexatin and formetanate in *Tetranychus urticae* Koch (Acarina: Tetranychidae). J. Econ. Entomol. 77: 574–578.

Croft, B. A., S. C. Hoyt, and P. H. Westigard. 1987. Integrated mite control revisited: Organotin and acaricide resistance management. J. Econ. Entomol. 80: 304–311.

Croft, B. A., E. C. Burts, H. E. van de Baan, P. H. Westigard, and H. W. Riedl. 1989. Local and regional resistance to fenvalerate in *Psylla pyricola* Foerster (Homoptera: Psyllidae) in western North America. Can. Entomol. 121: 121–129.

Daly, J. C., and J. A. McKenzie. 1987. Resistance management strategies in Australia: the "Heliothis" and "wormkill" programmes, pp. 951–959. *In* Proc. 1986 British Crop Prot. Soc. Bristol, U.K.

Dover, M. J., and B. A. Croft. 1984. Getting tough: Public policy and the management of pesticide resistance. World Res. Inst. Study 1. 80 pp.

Dover, M. J., and B. A. Croft. 1986. Pesticide resistance and public policy. Bioscience 36: 78–85.

Edge, V. E., and D. G. James. 1986. Organotin resistance in *Tetranychus urticae* (Acari: Tetranychidae) in Australia. J. Econ. Entomol. 79: 1477–1483.

Flexner, J. L. 1987. Organotin resistance in *Tetranychus urticae* on pear: Components and their integration for resistance management. Ph.D thesis, Oregon St. Univ. Corvallis, OR. 103 pp.

Flexner, J. L., K. M. Theiling, B. A. Croft, and P. H. Westigard. 1989. Fitness and immigration: Factors affecting reversion of organotin resistance in the twospotted spider mite (Acari: Tetranychidae) J. Econ. Entomol. 82: 996–1002.

Follett, P. A., B. A. Croft, and P. H. Westigard. 1985. Regional resistance to insecticides in *Psylla pyricola* from pear orchards in Oregon. Can. Entomol. 117: 565–573.

Frisbie, R. F. and P. L. Adkisson. 1986. Eds. Integrated pest management of major agricultural systems. CIPM IPM Project Rept. Texas Agric. Exper. Sta. Pub. MP-1616.

Georghiou, G. P. 1981. The occurrence of resistance to pesticides in arthropods: An index of cases reported through 1980. FAO Rept. Rome 129 pp.

Georghiou, G. P., and C. E. Taylor. 1976. Pesticide resistance as an evolutionary phenomenon, pp. 759–785. *In* Proc. 15th Int. Cong. Entomology.

Georghiou, G. P., and C. E. Taylor. 1977a. Genetic and biological influences in the evolution of insecticide resistance. J. Econ. Entomol. 70: 319–323.

Georghiou, G. P., and C. E. Taylor. 1977b. Operational influences in the evolution of insecticide resistance. J. Econ. Entomol. 70: 653–658.

Hoy, M. A. 1985. Recent advances in genetics and genetic improvement of the Phytoseiidae. Annu. Rev. Entomol. 30: 345–370.

Keiding, J. 1986. Prediction or resistance risk assessment. pp. 279–297. *In* Pesticide resistance: strategies and tactics for management, National Academy of Sciences, Washington, D.C.

Liu, M. Y., Y. J. Tzeng, and C. N. Sun. 1981. Diamondback moth resistance to several synthetic pyrethroids. J. Econ. Entomol. 74: 393–396.

Luttrell, R., and R. T. Roush. 1987. Strategic approaches to avoid or delay development of resistance to insecticides, pp. 31–33. *In* Proc. 1987 Beltwide Cotton Production Research Conferences, Dallas, Tex. National Cotton Council of America.

National Academy of Sciences (NRC, Board on Agric.) 1986. Pesticide resistance: strategies and tactics for management. National Academy of Sciences, Washington, D.C.

Plapp, F. W., Jr., and C. Campanhola. 1986. Synergism of pyrethroids by chlordimeform against susceptible and resistant *Heliothis*, pp. 167–169. *In* Proc. 1986 Beltwide Cotton Production Research Conferences, Las Vegas, Nev. National Cotton Council of America, Memphis, Tenn.

Pree, D. J., D. B. Marshall, and D. E. Archibald. 1986. Resistance to pyrethroid insecticides in the spotted tentiform leafminer, *Phyllonorycter blancardella* (Lepidoptera: Gracillariidae), in southern Ontario. J. Econ. Entomol. 79: 318–322.

Riedl, H., P. H. Westigard, R. S. Bethell, and J. E. DeTar. 1981. Problems with chemical control of pear psylla. Calif. Agric. 35: 7–9.

Roush, R. T., and B. A. Croft. 1986. Experimental population genetics and ecological studies of pesticide resistance in insects and mites, pp. 257–270. *In* Pesticide resistance: strategies and tactics for management. National Academy of Sciences, Washington, D.C.

Roush, R. T., and R. G. Luttrell. 1987. The phenotypic expression of resistance in *Heliothis* and implications for resistance management, pp. 220–224. *In* Proc. 1987 Beltwide Cotton Production Research Conferences, Dallas, Tex. National Cotton Council of America, Memphia, Tenn.

Roush, R. T., and J. McKenzie. 1987. Ecological genetics of insecticide and acaracide resistance. Annu. Rev. Entomol. 32: 361–380.

Roush, R. T., and G. L. Miller. 1986. Considerations for design of insecticide resistance monitoring programs. J. Econ. Entomol. 79: 293–298.

Scott, J. G., R. T. Roush, and D. A. Rutz. 1989. Insecticide resistance of house flies from New York dairies (Diptera: Muscidae). J. Agric. Entomol. 6: 53–64.

Sutherst, R. W., and H. N. Comins. 1979. The management of acaricide resistance in the cattle tick (*Boophilus microplus*) (Canestrini) (Acari: Ixodidae) in Australia. Bull. Entomol. Res. 69: 519–537.

Tabashnik, B. E. 1986a. Evolution of pesticide resistance in predator-prey systems. Bull. Entomol. Soc. Amer. 32: 156–161.

Tabashnik, B. E. 1986b. Model for managing resistance to fenvalerate in the diamondback moth (Lepidoptera: Plutellidae). J. Econ. Entomol. 79: 1447–1451.

Tabashnik, B. E. 1989. Managing resistance with multiple pesticide tactics: theory, evidence, and recommendations. J. Econ. Entomol. 82: 1263–1269.

Tabashnik, B. E., and B. A. Croft 1982. Managing pesticide resistance in crop-arthropod complexes: interactions between biological and operational factors. Environ. Entomol. 11: 1137–1144.

Tabashnik, B. E., and B. A. Croft. 1985. Evolution of pesticide resistance in apple pests and their natural enemies. Entomophaga 30: 37–49.

Tabashnik, B. E., N. L. Cushing, and M. W. Johnson. 1987. Diamondback moth (Lepidoptera: Plutellidae) resistance to insecticides in Hawaii: Intra-island variation and cross-resistance. J. Econ. Entomol. 80: 1091–1099.

Taylor, C. E. 1983. Evolution of resistance to insecticides: The role of mathematical models and computer simulations, pp. 163–173. *In* G. P. Georghiou and T. Saito (eds.), Pest resistance to pesticides. Plenum, New York.

Vijverberg, H. P. M., J. M. van der Zalm, R. G. D. M. van Kleef, and J. van der Bercken. 1983. Temperature- and structure-dependent interaction of pyrethroids with the sodium channels in frog node of ranvier. Biochimica et Biophysica Acta 728: 73–82.

Waddill, V. H. 1978. Contact toxicity of four synthetic pyrethroids and methomyl to some adult insect parasites. Fla. Entomol. 61: 27–30.

Index